PRECALCULUS I
College Algebra

NDSU

Pages 143–159, 165–182, 185–205, 211–226, 233–240, 241–255, 259–275, 281–293, 299–316, and 327–333 from *Precalculus: Functions and Graphs,* Eighth Edition by M.A. Munem and J.P. Yizze. Copyright © 2006 by Kendall Hunt Publishing Company. Used with permission.

Cover image © 2014, Shutterstock, Inc.

www.kendallhunt.com
Send all inquiries to:
4050 Westmark Drive
Dubuque, IA 52004-1840

ISBN: 978-1-4652-4762-9

Printed in the United States of America

Contents

Lines, Linear Models, and Graphing

CHAPTER CONTENTS

Overview

It is often said that a picture is worth a thousand words, and this is especially true in mathematics. Using graphs to demonstrate the relationship between two quantities is common practice in many areas, and perhaps the simplest such graph is that of a line. Mathematical models of linear growth and decay are used by federal policy makers to project the increase in the number of welfare recipients, biologists to explore the destruction of our natural environment, and business managers to project weekly gains in revenue. We begin a rather lengthy process of describing relationships between quantities with a discussion of linearity and graphing.

OBJECTIVES

1. Identify Total Change and Average Rate of Change
2. Characterize Linear Relationships
3. Determine Formulas That Model Linear Behavior

1.1 Linearity

In order to truly understand how one quantity relates to another, one must be familiar with the concept of change. There are two measures of change that we will discuss in this section, *total change* and *average rate of change*.

Total Change and Average Rate of Change

Total change is simply the complete change that has occurred in some quantity over a fixed reference period (time, for instance). Algebraically, it is the difference between the initial value and ending value of the quantity over that reference period. Geometrically, it is the vertical difference between the first and last points on a graph representing the values of the quantity over that reference period. For example, if a plant was two inches high at the beginning of the year and eight inches high at the end of the year, then we can say that the total change in height is eight minus two, or six inches over the course of the year.

Given two quantities, say x and y, that are related in some way, we often use the symbol Δy (read "delta y") to indicate the total change in the values of y given a corresponding total change in the values of x, denoted by Δx.

Also, given two related quantities, it is often desirable to compute the *rate* at which one is changing with respect to the other. In particular, we are interested in the **average rate of change** of one quantity with respect to the other. If x and y are two related quantities, and Δy is the total change in values of y corresponding to a total change Δx in values of x, then the corresponding average rate of change of y with respect to x is the quantity $\frac{\Delta y}{\Delta x}$. Notice that if units are attached to x and y, then this average rate of change has units $\frac{\text{units of } y}{\text{units of } x}$, or (units of y) per (units of x).

For example, if the total change in height of the plant was six inches over the past year, then we could say that the average rate of growth over the past year was 6 inches / 1 year or 6 inches per year. We could also say that the average rate of growth each month was 6 inches / 12 months or ½ inch per month, and so on.

If x and y are two related quantities, then as x changes from $x = a$ to $x = b$, we summarize the ideas of total change and average rate of change as follows:

Total Change and Average Rate of Change

- total change in $y = \Delta y =$ (value of y when $x = b$) $-$ (value of y when $x = a$)
- $\text{average rate of change of } y \text{ with respect to } x = \frac{\text{total change in } y}{\text{total change in } x} = \frac{\Delta y}{\Delta x}$

To illustrate this topic, we will first explore population growth projections used by city planners.

EXAMPLE 1 Population Growth

Suppose that a town of 15,000 people increases in population by 500 people per year.

(a) Create a table that shows the population of the town over fifteen years in three-year increments.

(b) Plot the population over time as found in (a).

(c) Find a formula for population in terms of time.

Solution We let P represent population and t represent number of years, with $t = 0$ corresponding to the point in time when $P = 15{,}000$.

(a)

Years (t)	Population (P)
0	15,000
3	16,500
6	18,000
9	19,500
12	21,000
15	22,500

(b) Plotting P in terms of t, we have the graph shown in Figure 1.

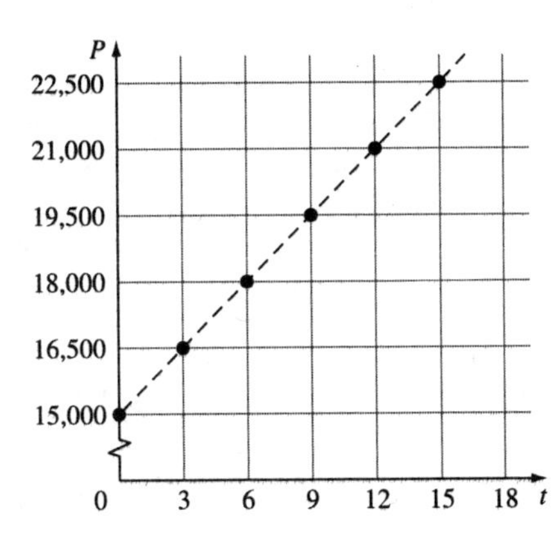

Figure 1

(Note: The dashed line is not part of the graph. It is just to illustrate that the points appear to lie on a line.)

(c) $P = 15{,}000 + 500t$ ❖

In the previous example, the average rate of change of 500 people per year remains constant, irrespective of pairs of distinct years for which this is computed (calculate the average rate of change over different periods of time, not all the same length). When this is the case, the relationship between the two quantities is called **linear**. Graphically, as you might expect, this relationship is illustrated by a line. So, in the last example the population P is linearly related to time t. We have included a line through the plotted points as an illustration (see Figure 1). Any equation, graph, or table of values that describes two quantities that are related linearly is said to be a **linear model**.

EXAMPLE 2 **Creating Linear Models**

A manufacturer of pianos has *fixed* costs (for building, machinery, recycling, etc.) of $1200 each day. These are expenses that must be paid regardless of the number of pianos produced. The *variable costs* (for material, labor, shipping, etc.) for making and selling a single piano amount to $1500. Let P represent the number of pianos produced in a day.

(a) Calculate the factory's total cost, C, to make 1, 5, 8, and 12 pianos.
(b) Find a formula for C in terms of P.
(c) What is the average rate of change of cost with respect to the number of pianos produced? How does this relate to the variable cost of producing a single piano?

Solution As given, daily fixed costs are $1200. To calculate the total production costs for a day when pianos are produced, we simply multiply the number of pianos by the variable cost per piano and add the fixed costs.

(a)

Pianos Produced (d)	Total Cost (C)
1	15(1) + 1200 = $1215
5	15(5) + 1200 = $1275
8	15(8) + 1200 = $1320
12	15(12) + 1200 = $1380

(b) $C = 1500P + 1200$
(c) We will show that C is linearly related to P by demonstrating that the average rate of change of C with respect to P is always constant. Choose two arbitrary production levels, say $P = P_1$ and $P = P_2$ pianos. We calculate the average rate of change of C with respect to P as P changes from P_1 to P_2. Using our results from (b), we have

$$\frac{\Delta C}{\Delta P} = \frac{(1500P_2 + 1200) - (1500P_1 + 1200)}{P_2 - P_1}$$

$$\frac{\Delta C}{\Delta P} = \frac{1500P_2 + 1200 - 1500P_1 - 1200}{P_2 - P_1}$$

$$\frac{\Delta C}{\Delta P} = \frac{1500P_2 - 1500P_1}{P_2 - P_1}$$

$$\frac{\Delta C}{\Delta P} = \frac{1500(P_2 - P_1)}{(P_2 - P_1)}$$

$$\frac{\Delta C}{\Delta P} = 1500$$

So, the average rate of change of C with respect to P is always \$1500 per piano. Notice that this rate of change is exactly the variable cost associated with producing one piano. ❖

Remark: In the example above, the average rate of change that was calculated is often referred to as the **marginal cost** of producing one dozen rubber bands. If the situation was such that the average rate of change in cost with respect to dozens of rubber bands produced was not constant, we might still be interested in the marginal cost of producing one dozen rubber bands, but that quantity may change depending upon the production level where we consider how fast cost of production is changing.

The last two examples show some important characteristics of equations that model linear relationships between two quantities.

In Example 1,

$$\text{current population} = \begin{pmatrix} \text{average rate of} \\ \text{growth per year} \end{pmatrix} \times \begin{pmatrix} \text{number} \\ \text{of years} \end{pmatrix} + \text{initial population}$$

In Example 2,

$$\text{total costs} = \begin{pmatrix} \text{average rate of} \\ \text{increase per piano} \end{pmatrix} \times \begin{pmatrix} \text{number} \\ \text{of piano} \end{pmatrix} + \text{fixed costs}$$

In general, if x and y are two linearly related quantities such that when $x = 0$ we have $y = b$ (call this the *initial value of* y) and the average rate of change of y with respect to x is constant, say m, then for $\Delta x \neq 0$, we have

$$m = \frac{\Delta y}{\Delta x}$$

So, given any value of x with corresponding value of y, and letting

$$\Delta x = x - 0 \text{ and } \Delta y = y - b$$

we have

$$m = \frac{\Delta y}{\Delta x} = \frac{y - b}{x - 0} = \frac{y - b}{x}$$

Solving for y gives

$$y = mx + b$$

In other words,

$$\text{value of } y = (\text{average rate of change of } y \text{ with respect to } x) \times (\text{value of } x) + (\text{initial value of } y)$$

This is a way of describing linearly related quantities by an equation (formula), provided that we know the average rate of change and an initial value. In the next section we expand upon this idea.

PROBLEM SET 1.1

1. Which of the following tables represents linearly related quantities? Check by determining if equal increments in x-values correspond to equal increments in y-values.

(a)

x	y
0	10
3	15
6	22
9	28

(b)

x	y
0	50
200	100
600	150
1200	200

(c)

x	y
0	10
10	20
20	25
30	35

(d)

x	y
−3	10
−1	2
0	−2
3	−14

2. The table below shows the cost of selling various amounts of iced tea per day.

x (cups/day)	0	5	10	50	100	200
C (dollars)	75.00	77.50	80.00	100.00	125.00	175.00

(a) Explain why the relationship between C and x appears to be linear.
(b) Using the values in the table, make a plot of C against x.
(c) Find the average rate of change in cost with respect to the number of cups per day. Label your response with appropriate units. Explain what this means in the context of the given situation.
(d) How can it cost $75 to serve zero cups of iced tea?

3. The population, P, in millions, of a country in year t is given by the formula

$$P = 14 + 0.2t.$$

(a) Construct a table of values for P for $t = 0, 5, 10, 15, 20, 25$.
(b) Make a plot of P against t using the results from part **(a)**.
(c) What is the country's initial population?
(d) Find the average rate of change of the population in millions of people per year.

4. An office supply company produces staplers. The company has fixed monthly costs of $2000 and it costs $3.75 to produce each stapler. Write a formula that expresses the company's total cost, C, of producing x staplers in a month.

5. In 2005, the population of a town was 4650 and was growing by 23 people per year. Find a formula for P, the town's population, in terms of t, the number of years since 2005.

6. A new boat was purchased for $12,500. The boat's value depreciates linearly to $8600 in three years. Write a formula that expresses the boat's value, V, in terms of its age, t, in years.

7. Tuition cost, T, in dollars, for part-time students at Northeast Bible College is given by $T = 3000 + 200c$, where c represents the number of credits taken.

(a) Find the tuition cost for six credits.
(b) How many credits were taken if the tuition was $6200?
(c) Make a table showing costs for taking from four to sixteen credits. For each value of c, give the tuition cost, T.
(d) Which of these values of c has the smallest cost per credit?
(e) What does the 3000 represent in the formula for T?
(f) What does the 200 represent in the formula for T?

8. Survivor Sanctuary, a hidden island in the Pacific Ocean, experienced approximately linear population growth from 1980 to 2000. On the other hand, Afghanistan, a country in Asia, was torn by warfare in the 1980's and did not experience linear nor near-linear growth.

Year	1980	1985	1990	1995	2000
Pop. of Place A (millions)	16	14	16	15.8	15.9
Pop. of Place B (millions)	12.2	13.5	14.7	15.8	17

(a) Which of these two places is Place A and which is Place B? Explain.
(b) What is the approximate average rate of change in population with respect to time for the population that is growing approximately linearly?
(c) Using average rate of change, estimate the population of Survivor Sanctuary in 1988.

9. Outside the US, temperature readings are usually given in degrees Celsius; inside the US, they are often given in degrees Fahrenheit. The exact conversion from Celsius, C, to Fahrenheit, F, is given by the formula

$$F = \frac{9}{5}C + 32.$$

An approximate conversion is obtained by doubling the Celsius temperature and adding 30 to get the equivalent Fahrenheit temperature.

(a) Write a formula using C and F to express the approximate conversion.

(b) How far off is the approximation if the Celsius temperature is $-10°$, $0°$, $10°$, $25°$?

(c) For what temperature (in degrees Celsius) does the approximation agree with the actual formula?

10. A woodworker goes into business selling birdhouses. His start-up costs, including tools, plans, and advertising, total \$2000. Labor and materials for each birdhouse cost \$35.

(a) Calculate the woodworker's total cost, C, to make 1, 3, 10, 15, and 20 birdhouses. Sketch a graph of C versus n, the number of birdhouses that he makes.

(b) Find a formula for C in terms of n.

(c) What is the average rate of change of C with respect to n? What does this rate indicate about the woodworker's expenses?

11. A company has found that there is a linear relationship between the amount of money that it spends on marketing a product on the internet and the number of units that it sells. If the company spends no money on marketing, it sells 400 units. For each \$5000 spent on marketing, an additional 30 units are sold.

(a) If x is the amount of money that the company spends on marketing (in dollars), then find a formula for y, the number of units sold, in terms of x.

(b) How many units does the firm sell if it spends \$25,000 on marketing? \$40,000?

(c) How much marketing money must be spent to sell 1000 units?

(d) What is the average rate of change of y with respect to x? Give an interpretation that relates units sold and marketing costs.

12. At sea level, water boils at $212° F$. At an altitude of 1100 feet, water boils at $210° F$. The relationship between the boiling point and altitude is linear.

(a) Find the average rate of change in boiling point with respect to altitude.

(b) Find a formula that gives the boiling point of water in terms of altitude.

(c) Find the boiling point of water in each of the following cities: Minneapolis, MN (550 ft.); St. Joseph, MO (1300 ft.); Gillette, WY (3120 ft.)

13. A plane costs \$100,000 new. After 10 years, its value is \$25,000. Assuming linear depreciation (value decreases linearly with time), find the value of the plane when it is 8 years old and when it is 12 years old.

14. The profit, P, (in thousands of dollars) on x thousand units of a specialty item is given by $P = 0.8x - 16.5$. The cost, C, (in thousands of dollars) of manufacturing x thousand units is given by $C = 0.9x + 16.5$.

(a) Find a formula that gives the revenue R from selling x items, where $P = R - C$.

(b) How many items must be sold for the company to *break even* (zero profit)?

15. A doctor has a "rule of thumb" for determining the target weights for her patients, based on their heights. For women, the target weight is 100 lbs plus 5 lbs for each inch over 5 ft. in height. For men, it is 106 lbs plus 6 lbs for each inch over 5 ft. in height.

(a) Fill in the following tables:

Women's Target Weight

inches over 5 ft.	*Weight (lbs.)*
0	
1	
2	
4	
6	
12	
15	

Men's Target Weight

inches over 5 ft.	*Weight (lbs.)*
0	
1	
2	
4	
6	
12	
15	

(b) Find a formula that gives a woman's target weight, y, in terms of the number of inches, x, by which her height exceeds 5 ft.

(c) What is the target weight for a woman who is 69 inches tall?

(d) Do parts **(b)** and **(c)** for men.

16. The first edition of the textbook *Finite Math For Fun* was 372 pages long. By the time the ninth edition came along, the text had grown to 596 pages in length. Assume that the length L of the book is linearly related to the edition number n.

(a) Write an equation that gives L in terms of n.

(b) If the length trend continues, what edition of the book will be the first to have 1000 pages?

17. According to production records, on a certain production line the number of defective radio components produced is linearly related to the total number of these components produced. Suppose that one day 10 components out of a total of 300 produced are defective, and on another day 15 components out of a total of 425 produced are defective. How many defective components are expected on a day when a total of 500 components are produced?

18. A certain car can be rented according to one of two different rates:

- \$40 per day and \$0.30 per mile driven
- \$30 per day and \$0.50 per mile driven

(a) For each rate, write an equation that gives the cost, C (in dollars), of driving x miles in a day.

(b) Which rate is the least expensive if only 30 miles is to be driven on a single day?

19. Suppose that the cost of an airline ticket is linearly related to the distance traveled. The cost of a 200-mile flight is \$76 and the cost of a 350-mile flight is \$100. Let C be the cost (in dollars) of a flight of x miles.

(a) Write an equation that gives C in terms of x.

(b) Determine the cost of a 275-mile flight.

(c) How many miles of flight result in a cost of \$524?

20. The height, h, of a falling sheet of paper, in feet above the ground, t seconds after it is dropped is given by $h = -1.8t + 9$.

(a) How fast is the piece of paper falling?

(b) How far above ground is the paper after 4 seconds?

(c) How long will it take for the paper to reach the ground?

21. A soft-drink manufacturer can produce 1000 cases of soda in a week at a total cost of \$6000, and 1500 cases of soda in a week at a total cost of \$8500. Determine the manufacturers weekly fixed costs and weekly variable costs.

22. Suppose that a new tractor is worth \$120,000 right now and is expected to have a useful life of 20 years, at which time the tractor will have a scrap value of \$6000. Suppose that the value of the tractor decreases at the same rate each year. Calculate the rate at which the value decreases and also determine how long into its useful life the tractor is if its value has dropped below \$80,000.

Interlude: Progress Check 1.1

Several student organizations on campus are selling tickets to a large dance to be held following the Homecoming football game. It has been determined that the cost, C, to put on the event is linearly related to the number of tickets sold, t. The revenue, R, from ticket sales is also linearly related to t. The graphs shown in Figure 2 have been generated to show these relationships.

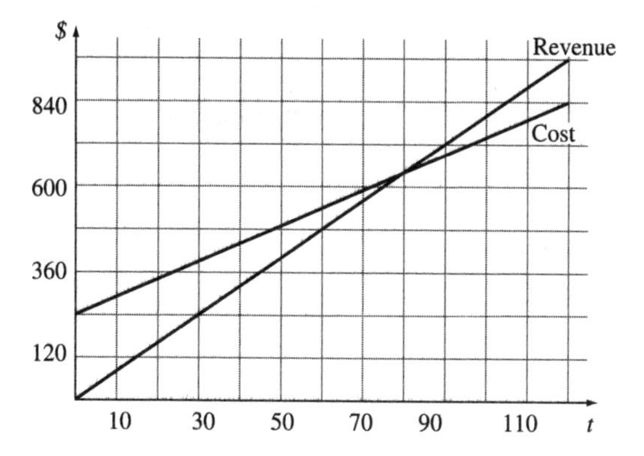

Figure 2

1. Calculate the average rate of change of C with respect to t. Calculate the average rate of change of R with respect to t (fill in the blanks). Label both with appropriate units.

$\frac{\Delta C}{\Delta t} =$ ______________________________

$\frac{\Delta R}{\Delta t} =$ ______________________________

2. Find formulas for both C and R in terms of t (fill in the blanks).

$C =$ ______________________________

$R =$ ______________________________

3. Estimate the coordinates of the point on the graph where the two lines meet. What is the significance of that point for event organizers? Explain in regards to number of tickets sold.

4. Let P denote profit realized by the sale of t tickets. Find a formula for P in terms of t (fill in the blank). Sketch a graph of P against t on the provided grid. Put scale on the vertical axis.

$P =$ ______________________________

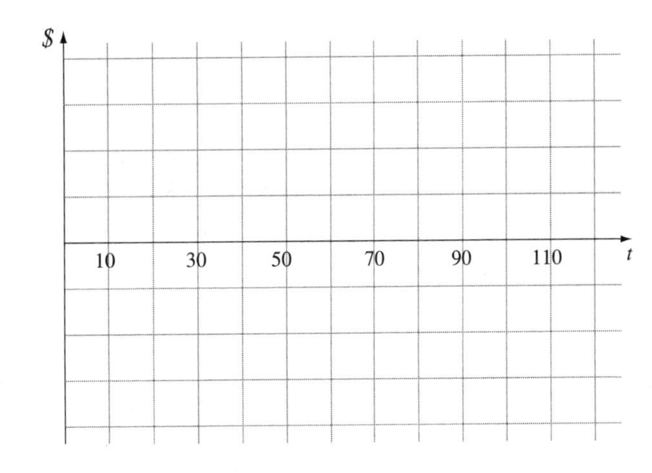

OBJECTIVES

1. Define Slope
2. Calculate and Interpret Slope
3. Utilize Slope-Intercept and Point-Slope Forms of Linear Equations
4. Solve Applied Problems

1.2 Lines: A Quick Review

In the previous section, we saw that if two quantities are linearly related, then the relationship can be illustrated graphically with a line. We also saw that average rate of change is one way to measure the rate at which one quantity is changing with respect to the other. In this section, we generalize the idea of average rate of change with the notion of *slope of a line.*

Slope of a Line

When moving between point S and point T on a line lying in the xy-plane, as shown in Figure 1, we move vertically (parallel to the y-axis) and horizontally (parallel to the x-axis).

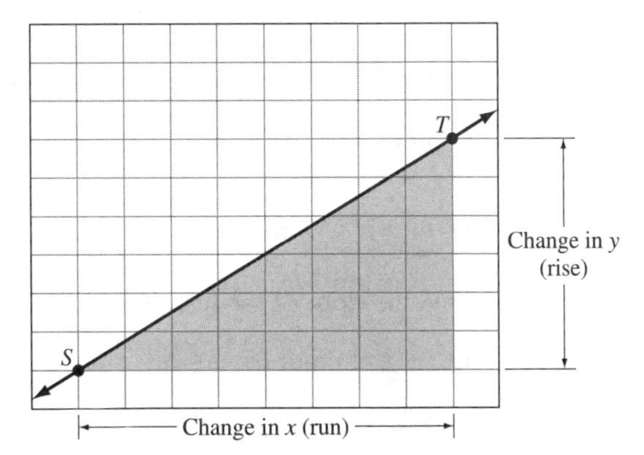

Figure 1

Evaluating the ratio

$$\frac{\text{change in } y}{\text{change in } x}$$

gives a measure of the steepness of the line. The larger the magnitude of this ratio, the steeper the line. It also indicates if the line is climbing from left to right (positive ratio) or falling from left to right (negative ratio). This ratio is called the **slope** of the line. A common interpretation of slope is "rise over run", as illustrated in Figure 1.

Consider two points in the xy-plane with coordinates (x_1, y_1) and (x_2, y_2). We can then say that the change in x, or "run", is the difference between x_1 and x_2, and the change in y, or "rise", is the difference between y_1 and y_2. There is a correspondence in change, in that as x_1 changes to x_2, y_1 changes to y_2, and vice versa. Consequently, we define the slope, m, of the line through these two points, assuming $x_1 \neq x_2$, as follows:

Slope of a Line

The slope, m, of a nonvertical line through the points (x_1, y_1) and (x_2, y_2) is

$$m = \frac{\text{rise}}{\text{run}} = \frac{\text{change in } y}{\text{change in } x} = \frac{\Delta y}{\Delta x} = \frac{y_2 - y_1}{x_2 - x_1}$$

EXAMPLE 1 Calculating Slopes

(a) Find the slope of each of the lines in Figure 2. Assume that the lines lie in the xy-plane and that the gridlines each mark one unit.

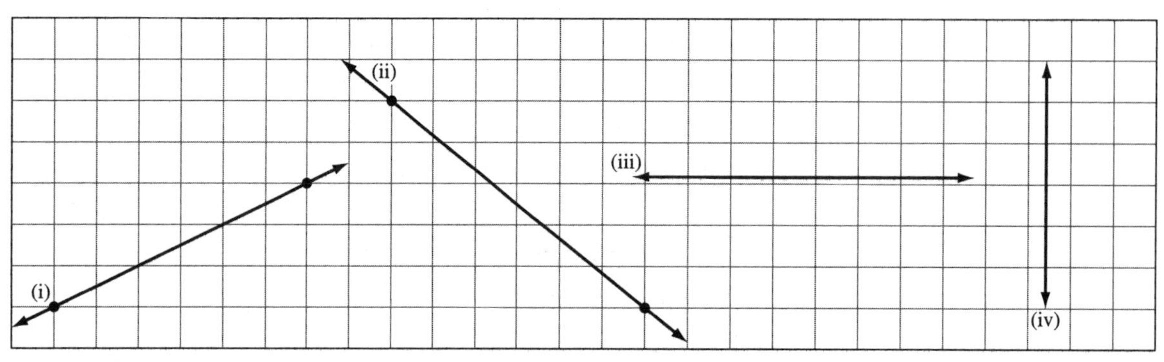

Figure 2

(b) Graph the line containing the point p with slope m.

i. $p(1, -3),\ m = -\frac{2}{5}$

ii. $p(-1, 4),\ m = \frac{3}{2}$

(c) Find the slope of the line through the given points in the xy-plane.

i. $(-2, 1)$ and $(5, 7)$

ii. $\left(\frac{1}{2}, 3\right)$ and $(2, -1)$

Solution (a) For i. and ii., we will find Δx and Δy using the marked points on the lines.

i. The line is rising from left to right, so the slope is positive. Also, $\Delta y = 3$ while $\Delta x = 6$.

Therefore, the slope is $\frac{\Delta y}{\Delta x} = \frac{3}{6} = \frac{1}{2}$.

ii. The line is falling from left to right, so the slope is negative. Also, $\Delta y = -5$ while $\Delta x = 6$.

Therefore, the slope is $\frac{\Delta y}{\Delta x} = \frac{-5}{6} = -\frac{5}{6}$.

iii. The line is horizontal. Consequently, $\Delta y = 0$ between any two distinct points (and $\Delta x \neq 0$). Therefore, the slope is $\frac{\Delta y}{\Delta x} = \frac{0}{\Delta x} = 0$.

iv. The line is vertical. Therefore $\Delta x = 0$ regardless of the points at which we choose to look. Since $\frac{\Delta y}{\Delta x}$ is undefined if $\Delta x = 0$, the slope of this line is undefined.

(b) See Figure 3.

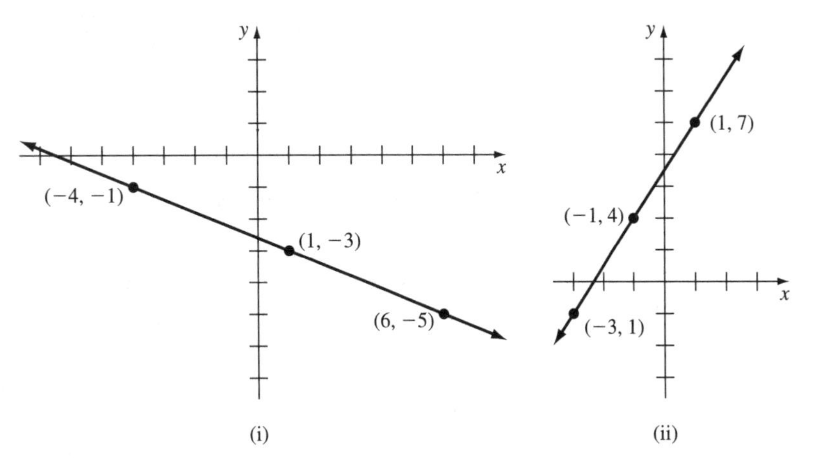

Figure 3

(c) Slope, m, is given by $m = \dfrac{y_2 - y_1}{x_2 - x_1}$.

i. The slope of the line through $(-2, 1)$ and $(5, 7)$ is

$$m = \frac{7-1}{5-(-2)} = \frac{6}{7}$$

ii. The slope of the line through $\left(\frac{1}{2}, 3\right)$ and $(2, -1)$ is

$$m = \frac{-1-3}{2-\frac{1}{2}} = \frac{-4}{\frac{3}{2}} = -4\left(\frac{2}{3}\right) = -\frac{8}{3}$$ ❖

It should be noted here that, given any nonvertical line, any two distinct points on the line can be used to calculate the slope of the line. It can be shown, using similar triangles, that we will always get the same value for slope, regardless of which two points are used. More relevant to our previous discussion, though, is that the slope of a line in the *xy*-plane is the average rate of change of *y* with respect to *x*, which we have already shown is constant when the relationship between *x* and *y* is linear.

Consider the graphs shown in Figure 4. The first graph contains *parallel* lines L_1 and L_2. The second graph contains *perpendicular* lines L_3 and L_4.

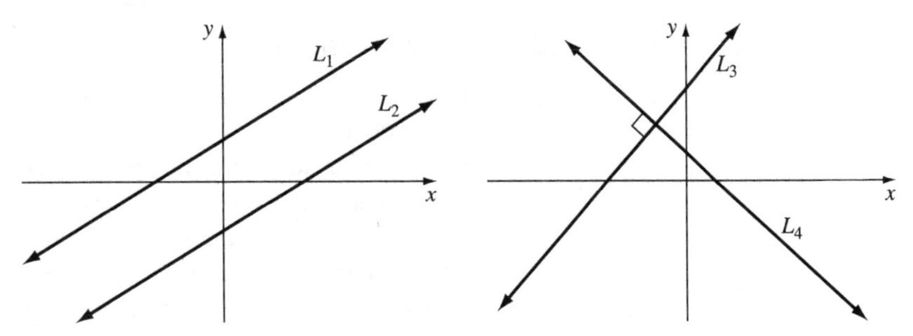

Figure 4

Two lines are parallel if they have the same slope. It can be shown that the slopes of two perpendicular lines are negative reciprocals of each other; that is, if a line L_1 that is neither horizontal nor vertical has slope m and L_2 is perpendicular to L_1, then

$$\text{slope of } L_2 = -\frac{1}{m}.$$

Equations for Lines

We now turn our attention to an algebraic treatment of lines, which means we will consider equations that are related to lines. **The graph of an equation** in the variables x and y consists of all points in the xy-plane whose coordinates (x, y) satisfy the equation. So, when we speak of **an equation for a line**, we mean an equation whose graph is the given line. We have already encountered one such type of equation in the previous section when we investigated equations of the form $y = mx + b$.

Suppose that L is a nonvertical line in the xy-plane that passes through the point $(0, b)$ and has slope m. If we let (x, y) denote the coordinates of any other point on the line, then we know that

$$m = \frac{\Delta y}{\Delta x} = \frac{y - b}{x - 0}$$

Solving for y gives

$$y = mx + b$$

This gives us one way of writing an equation whose graph is the line through the point $(0, b)$ with slope m. We call this equation the **slope-intercept form equation** for that line.

Slope-Intercept Form (Equation for a Line)

Given a nonvertical line in the xy-plane with slope m that passes through the point $(0, b)$, the slope-intercept form equation for the line is

$$y = mx + b$$

The number b is called the **y-intercept** of the line.

Notice that this equation is exactly the same as the equation we developed at the end of the last section, given two linearly related quantities x and y with constant average rate of change of y with respect to x being m.

EXAMPLE 2 Using Slope-Intercept Form

(a) Write the equation $3y - 2x = 6$ in slope-intercept form and sketch a graph of the equation.
(b) Write an equation for the line with slope ½ that passes through the point (2, 4).
(c) The math department buys a new calculator for \$130. Three years later, its value is \$70. Assuming *linear depreciation* (calculator is losing value in a linear fashion), what is the value of the calculator 6 months after it is purchased? How about after one year? What does this mean graphically?

Solution (a) Writing the given equation in slope-intercept form, we have

$$\begin{aligned} 3y - 2x &= 6 \\ 3y &= 2x + 6 \\ y &= \frac{2x+6}{3} \\ y &= \frac{2}{3}x + 2 \end{aligned}$$

Figure 5 shows the graph of the equation $3y - 2x = 6$.

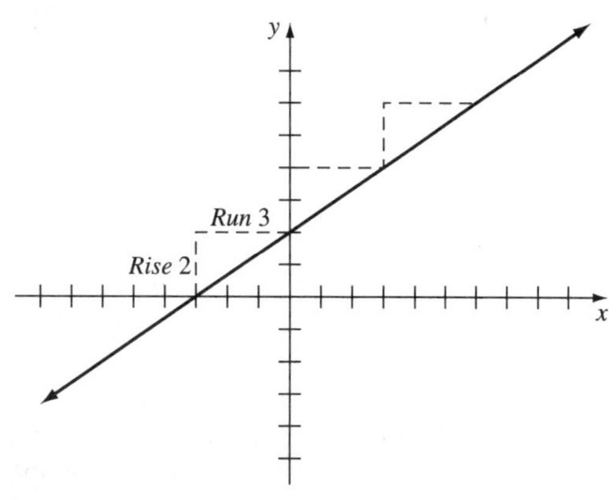

Figure 5

(b) We will write an equation for the line in slope-intercept form. Since $m = \frac{1}{2}$ we have

$$y = \frac{1}{2}x + b$$

for some number b. Now, we will find that number b, the y-intercept. Notice that we can do so by substituting in the x and y values of 2 and 4, respectively. We get

$$\begin{aligned} 4 &= \frac{1}{2}(2) + b \\ 4 &= 1 + b \\ 3 &= b \end{aligned}$$

Therefore, an equation for the line with slope ½ that passes through the point (2, 4) is

$$y = \frac{1}{2}x + 3$$

(c) Let x represent the number of years since the calculator was purchased, and let y represent the corresponding value of the calculator. Since the values of x and y are related in a linear fashion, we can represent the relationship between these two quantities by an equation of the form $y = mx + b$. Since we know that $y = 130$

when $x = 0$, we have $y = mx + 130$ for some number m. We note that when $x = 3, y = 70$. So,

$$m = \frac{\Delta y}{\Delta x} = \frac{70 - 130}{3 - 0} = -20$$

Consequently, $y = -20x + 130$, and

- after six months (½ year), the value is $-20(½) + 130 = \$120$
- after one year, the value of the calculator is $-20(1) + 130 = \$110$.

The number m calculated above is the average rate of change of y with respect to x. The equation $y = -20x + 130$ indicates that a graph of the value of the calculator (y) against years since purchase (x) is a line with slope -20 and y-intercept 130. Also, the points $\left(\frac{1}{2}, 120\right)$ and (1, 110) lie on that line.

EXAMPLE 3 More on Intercepts

This will be discussed in more detail in Section 1.4, but if the graph of an equation in the xy-plane intersects the x-axis at the point $(a, 0)$, then we say that the number a is an x-**intercept** for the graph. Clearly, the graph of a non-horizontal line in the xy-plane can have at most one x-intercept.

For instance, refer to Figure 5, which shows the graph of the line given by the equation $3y - 2x = 6$. From the graph, it appears that this line has x-intercept -3. We can verify this algebraically, as the point of intersection with the x-axis has y-coordinate $y = 0$. So, we have

$$3(0) - 2x = 6$$
$$-2x = 6$$
$$x = -3$$

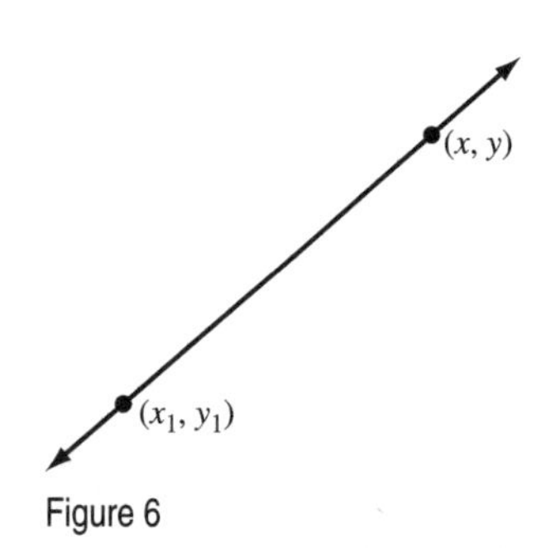

Figure 6

While slope-intercept form is a common way to write an equation for a line, it is not the only way. Consider the line in Figure 6, where (x, y) represents an arbitrary point on the line in the coordinate plane, and (x_1, y_1) is some fixed point on the line.

The slope of this line is $m = \frac{\Delta y}{\Delta x} = \frac{y - y_1}{x - x_1}$

Multiplying both sides of the above equation by $x - x_1$ gives

$$y - y_1 = m(x - x_1).$$

This equation is called the **point-slope form equation** for the given line.

Point-Slope Form (Equation for a Line)

Given a nonvertical line in the xy-plane with slope m that passes through the point (x_1, y_1), the point-slope form equation for the line is

$$y - y_1 = m(x - x_1)$$

We most often use the point-slope form of an equation for a line in the following situations:

- When we are given the slope and a point on the line. Often we then rewrite the equation in slope-intercept form.
- When we are given two points on the line. Since only two points are needed to determine the slope of a line, we are able to find the slope. Our task is then to use the slope that we find along with one of the points (it doesn't matter which) to find the equation for the line. Again, we will most likely find it useful to then write the equation in slope-intercept form.

EXAMPLE 4 Using Point-Slope Form

(a) Find the point-slope form equation for the line with slope 3 that passes through the point $(-5, 1)$, and then find slope-intercept form equation for the line.

(b) The table below gives the velocity, v, of a softball t seconds after being thrown straight up in the air. Find a formula for v in terms of t. Interpret the value of v predicted when $t = 0$ and the average rate of change.

Velocity of a softball t seconds after being thrown in the air

t (sec)	1	2	3	4
v (ft/sec)	58	26	-6	-38

(c) Find equations for each line shown in Figure 7 and interpret the slopes in terms of temperature.

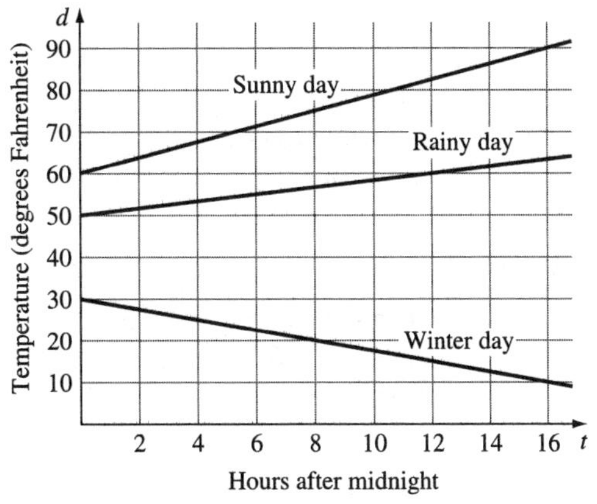

Figure 7

(d) Sales of a cosmetics company increased linearly from \$550,000 in 1992 to \$2.5 million in 2002. Find an equation that expresses the sales, y, in year x, where $x = 0$ corresponds to 1992, and use the equation to estimate the sales in 1999.

Solution (a) Since the slope and a point on the line are given, we use the point-slope form of an equation for a line, $y - y_1 = m(x - x_1)$. Using 3 for the slope and the point $(-5, 1)$ for (x_1, y_1) gives

$$y - 1 = 3(x - (-5))$$
$$y - 1 = 3(x + 5)$$
$$y - 1 = 3x + 15$$
$$y = 3x + 16$$

(b) Notice that in the data table, for each second of time that passes by, the velocity decreases by 32 feet per second. Since the rate of decrease in velocity is constant, we can model the behavior by a linear equation using a slope of -32 (this constant rate of decrease is the slope). We are also given four data points in the table for (t, v): $(1, 58)$, $(2, 26)$, $(3, -6)$ and $(4, -38)$. Since we only need one point to use the point-slope form of equation for a line, any point can be chosen. We arbitrarily choose $(1, 58)$, obtaining

$$v - 58 = -32(t - 1)$$
$$v - 58 = -32t + 32$$
$$v = -32t + 90$$

So, the softball started out (at time $t = 0$) traveling 90 feet per second and its velocity is decreasing at a rate of 32 feet per second each second.

(c) On a typical sunny day, the temperature starts out at 60 degrees and has a constant rate of increase (slope) of 30 degrees/16 hours = 15/8 or 1.875 degrees per hour. Therefore, an equation of the line is

$$d = \frac{15}{8}t + 60$$

On a rainy day, the temperature starts out at 50 degrees and has a constant rate of increase (slope) of 10 degrees/12 hours = 5/6 (about 0.83) degrees per hour. Therefore, an equation of the line is

$$d = \frac{5}{6}t + 50$$

On a winter day, the temperature starts out at 30 degrees and has a constant rate of decrease (slope) of -10 degrees/8 hours = $-5/4$ or -1.25 degrees per hour. Therefore, an equation of the line is

$$d = -\frac{5}{4}t + 30$$

Notice that slope here is negative, indicating that the temperature is decreasing at the described rate throughout the day.

(d) If we set 1992 as year $x = 0$ and put the sales y in thousands of dollars, then we have the points $(0, 550)$ and $(10, 2500)$ on the line that is the graph of y against x. Given these two points, we are now able to find an equation of the line modeling the data using the point-slope form. First, we find the slope, m.

$$m = \frac{y_2 - y_1}{x_2 - x_1} = \frac{2500 - 550}{10 - 0} = \frac{1950}{10} = 195$$

The slope indicates that sales are increasing at a rate of \$195,000 per year. We could use either of the points with the slope that we found to find the equation of the line and we would get the same result. However, since we have an initial value for y, we see that

$$y = 195x + 550$$

Since 1999 is 7 years after our beginning year of 1992, we find the value of y when $x = 7$, yielding

$$y = 195(7) + 550$$

$$y = 1915$$

We opted to let y represent thousands of dollars in sales, so the estimated sales in 1999 are \$1,915,000. ❖

From the slope-intercept form of equation for a line, we see that a horizontal line (slope $m = 0$) that passes through the point $(0, b)$ has an equation of the form $y = b$. This reinforces the fact that every point on the line has y-coordinate b, irrespective of x-coordinate.

Similarly, each point of a vertical line that passes through the point $(a, 0)$ has x-coordinate a, irrespective of y-coordinate. Consequently, the graph of the equation $x = a$ is exactly that line. Notice that no slope-intercept (or point-slope) form of equation for a vertical line exists, as the slope of a vertical line is undefined.

We briefly summarize some basic equations for graphs of lines in the xy-plane.

Equations for Lines

- Vertical line through $(a, 0)$: $x = a$
- Horizontal line through $(0, b)$: $y = b$
- Slope-intercept form: $y = mx + b$
- Point-slope form: $y - y_1 = m(x - x_1)$

As we have seen, the defining characteristic of linearly related quantities is a constant average rate of change. Graphically, this resulted in using lines to describe how the quantities were related. However, many situations we encounter show a greater degree of variability in that two quantities of interest may be related in a nonlinear fashion. In the next section, we investigate the notion of using lines and linear equations to *approximately* represent nonlinear relationships.

PROBLEM SET 1.2

For problems 1–5 determine whether each statement is true or false. If it is false, correct the statement to make it true.

1. The slope of a line is another way of describing the average rate of change.
2. The graph of $2y - 8 = 3x$ has a y-intercept of 4.
3. The lines $3x + 4y = 12$ and $4x + 3y = 12$ are perpendicular.
4. Slope is not defined for horizontal lines.
5. The line $3x - y = 4$ rises more steeply from left to right than the line $4x - y = -1$.

Graphing a linear equation is often easier if the equation is in slope-intercept form. If possible, rewrite the equations in problems 6–10 in this form.

6. $2x + 9y = 18$
7. $y - 0.5 = 3(x - 0.1)$
8. $7(x + y) = 6$
9. $4x - 2y + 1 = 0$
10. $6y = 18$
11. Find equations for the lines with the indicated properties. Use slope-intercept form.
    - **(a)** Slope 5 and y-intercept 6
    - **(b)** Slope -3 and x-intercept 8
    - **(c)** passes through the points $(-1, 6)$ and $(3, -2)$
    - **(d)** has x-intercept 5 and y-intercept -7
    - **(e)** slope ¾ and passes through the point $(-2, 12)$
    - **(f)** has x-intercept -4 and is parallel to the line with equation $3x-8y=36$
12. Figure 8 shows five different line segments. Without using a calculator, match each line segment to one of the following equations.

$$y_1 = 15 + 2x \qquad y_4 = 70 - x$$
$$y_2 = 15 + 4x \qquad y_5 = 70 - 2x$$
$$y_3 = 2x - 20$$

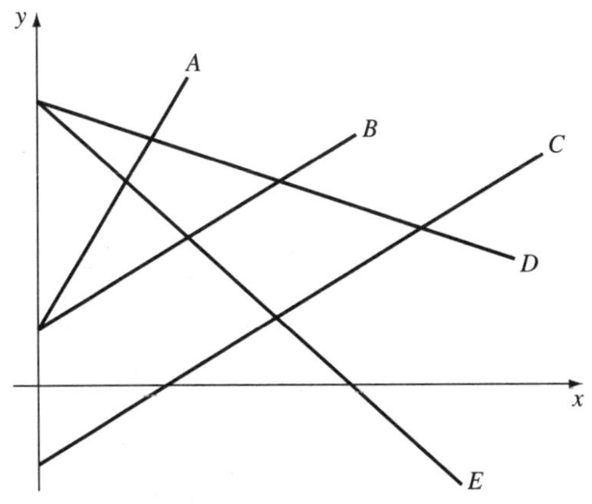

Figure 8 (not drawn to scale)

13. Write equations for the lines in the xy-plane with the indicated characteristics.
    - **(a)** vertical, passes through $(-4, 0)$
    - **(b)** horizontal, passes through $(0, 9)$
    - **(c)** vertical, intersects a horizontal line at $(0, 3)$
    - **(d)** horizontal, intersects a vertical line at $(-1, 5)$
14. Write equations for the lines with the indicated characteristics. Use point-slope form.
    - **(a)** passes through the points $(5, -1)$ and $(-5, 5)$
    - **(b)** passes through the points $(1, 1)$ and $(6, -¾)$
    - **(c)** passes through $(8, -2)$ and is parallel to the line with equation $y = \frac{3}{4}x - 12$
    - **(d)** passes through $(-1, 1)$ and is perpendicular to the line with equation $y = -\frac{1}{8}x + 10$
15. Without using a calculator, match the following formulas to the lines shown in Figure 9.

$$y_1 = 3 + 3x$$
$$y_2 = -5 + 3x$$
$$y_3 = 3 + 4x$$
$$y_4 = 3 - 3x$$
$$y_5 = 3 - 4x$$

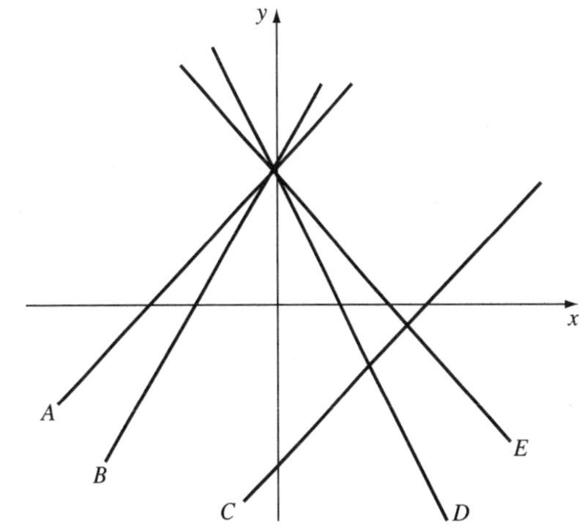

Figure 9 (not drawn to scale)

16. Use slopes to determine if the points $(9, 6)$, $(-1, 2)$, and $(1, -3)$ are the vertices of a right triangle.

17. Line l is given by $y = 5 - \frac{1}{3}x$ and point p has coordinates (4, 3).
   (a) Find an equation of the line containing p that is parallel to l.
   (b) Find an equation of the line containing p that is perpendicular to l.
   (c) Graph the equations from parts **(a)** and **(b)** on the same set of axes.

18. You need to rent a van and compare the charges of three different companies A, B, and C, in terms of x, the number of miles driven in one day. Company A charges 20 cents per mile plus $40 per day, company B charges 10 cents per mile plus $60 per day, and company C charges $110 per day with no mileage charges.
   (a) Find formulas for the cost of driving vans rented from companies A, B, and C, in terms of x.
   (b) Graph the cost for each company for $0 \leq x \leq 600$ and $0 \leq y \leq 150$. Put all three graphs on the same set of axes.
   (c) What do the slope and the y-intercept tell you in each situation?
   (d) Use the graph in part **(b)** to find the circumstances under which company A would be the cheapest. Do the same for company B and company C.

19. The cost of a Hotbox oven is $840 and it depreciates $40 each year. The cost of an Equator-bake oven is $1,400 and it depreciates $75 per year.
   (a) If a Hotbox and an Equator-bake are bought at the same time, when do the two ovens have equal value?
   (b) If the depreciation continues at the same rate, what happens to the values of the ovens in 20 years time?

20. John wants to buy a dozen muffins. The local bakery sells blueberry and poppy seed muffins for the same price.
   (a) Make a table of all the possible combinations of muffins if he buys a dozen, where b is the number of blueberry muffins and p is the number of poppy seed muffins.
   (b) Find a formula for p in terms of b.
   (c) Graph your equation from above.

21. A theme park manager graphed weekly profits in terms of the number of visitors, and found that the relationship was linear. One week the profit was $990,000 when 132,400 visitors attended. Another week 174,525 visitors produced a profit of $1,327,000.
   (a) Find an equation for weekly profit, p, in terms of the number of visitors, n.
   (b) Interpret the slope and the p-intercept.
   (c) What is the number of visitors for which there is zero profit?
   (d) Find a formula for the number of visitors in terms of profit.
   (e) If the weekly profit was $125,000, how many visitors attended the theme park?

22. According to one economic model, the demand for gasoline in terms of price is a linear equation. If the price of gasoline is $p = \$1.30$ per gallon, then the quantity demanded in a fixed period is $q = 50$ gallons. If the price rises to $1.55 per gallon, then the quantity demanded falls to 40 gallons in that period.
   (a) Find a formula for q in terms of p.
   (b) Explain the economic significance of the slope used to generate your formula.
   (c) Explain the economic significance of the q-axis and p-axis intercepts.

Interlude: Progress Check 1.2

Sandy has been paying down her credit card balance over the last several months. The accompanying table shows her card balance when she received her monthly statement over that period of time. Here, $m = 1$ corresponds to January.

Months (m)	1	2	3	4	5	6
Balance (B)	$9650	$9494.75	$9187.20	$8726.50	$8094.25	$7354.40

1. Explain why m and B are not linearly related.

2. On the scatterplot below (from the data above), sketch two lines, one through the first and last points on the plot and one through the last two points on the plot. Calculate the slope of each line.

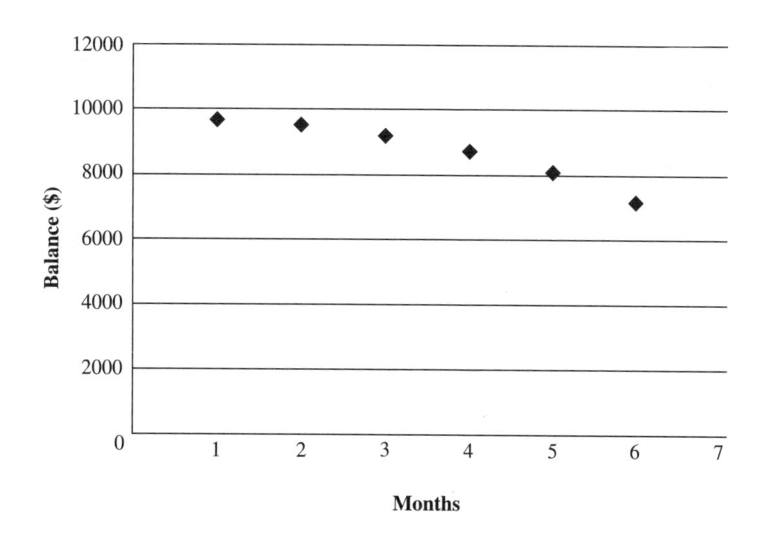

3. Interpret the slopes you found in **2**. in terms of the credit card balance and time. Label each appropriately.

4. Find equations for the lines from **2**. Use each equation to predict the credit card balance 8 months into the year. Which prediction do you think is more accurate? Explain your choice.

OBJECTIVES

1. Determine Whether or Not a Line Properly Fits Given Data
2. Utilize Graphing Calculators to Create: Scatterplots, Lines of Best Fit, Linear Models, and Correlation Coefficients
3. Properly Describe and Apply Displayed Results

1.3 Linear Models

Consider the following questions.

"Do large doses of vitamin C decrease the number of cases of colds?"

"What effect does the number of sports articles in the newspaper have on daily readership?"

"How much does an increase in wages affect overall company satisfaction?"

Each of the questions above refers to the relationship between two quantities. In order to answer these questions, sufficient data must be made available to construct mathematical models that accurately represent the data. In this section, we will consider those applications where the relationship between two quantities can be reasonably modeled by a linear equation. Real data may not perfectly conform to a linear relationship, but none-the-less we may still be able to use it to help analyze the data. Other mathematical models will follow in later sections.

Lines of Best Fit

Consider the following table of data that shows the average salary of professional baseball players from 1991 to 1996.

Year	1991	1992	1993	1994	1995	1996
Salary (millions)	\$0.85	\$1.03	\$1.08	\$1.17	\$1.11	\$1.12

We will let t represent the year, with $t = 0$ corresponding to 1990. Then we plot the points (1, 0.85), (2, 1.03), and so on. Plotting the points creates a *scatterplot* (or *scattergram*) with which we can examine the data and potentially identify an appropriate mathematical model that reasonably *approximates* the given data. Figure 1 gives a scatterplot of the average salary data.

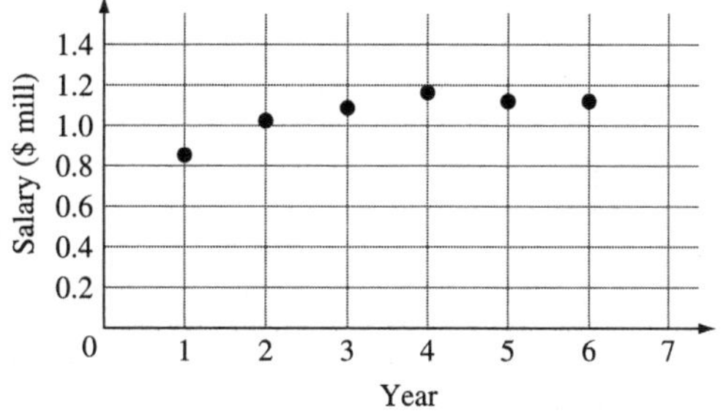

Figure 1

The plotted points do not fall on a straight line. However, since the points do seem to be reasonably close to falling on a line, we may be able to "fit" a line to the scatterplot that will do an acceptable job of approximating how the average salary has changed over the indicated time period. Below are three attempts to "fit" a line to the given points. An equation for the line has been given in each case.

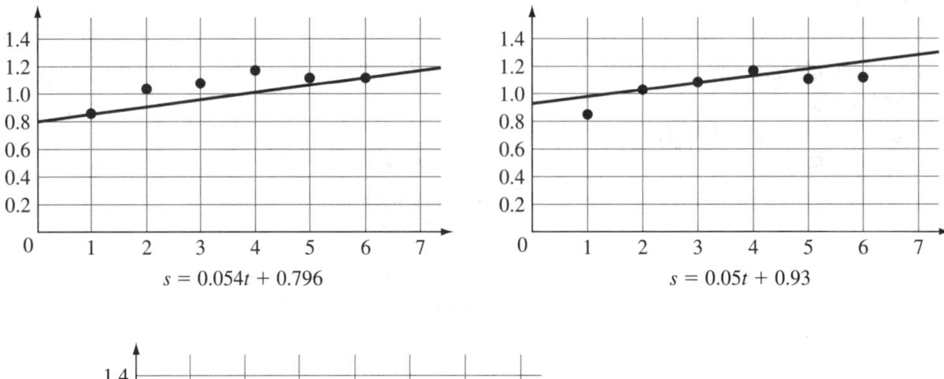

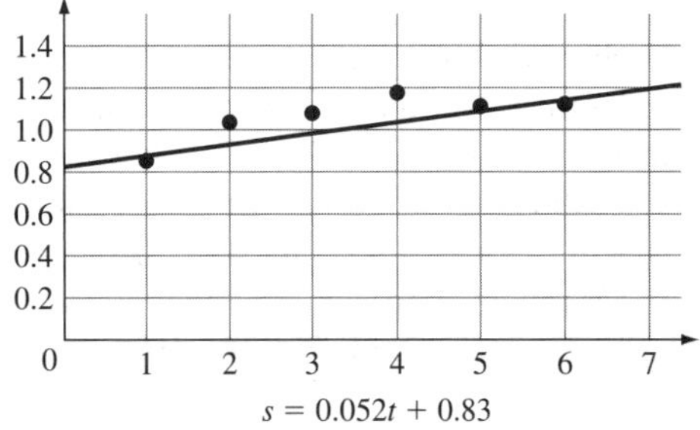

Fitting a line to a plotted set of data is called *linear regression*. There are several ways to find such a line. From inspection, it is likely quite clear how the first two lines were found. The third line is "in between" the first two, with a general goal of being "close" to each point, even if it doesn't go through any point. A very natural question to consider is which of those three lines is the "best fit", or, perhaps, if there is some other line that is a "better fit".

Consider the following table, which shows the average annual salaries predicted by the three linear models above as well as the average salaries actually recorded.

Year (t)	Actual s	$s = 0.054t + 0.796$	$s = 0.05t + 0.93$	$s = 0.052t + 0.83$
1	0.85	0.85	0.98	0.882
2	1.03	0.904	1.03	0.934
3	1.08	0.958	1.08	0.986
4	1.17	1.012	1.13	1.038
5	1.11	1.066	1.18	1.09
6	1.12	1.12	1.23	1.142

The question of "best fit" is often very difficult and is highly dependent on the proposed use of the resulting linear model. One mathematical way to decide which line is "best" is to sum the measured errors between the proposed linear model and the actual values displayed in the data. Whichever model produces the smallest sum of errors could be considered the "best" for the given data. Notice what happens when this approach is applied to two of the models proposed above.

Year (t)	Actual s	Error via $s = 0.05t + 0.93$	Error via $s = 0.052t + 0.83$
1	0.85	$0.98 - 0.85 = 0.13$	$0.882 - 0.85 = 0.032$
2	1.03	$1.03 - 1.03 = 0$	$0.934 - 1.03 = -0.096$
3	1.08	$1.08 - 1.08 = 0$	$0.986 - 1.08 = -0.094$
4	1.17	$1.13 - 1.17 = -0.04$	$1.038 - 1.17 = -0.132$
5	1.11	$1.18 - 1.11 = 0.07$	$1.09 - 1.11 = -0.02$
6	1.12	$1.23 - 1.12 = 0.11$	$1.142 - 1.12 = 0.022$
		Sum of errors: 0.27	Sum of errors: -0.288

How should we interpret these results? Positive and negative errors may cancel each other out and become more misleading than helpful, as in each case above the *magnitude* of the sum of the errors is very close, even though the model $s = 0.052t + 0.83$ does not agree with a single actual value of s. Instead, we might opt to use the *sum of the squares* of the errors, as this will eliminate negative values that may wind up concealing the size of the error. In addition, this approach will emphasize the large errors and minimize the smaller errors (because the square of the error is less than the error if the error is less than one). Graphically, this amounts to summing the squares of the vertical distances between the data points and the proposed line of best fit.

Year (t)	Actual s	Squared error via $s = 0.05t + 0.93$	Squared error via $s = 0.052t + 0.83$
1	0.85	$(0.98 - 0.85)^2 = 0.0169$	$(0.882 - 0.85)^2 = 0.001024$
2	1.03	$(1.03 - 1.03)^2 = 0$	$(0.934 - 1.03)^2 = 0.009216$
3	1.08	$(1.08 - 1.08)^2 = 0$	$(0.986 - 1.08)^2 = 0.008836$
4	1.17	$(1.13 - 1.17)^2 = 0.0016$	$(1.038 - 1.17)^2 = 0.017424$
5	1.11	$(1.18 - 1.11)^2 = 0.0049$	$(1.09 - 1.11)^2 = 0.0004$
6	1.12	$(1.23 - 1.12)^2 = 0.0121$	$(1.142 - 1.12)^2 = 0.000484$
		Sum of squares of errors: 0.0355	Sum of squares of errors: 0.037384

Using this approach, it appears that the second model is a better fit than the third, although the difference in sums of squared errors is very slight. Moreover, although it is beyond the scope of this text, it can be shown for that any given set of data there exists exactly one linear model such that the sum of the squares of the errors is as small as possible. This model is called the *least-squares regression line*, and most graphing calculators have programs for finding this line. In addition, most graphing calculators are able to produce various types of data plots, including scatterplots (see your owner's manual for details).

EXAMPLE 1 **Using a Graphing Calculator to Find Linear Models**

Twenty people were randomly selected for a survey that asked for their education level and the number of books they read each year. The results are as follows:

Education Level	Number of Books
10	2
11	2
11	4
12	4
13	5
13	4
13	6
13	6
14	7
15	8
15	9
16	10
17	12
17	13
17	15
19	15
19	14
19	16
19	18
21	20

Note: Education Levels are defined to be ...

HS Graduate: 13
2 Year College: 15
4 Year College: 17
Masters: 19
Doctorate: 21

(a) Make a scatterplot of the data, with education level on the horizontal axis and number of books read on the vertical axis.

(b) Let x represent the education level and let y represent the number of books read. Without graphing, which of the following models seems to make the most sense for this data? Why?

i. $y = x + 5$

ii. $y = \frac{3}{2}x - 15$

iii. $y = \frac{1}{2}x - 10$

iv. $y = 2x - 20$

(c) Find the equation for the least-squares regression line for this data.
(d) Interpret the meaning of the slope of the least-squares regression equation.
(e) What are some of the limitations of this model, if any?

Solution (a)

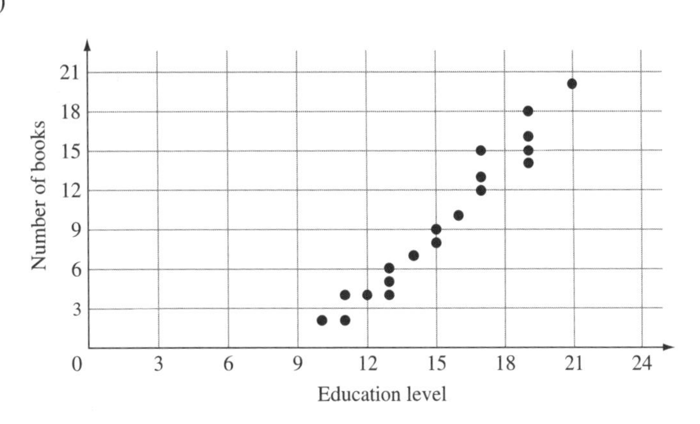

(b) Equation ii) seems to be the most reasonable. A rough estimate of the slope of a line of best fit is between one and two.
(c) Following the directions for linear regression on your specific calculator should yield the equation $y = 1.685x - 16.119$, rounding coefficients to three decimal places.
(d) The slope of 1.685 indicates that for every one-unit increase in education level, an additional 1.685 books are read annually.
(e) The model does not seem to be of much value for education levels below 10. Also, while the data does show a linear trend, it does not show that there is a *causal* relationship between education level and reading patterns. Varied results may occur with gender, geographical location, career path, and so on. As seen above, the y-intercept does not have any practical interpretation. ❖

Correlation

Most graphing calculators also calculate a number called the **correlation coefficient** when calculating the least-squares regression line. This number is typically denoted by r, and $-1 \leq r \leq 1$. The correlation coefficient is a measure of the strength of the linear relationship between the two quantities involved. When $|r|$ is close to 1, the quantities have a very strong linear relationship, whereas if r is close to zero, there is little or no linear relationship present. Moreover, when $r = -1$ there is a *perfect negative correlation*; so the data points lie on a line with negative slope.

If $r = 1$, there is a *perfect positive correlation*; so the data points lie on a line with positive slope. Figure 2 shows some scatterplots and information on the correlation coefficient associated with the least-squares regression line.

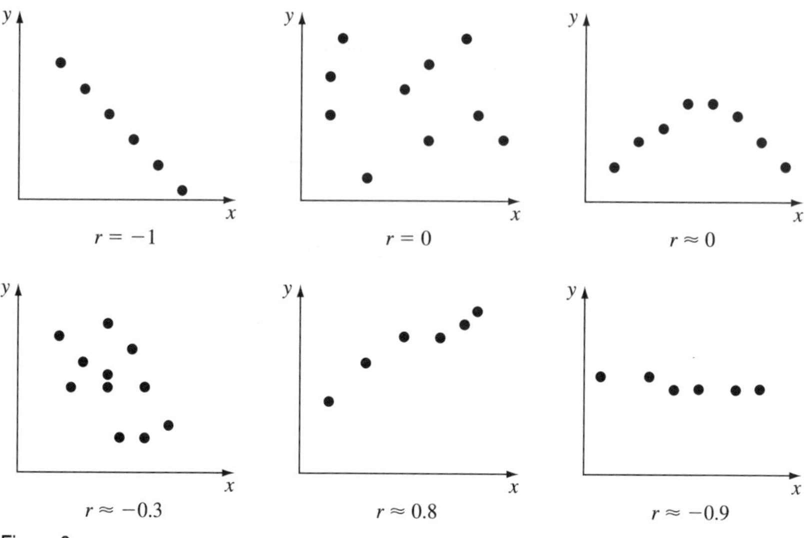

Figure 2

As noted before, even if there may be a linear relationship between two quantities, and even if the relationship is very strong, correlation and causation are quite different.

PROBLEM SET 1.3

In each of the problems where you are asked to find an equation for the least-squares regression line, round coefficients to three decimal places.

1. Table 1 shows the total retail e-commerce sales in the United States for the second quarter of various years (as reported by the US Census Bureau, 2007). (see Table 1).

TABLE 1

Year	2000	2004	2007
$Billions	90	125	200

(a) Make a scatterplot of this data. Plot year on the horizontal axis.
(b) Use a calculator to find the equation of the least-squares regression line.
(c) Interpret the slope of the regression equation.

2. A nature conservationist traced 150 otters that were born in 2002. The number of these otters, y, still alive in each subsequent year is recorded in Table 2.

TABLE 2

Year	2002	2003	2004	2005	2006	2007	2008	2009	2010
Otters	150	148	139	108	74	50	37	26	10

(a) Make a scatterplot of this data. Let x represent years, with $x = 0$ corresponding to 2000.
(b) Draw a good fitting line that you think best fits the data and estimate its equation.
(c) Use a calculator to find the equation of the least-squares regression line.
(d) Interpret the slope and each intercept of the line.

3. Table 3 shows the IQ of ten students and the number of hours of videogames each student played per week.

TABLE 3

IQ	115	107	120	150	100	145	128	108	119	112
Hours of VG	12	11	9	4	14	3	6	10	8	5

Obtained from "Functions Modeling Change: A Preparation for Calculus" by Connally, Hughes-Hallet, Gleason, et al., 2000, page 66.

(a) Make a scatterplot of this data. Plot IQ on the horizontal axis.
(b) Make a rough estimate of the correlation coefficient.
(c) Use a calculator to find the equation of the least-squares regression line and the correlation coefficient.

4. The data on preferred lead foot strength (lbs per square inch) in Table 4 was collected from high school seniors using a starting block strength meter.

TABLE 4

Preferred foot	**27**	**26**	**42**	**18**	**38**	**45**	**27**	**58**	**51**	**23**
Non-preferred foot	24	26	43	22	40	45	26	46	46	22
Preferred foot	**51**	**53**	**48**	**44**	**37**	**25**	**24**	**29**	**32**	**31**
Non-preferred foot	47	47	41	44	33	20	27	30	29	29

(a) Make a scatterplot of this data treating the strength of the preferred foot as the independent quantity, x, and the strength of the non-preferred foot as the dependent quantity, y.
(b) Draw a line on your scatterplot that is a good fit for the data and use it to find an approximate linear equation.
(c) Use a calculator to find the equation of the least-squares regression line.
(d) What would the predicted foot strength in the non-preferred foot be for a student with preferred foot strength of 37?
(e) Why is the correlation coefficient, r, both positive and close to 1?
(f) Why do the points tend to "cluster" into two groups?

5. In baseball, Henry Aaron holds the record for the greatest number of home runs hit in the major leagues. Table 5 shows his cumulative yearly record from the start of his career in 1954 until 1973.

TABLE 5

Year	1954	1955	1956	1957	1958	1959	1960	1961	1962	1963	1964	1965	1966	1967	1968	1969	1970	1971	1972	1973
Home runs	13	40	66	110	140	179	219	253	298	342	366	398	442	481	510	554	592	639	673	713

Adapted from "Graphing Henry Aaron's Home-Run Output" by H. Ringel, The Physics Teacher, January 1974, page 43.

(a) Plot Aaron's cumulative number of home-runs, H, on the vertical axis, and the time, t, in years, along the horizontal axis, where $t = 0$ corresponds to 1954.

(b) Draw a straight line that you believe fits the data well and estimate its equation.

(c) Use a calculator to find the equation of the least-squares regression line. What is the correlation coefficient? What does this tell you?

(d) What does the slope of the regression line mean in terms of Henry Aaron's home-run record?

(e) From your answer to part **(c)**, how many home-runs do you estimate Henry Aaron hit in each of the years 1974, 1975, 1976, and 1977? If you were told that Henry Aaron retired at the end of the 1976 season, would this affect your answers?

6. Table 6 shows men's and women's world records for swimming distances from 50 meters to 1500 meters.

TABLE 6

Distance (meters)	50	100	200	400	800	1500
Men (seconds)	21.81	48.21	106.69	223.80	466.00	881.66
Women (seconds)	24.51	54.01	116.78	243.85	496.22	952.10

Data from "The World Almanac and Book of Facts: 1996" edited by R. Famighetti, Funk and 'Wagnalls, New Jersey, 1995.

(a) What values would you add to Table 6 to represent the time taken by both men and women to swim 0 meters?

(b) Plot the women's times against distance, with time t in seconds on the vertical axis and distance d in meters on the horizontal axis. It is claimed that a straight line models this behavior well. What is the equation for that line? What does its slope represent?

(c) On the same graph, plot the men's times against distance and find the equation of the straight line that models this behavior well. Is this line steeper or flatter than the women's line? What does that mean in terms of swimming? What are the values of the vertical intercepts? What practical interpretation do these values have?

(d) On another graph, plot the women's times against the men's times, with women's times, w, on the vertical axis and men's times, m, on the horizontal axis. It should look linear. How could you have predicted this linearity from the equations you found in part **(b)** and **(c)**? What is the slope of this line and how can it be interpreted?

(e) A newspaper reporter claims that the women's records are about 8% slower than the men's. Do the facts support this statement? What is the value of the vertical intercept? What practical interpretation does this value hold?

7. Fran exercised for 30 minutes and then measured her ten-second pulse count at one-minute intervals as she rested. The data is shown in Table 7, where time is in minutes after exercise.

TABLE 7

Time (min)	0	1	2	3	4	5
Pulse	26	22	19	16	14	14

(a) Make a scatterplot of this data.

(b) Because the pulse values are the same after 4 and 5 minutes, this data is certainly not linear. Are there values of t for which the least-squares regression line is a good model for the data? If so, what values?

(c) Compare the r-values (correlation coefficients) when leaving out the last data point to when keeping it in the data set.

8. Record the height and arm-span length of a least 10 people who live with you, sit next to you in class, or work with you (i.e., people who are willing to be measured) and make a data table.

(a) Make a scatterplot of this data, with height on the y-axis and arm-span on the x-axis.

(b) If the data appears to be linear, sketch a line that fits the data and estimate a linear equation for it.

(c) Use a graphing calculator to find the equation of the least-squares regression line.

(d) Find and interpret r, the correlation coefficient.

9. Similar to Problem 7, collect your own data in a table like Table 7.

   (a) Make a scatterplot of this data, with pulse on the y-axis and time on the x-axis.

   (b) If the data appears to be linear, sketch a line that fits the data and estimate a linear equation for it.

   (c) Use a graphing calculator to find the equation of the least-squares regression line.

   (d) Find and interpret r, the correlation coefficient.

10. An initial investment of $13,000 earns interest monthly as summarized in Table 8.

   (a) Let V be the value of the investment after t months. Make a similar table showing t and V.

   (b) Plot the data points you got from part **(a)**.

   (c) Find the equation of the least-squares regression line with the data points from part **(a)**.

   (d) Use the equation of the line to predict the value of the investment after 12 and 60 months. Are these reasonable predictions?

TABLE 8

Month	1	2	3	4	5	6	7	8
Interest	$92.36	$93.18	$93.99	$94.82	$95.64	$96.48	$97.30	$98.11

Interlude: Progress Check 1.3

Consider the table below, which shows the total highway registered motorcycles in the the U.S. reported in the *2012 Statistical Abstract of the United States*.

Year	2004	2005	2006	2007	2008	2009
Thousands of motorcycles	5781	6227	6679	7138	7753	7930

1. Graph the data, with year represented on the horizontal axis and thousands of motorcycles represented on the vertical axis (let $x = 0$ correspond to 2004). Do the data points appear to *almost* lie on a straight line?

2. Find equations for three lines that come "close" to fitting the data points. **Do not** attempt to find the least-squares regression line. Indicate how you arrived at the equations. Which of the three do you believe is the "best fit"? Explain your choice.

3. Use a calculator to find an equation for the least-squares regression line for this data and graph this line with your data points. Does the line appear to fit the data reasonably well? How does it compare to the lines you found above?

4. Use the equation for the least-squares regression line to predict the number of highway registered motorcycles in 2010 and in 2015. Do the same for 2000. Do you think those predictions are reasonable? Why or why not?

OBJECTIVES

1. Graph Equations by Point-Plotting
2. Determine Intercepts and Symmetry of a Graph
3. Graph Equations Using Graphing Calculators
4. Solve Problems Using Graphs

1.4 Graphing Equations

As we have already seen, illustrations such as scatterplots and lines of best fit play an important role in helping us investigate the relationship between two quantities. In the case where the relationship between two quantities is described by a two-variable equation, it is often desirable to represent this relationship geometrically with a **graph**. We restate a definition from earlier.

Graph of an Equation

The graph of an equation in the variables x and y consists of all points in the xy-plane whose coordinates (x, y) satisfy the equation.

Graphing by Point-Plotting

A common technique for obtaining a sketch of the graph of an equation in two variables is to first plot several points that lie on the graph and then connect the points with a smooth curve. This is called, naturally, *point-plotting*.

EXAMPLE 1 Graphing an Equation by Plotting Points

Sketch the graph of $y = x^2$.

Solution Since the given equation clearly shows how values of y are related to values of x, it seems reasonable to start by assigning several different numbers to x and then find the corresponding values of y to get points that lie on the graph. We then plot these points and connect them with a smooth curve. See Figure 1.

x	y	Point on Graph
-3	9	$(-3, 9)$
-2	4	$(-2, 4)$
-1	1	$(-1, 1)$
0	0	$(0, 0)$
1	1	$(1, 1)$
2	4	$(2, 4)$
3	9	$(3, 9)$

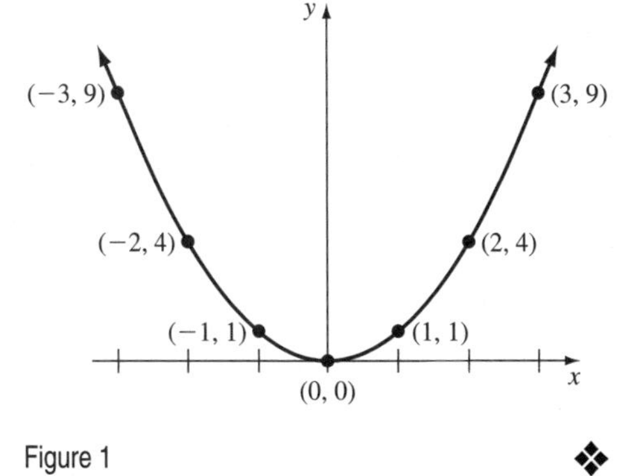

Figure 1 ❖

Notice that the sketch doesn't show *all* points on the graph, but it does establish a continuing pattern. If a sketch of the graph of an equation shows enough of the graph so that the viewer is able to "see" the rest of the graph as a continuation of an established pattern, we often call the sketch *complete*. So, when seeking such a sketch, one approach that might be taken is to plot a sufficient number of points so that a pattern becomes evident and then connect the points by a smooth curve. However, it is not always clear how many points are sufficient. Some knowledge about the given equation and what characteristics to expect the graph of the equation to have is certainly helpful. For instance, we know that the graph of any equation of the form $y = mx + b$ is a straight line. Often a graphing utility can be very helpful as it has the ability to plot points very rapidly and in large quantity. For now, we will investigate some properties of the graph of an equation that can be obtained from an algebraic analysis of the equation. Later, we will return to the idea of using a graphing utility to obtain a sketch of the graph of an equation.

Intercepts and Symmetry

The points, if any, where a graph touches or crosses one of the coordinate axes are called **intercepts**. The x-coordinate of a point where a graph touches or crosses the x-axis is called an x-**intercept**. Similarly, the y-coordinate of a point where a graph touches or crosses the y-axis is called a y-**intercept**. See Figure 2.

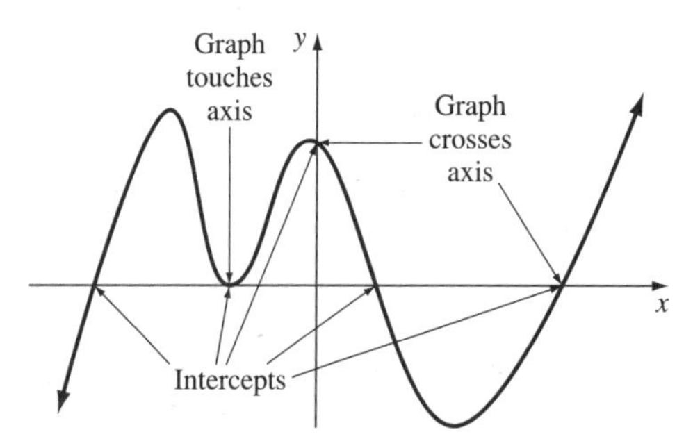

Figure 2

EXAMPLE 2 Finding Intercepts

Identify the x- and y-intercepts for the graph given in Figure 3.

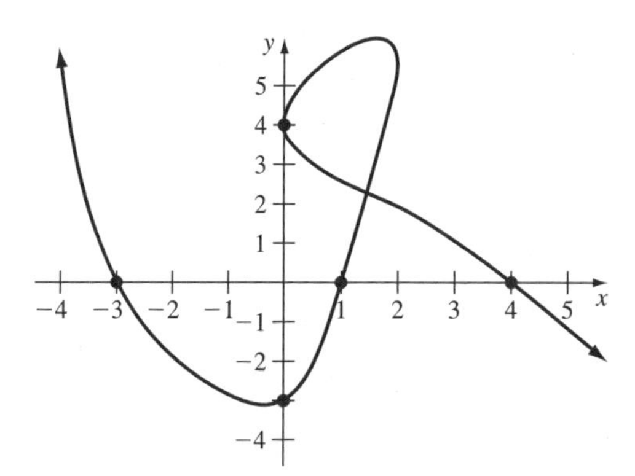

Figure 3

Solution From the graph, we see that there are two y-intercepts: -3 and 4. Similarly, there are three x-intercepts: -3, 1, and 4. ❖

The intercepts of the graph of an equation can often be located algebraically by using the fact that points on the x-axis have y-coordinates of zero and points on the y-axis have x-coordinates of zero.

Finding Intercepts

Given an equation in the variables x and y:

- To find the x-intercepts, if there are any, for its graph, set y equal to zero in the equation and solve for x.
- To find the y-intercepts, if there are any, for its graph, set x equal to zero in the equation and solve for y.

EXAMPLE 3 Finding Intercepts Algebraically

Find the intercepts of the graph of the equation $x - y^2 = -4$ and sketch its graph.

Solution We find the x-intercept(s) by setting y equal to zero, obtaining

$$x - 0^2 = 4$$
$$x = -4$$

Thus -4 is an x-intercept. Similarly, we find the y-intercept(s) by setting x equal to zero, obtaining

$$0 = y^2 - 4$$

Factoring and using the zero-product property easily solves this quadratic equation.

$$0 = (y - 2)(y + 2)$$
$$y - 2 = 0 \quad \text{or} \quad y + 2 = 0$$
$$y = 2 \quad \text{or} \quad y = -2$$

So, 2 and -2 are y-intercepts. We finish by plotting points and connecting the points with a smooth curve as in Example 1. Here, it is easier to select values of y and compute corresponding values of x. See Figure 4.

x	y	Point on Graph
5	-3	$(5, -3)$
0	-2	$(0, -2)$
-3	-1	$(-3, -1)$
-4	0	$(-4, 0)$
-3	1	$(-3, 1)$
0	2	$(0, 2)$
5	3	$(5, 3)$

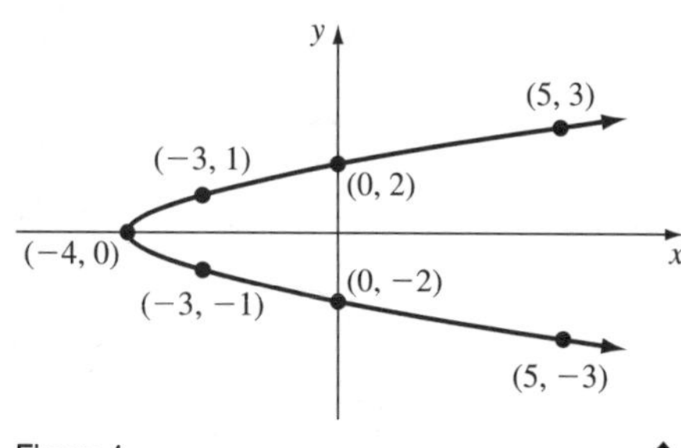

Figure 4

❖

Depending on the complexity of the equation involved, determining intercepts can be very challenging. For instance, Figure 5 shows the graph of the equation $y = 0.25x^3 - 7x + 12$ (not drawn to scale).

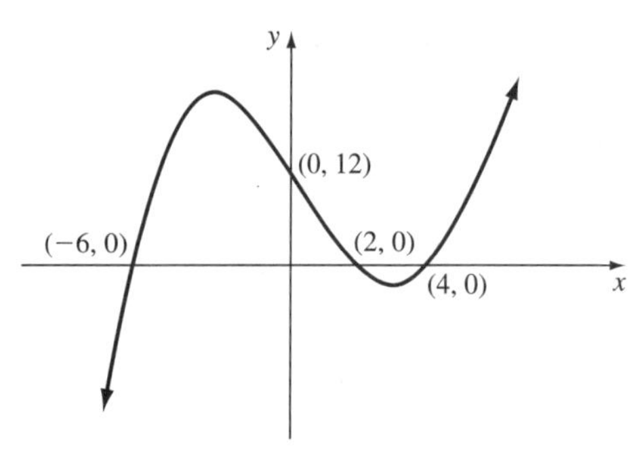

Figure 5

While it is easy to ascertain that the graph has an intercept at (0, 12), it is not readily apparent from looking at the equation

$$0 = 0.25x^3 - 7x + 12$$

that the graph also has intercepts at $(-6, 0)$, (2, 0), and (4, 0). *Cubic equations* are generally not easy to work with in regards to finding solutions using algebraic methods. In cases like this, and in fact, in many applied settings, we often rely on reasonable approximations of the intercepts of graphs. Graphing utilities typically have programs for just that task.

The graphs in Figure 1 and Figure 4 display readily evident types of *symmetry*. If the graph of Figure 1 is viewed in two pieces, one to the left of the y-axis and one to the right, the two parts are "mirror images" of each other. A similar phenomenon appears in the graph in Figure 4, if the graph is looked at as two parts divided by the x-axis.

There are many types of symmetry that a graph may display. We are particularly interested in **symmetry with respect to the y-axis, symmetry with respect to the x-axis,** and **symmetry with respect to the origin**.

Symmetry

- A graph is said to **be symmetric with respect to the y-axis** if, for every point (x, y) on the graph, the point $(-x, y)$ also lies on the graph.
- A graph is said to **be symmetric with respect to the x-axis** if, for every point (x, y) on the graph, the point $(x, -y)$ also lies on the graph.
- A graph is said to **be symmetric with respect to the origin** if, for every point (x, y) on the graph, the point $(-x, -y)$ also lies on the graph.

Figure 6 illustrates symmetry with respect to the y-axis. This is the type of symmetry displayed by the graph of the equation $y = x^2$ (see Figure 1). Figure 7 illustrates symmetry with respect to the x-axis. This is the type of symmetry displayed by the graph of the equation $x - y^2 = -4$ (see Figure 4).

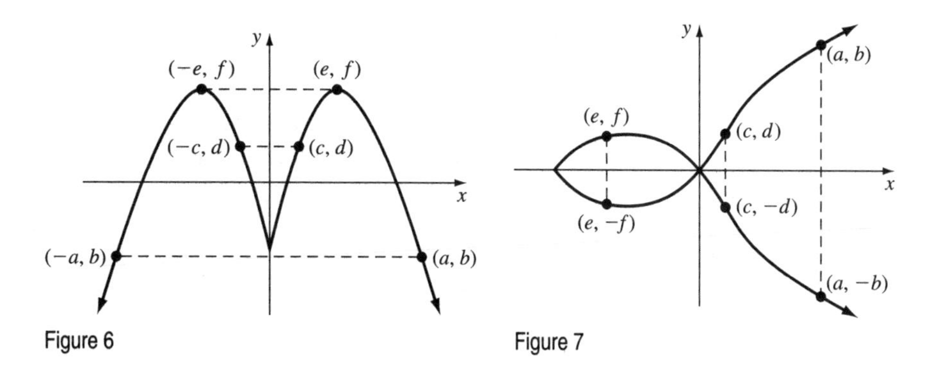

Figure 6

Figure 7

Figure 8 illustrates symmetry with respect to the origin.

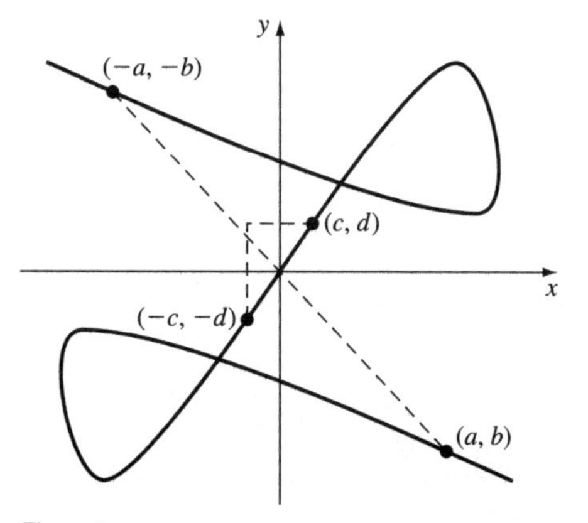

Figure 8

Symmetry with respect to the origin is often much more difficult to recognize than symmetry with respect to either coordinate axis. Notice, though, in the figure above, that symmetry with respect to the origin may be viewed as two consecutive reflections; looking at the portion of the graph above the x-axis, first reflect that portion across the y-axis, then follow with a reflection across the x-axis. See Figure 9.

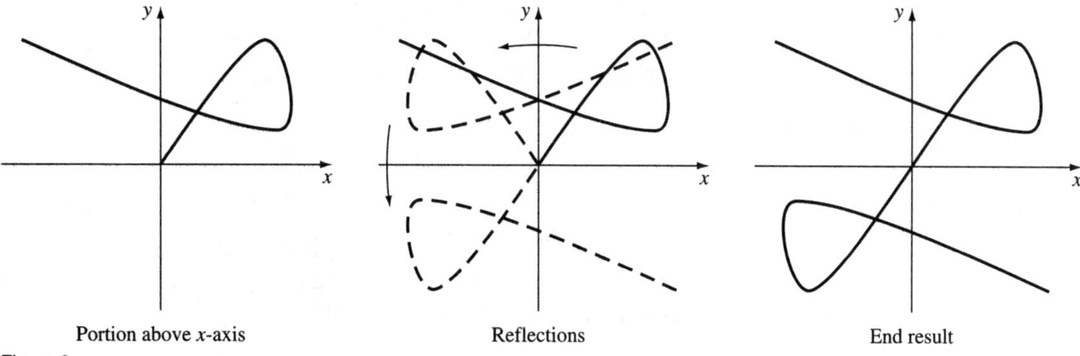

Figure 9

The definitions of the three types of symmetry we have considered suggest algebraic ways of testing an equation for the presence of one or more of these types of symmetry in its graph. For instance, when constructing the graph of $y = x^2$, a value of $x = 3$ corresponds to a value $y = 9$, while a value of $x = -3$ also corresponds to a value of $y = 9$. In fact, for any value of x, the resulting value of y found in the equation $y = x^2$ is precisely the same as when the negative of x is used to obtain y. That is,

$$y = (-x)^2 = x^2$$

We would say here that replacing x by $-x$ resulted in an *equivalent equation* (simplification will lead to the same expressions on both sides of the equation as in the original equation). Similar results occur when we look at equations whose graphs exhibit symmetry with respect to the x-axis or the origin. We summarize these below.

Algebraic Tests for Symmetry

The graph of an equation in the variables x and y exhibits symmetry with respect to

- the y-axis if replacing x by $-x$ in the equation results in an equivalent equation
- the x-axis if replacing y by $-y$ in the equation results in an equivalent equation
- the origin if replacing x by $-x$ and y by $-y$ in the equation results in an equivalent equation

EXAMPLE 4 Testing For Symmetry, Finding Intercepts, Sketching a Graph

Test the equation $4x^2 + y^2 = 16$ for the three types of symmetry, find all intercepts, and sketch its graph.

Solution First we test for symmetry with respect to the y-axis. Replacing x with $-x$ yields

$$4(-x)^2 + y^2 = 16$$
$$4x^2 + y^2 = 16$$

Thus the graph of the given equation is symmetric with respect to the y-axis. Similarly, we test for symmetry with respect to the x-axis by replacing y with $-y$ and obtain

$$4x^2 + (-y)^2 = 16$$
$$4x^2 + y^2 = 16$$

Consequently, the graph of the given equation is symmetric with respect to the x-axis as well. Testing for symmetry with respect to the origin, we replace both x and y with their negatives to obtain

$$4(-x)^2 + (-y)^2 = 16$$
$$4x^2 + y^2 = 16$$

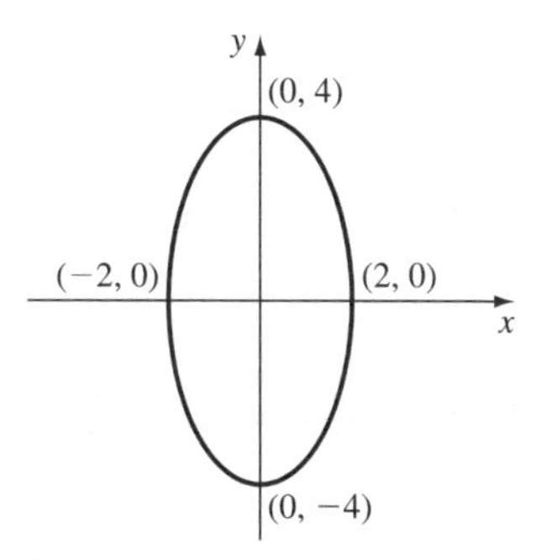

Figure 10

So, the graph of $4x^2 + y^2 = 16$ is also symmetric with respect to the origin. We leave it as an exercise to check that the graph of this equation has intercepts at the points $(-2, 0)$, $(2, 0)$, $(0, 4)$, and $(0, -4)$. A sketch of the graph is shown in Figure 10. ❖

Graphing Using Graphing Calculators

Although in Examples 1 and 2 we obtained sketches of graphs of equations by plotting points, this can be very time consuming, particularly if there is a need to plot a large number of points. Quite often we seek instead to obtain a graph using a graphing utility (grapher). Information we glean from a study of the equation may help determine an appropriate viewing window on the calculator in order to obtain a complete sketch. One thing to be careful of, however, is that most graphing utilities require expressing an equation in the form

$$y = \text{some expression involving only } x \text{ and constants}$$

in order to produce a graph. Appendix I contains exercises involving using some of the important features of most graphing calculators. Refer to your owner's manual for specific instructions on executing particular routines.

EXAMPLE 5 Graphing Using a Graphing Calculator

Graph the Equation $4x^2 + y^2 = 16$ using a graphing calculator.

Solution First we solve the equation for y in terms of x to obtain

$$y^2 = 16 - 4x^2$$

So, either

$$y = \sqrt{16 - 4x^2} \text{ or } y = -\sqrt{16 - 4x^2}$$

In order to graph the equation $4x^2 + y^2 = 16$ using a graphing calculator, we must actually enter two equations into the graphing equation menu. Based on the analysis done in Example 4, it is left as an exercise to find an appropriate viewing window to find a graph similar to that shown in Figure 10. ❖

EXAMPLE 6 **Graphing Using a Graphing Calculator**

Graph the equation $x + y^2 = 36$ using a graphing calculator, with window settings $-5 \le x \le 5$ and $-10 \le y \le 10$. Could this viewing window show a complete sketch of the graph of the given equation?

Solution First we solve the equation for y in terms of x, obtaining

$$y^2 = 36 - x$$

So,

$$y = \sqrt{36 - x} \quad \text{or} \quad y = -\sqrt{36 - x}$$

In order to graph the equation $x + y^2 = 36$ using a graphing utility, we must actually enter two equations into the graphing equation menu. The resulting sketch in the indicated viewing window should look very similar to Figure 11.

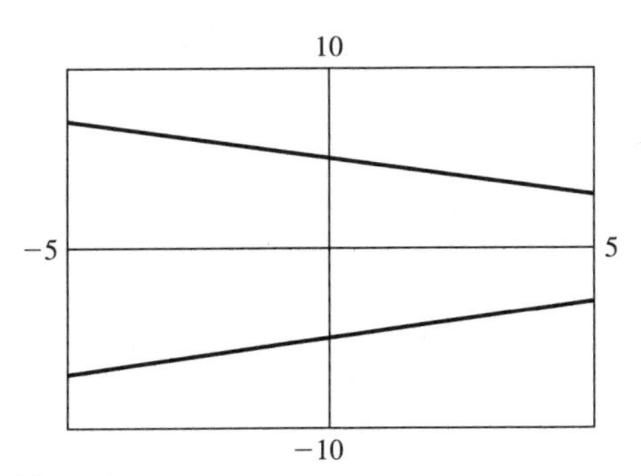

Figure 11

Notice, however, that if we look for x-intercepts of the graph, we solve the equations

$$0 = \sqrt{36 - x} \text{ and } 0 = -\sqrt{36 - x},$$

both of which yield the solution

$$x = 36$$

So, the graph of $y^2 = 16 - 4x^2$ must intersect the x-axis at the point (36, 0), which is not in the given window. Hence the sketch is not complete. ❖

Since graphs can be very useful, we will spend a great deal of time in subsequent chapters dealing with methods of analyzing special classes of equations so that we can obtain complete sketches of the corresponding graphs. While making observations about intercepts and symmetry is a good start, there are many other facts about specific types of equations that will be utilized in this endeavor.

Solving Problems Using Graphs

If we can obtain a complete graph of an equation in two variables, then it is frequently possible to answer concrete questions of interest about the relationship between the two variables, or at least give reasonable approximate answers.

EXAMPLE 7 **Solving an Inequality Graphically**

Use a graph to approximate solutions to the inequality $x^2 \le 2$.

Solution Consider the equation $y = x^2$. Since we seek all values of x such that $x^2 \le 2$, we can look at this problem graphically as a matter of locating the x-coordinates of all points of the graph of $y = x^2$ with y-coordinates less than or equal to 2. See Figure 12.

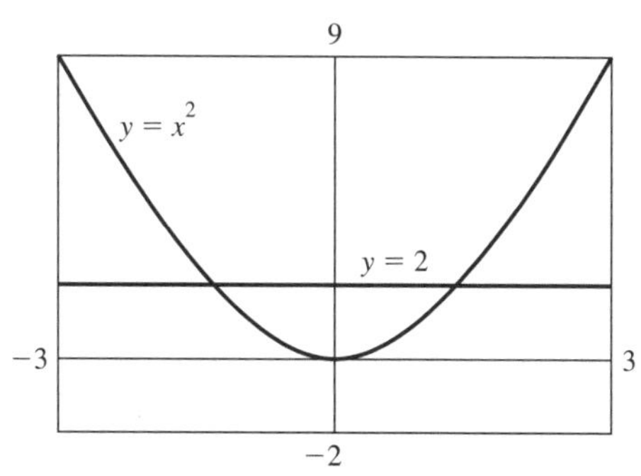

Figure 12

Using an intersection routine on a graphing calculator, we find that the x-coordinates of the two points of intersection (accurate to four decimal places) are

$$x \approx -1.4142 \quad \text{and} \quad x \approx 1.4142$$

Since all points with x-coordinates between those two values have corresponding y-coordinates that are less than 2, we conclude that the solution (accurate to four decimal places) is given by

$$-1.4142 \le x \le 1.4142$$

In interval notation, this corresponds to the closed interval $[-1.4142, 1.4142]$.

Solving the inequality

$$x^2 \le 2$$

is equivalent to solving the inequality

$$x^2 - 2 \le 0$$

Graphically, we could solve the second inequality by finding the x-coordinates of all points on the graph of $y = x^2 - 2$ with nonpositive y-coordinates. ❖

We conclude with an example that uses the graph of an equation modeling a relationship between two quantities to make a decision on how one quantity should be allocated in order to optimize the other.

EXAMPLE 8 Minimizing the Amount of Material Used To Make a Can

A typical soda can is made to hold 12 fluid ounces (about 355 cm^3) of liquid. Suppose that it is deemed desirable to construct a cylindrical can that will hold 355 cm^3 of liquid but uses the minimum possible amount of material. If the top, bottom and the side of the resulting can will be cut from a single sheet of aluminum, what would be the radius and height of the can that uses the least amount of aluminum? Assume here that there is no waste from the sheet of aluminum.

Solution This problem is really about minimizing the surface area of a right circular cylinder with a fixed volume. See Figure 13.

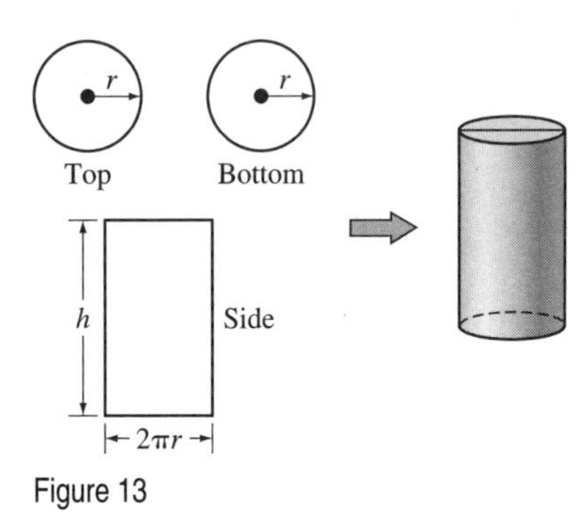

Figure 13

If we let S denote the surface area of the can with dimensions, in cm, as labeled in Figure 13, then

$$S = \begin{pmatrix}\text{area of top}\\ \text{and bottom}\end{pmatrix} + \begin{pmatrix}\text{area of}\\ \text{side piece}\end{pmatrix}$$
$$= 2\pi r^2 + 2\pi rh.$$

Since we also know that

$$\pi r^2 h = 355$$

we have

$$h = \frac{355}{\pi r^2}$$

Thus,

$$S = 2\pi r^2 + 2\pi r\left(\frac{355}{\pi r^2}\right)$$
$$= 2\pi r^2 + \frac{710}{r}$$

A sketch of the graph of the equation $S = 2\pi r^2 + \frac{710}{r}$ is shown in Figure 14. With the aid of a minimum-finding routine on a graphing calculator, we see that the approximate lowest point of the graph is (3.837, 277.545). This means that a radius of

$$r \approx 3.837 \text{ cm}$$

will yield the minimum possible surface area of the resulting can. This yields a height of

$$h \approx \frac{355}{\pi(3.837)^2} \approx 7.675 \text{ cm}$$

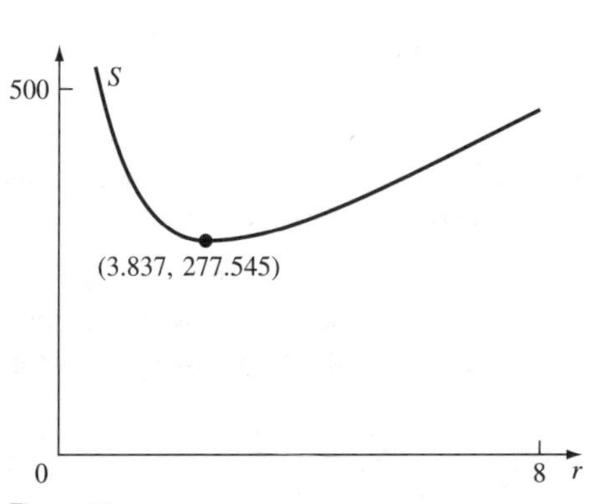

Figure 14

PROBLEM SET 1.4

1. For each of the following equations, determine if the point $(-1, 2)$ is on its graph.
   (a) $3x + y = 1$
   (b) $x^2 + y^2 = 5$
   (c) $x^3 + 7 = y^3$
   (d) $y = 2\sqrt{1 - x}$

2. Figure 15 shows the graph of some equation in the variables x and y. Use the graph for the problems that follow. Give exact answers whenever possible, your best approximations otherwise. Assume that gridlines each mark one unit on the coordinate axes.

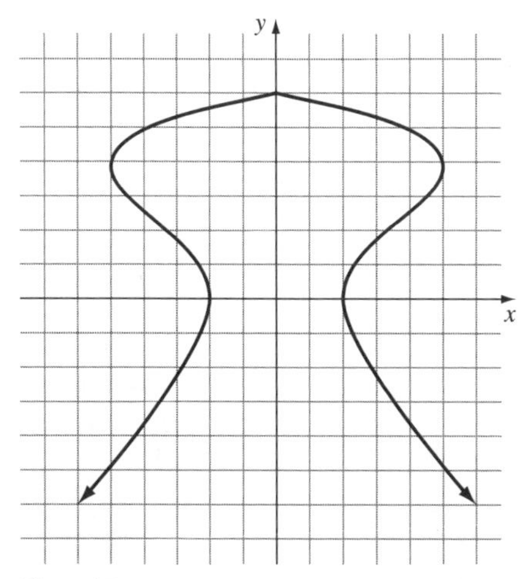

Figure 15

   (a) Find all values of x for which the equation $y = -1$ is true.
   (b) Give all of the intercepts of the graph.
   (c) Is the graph symmetric with respect to the y-axis? The x-axis? The origin?

In problems 3–7, test for the three types of symmetry, find all intercepts, and sketch a graph. Use a graphing utility and write down your window settings.

3. $x - y^2 = 3$
4. $x^2 + y^2 = 9$
5. $x^2y = 1 + x$
6. $y = 0.5x^2 - 2$
7. $y = 2 - 2x^3$

In problems 8 and 9, part of the graph of an equation is shown. Make a complete sketch of the graph of the equation if the graph is symmetric with respect to

(a) the y-axis
(b) the x-axis
(c) the origin

8.

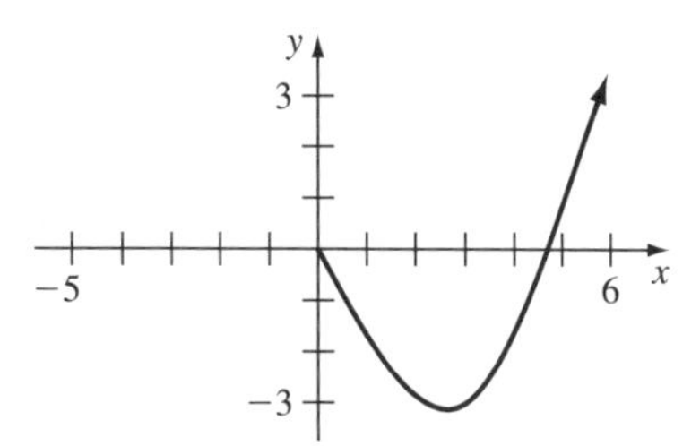

9.

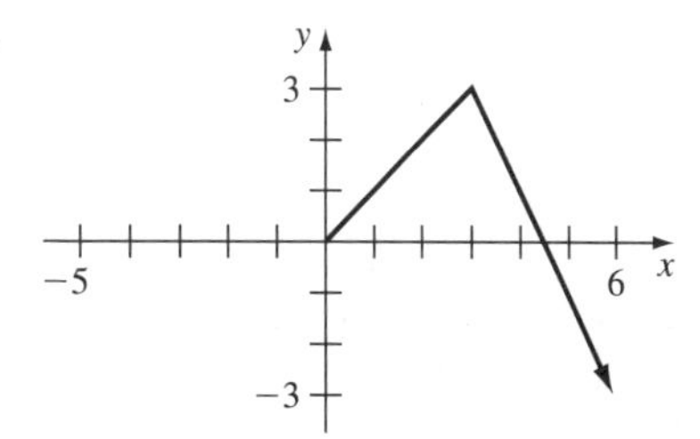

10. The equations $C = 380 + 28x$ and $R = x(150 - 2.5x)$ give the cost and revenue, respectively, in thousands of dollars, for manufacturing and selling x thousand VCRs, where $0 \leq x \leq 50$.
    (a) Sketch graphs of both equations on the same set of axes. Make sure your sketches are complete and relevant to the given information.
    (b) Determine the number of VCRs that must be produced and sold in order to generate the maximum possible revenue.
    (c) A production level where cost and revenue are equal is called a *break-even point*. Find all break-even points for the given cost and revenue information, to the nearest VCR.

11. An open box is to be made from a square sheet of tin 14 centimeters on a side by cutting small squares from each of the corners and turning up the edges (see the accompanying figure).

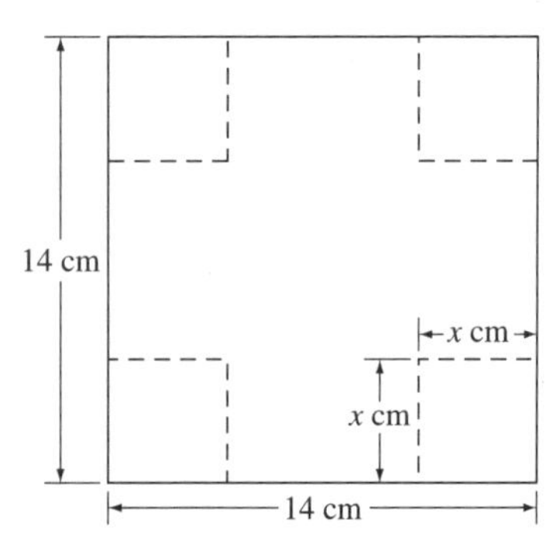

Figure 16

(a) Write an equation that gives the volume, V, of the box in terms of x (see Figure 16).

(b) What are the dimensions of the resulting box if its volume is to be maximized?

12. A closed box with a square base is to have a volume of 24 cubic feet. The material for the sides, top, and base costs \$0.25, \$0.50, and \$1 per square foot, respectively.

(a) Write an equation that gives the cost, C, of the box if the length of one side of the base of the box is x feet.

(b) Find the dimensions of the most economical box.

13. A closed box with a square base is to have a volume of 128 cubic inches.

(a) Let x denote the length, in inches, of one edge of the base of the box, and let S denote the amount of material necessary to construct the box. Find a formula for S in terms of x.

(b) Determine the dimensions of the box that can be constructed using the minimum amount of material. Round to two decimal places, if necessary.

14. A window in the shape of a rectangle surmounted (topped) by an equilateral triangle is to be cut through a wall but is subject to the condition that its perimeter cannot exceed 24 feet. What are the dimensions of the rectangle (accurate to two decimal places) if the window is to admit the maximum amount of light?

15. A manufacturer of hand-held calculators is confident that it can sell 3000 calculators per week at a price of \$12 each and believes that reducing the price by \$0.20 each will increase its weekly sales by 200. Assuming a maximum weekly capacity of 10,000 calculators, how many calculators should be manufactured to maximize weekly revenue?

16. United States Parcel Post regulations require that any package must have length plus girth of no more than 84 inches to be exempt from "Balloon" pricing. Given this, what are the dimensions (radius and height) of a cylindrical package of maximum volume that is still mailable by parcel post without being subject to "Balloon" pricing?

17. Use graphs to solve each equation or inequality. Round results to three decimal places and use interval notation when appropriate.

(a) $x^3 + 9 = 3x^2 + 6x$

(b) $2x^2 + 5x - 8 \leq 1$

(c) $x^2 = \sqrt{x+10}$

(d) $\sqrt{x^2+1} \leq \sqrt{x+9}$

18. Use graphs to solve each equation or inequality. Round results to three decimal places and use interval notation when appropriate.

(a) $|x^2-4| > 2$

(b) $\dfrac{3x+4}{x-1} \leq 1$

(c) $|x-8| = 100 - x^2$

(d) $0.003x^4 - 0.04x^3 - 0.201x^2 + 3.012x + 2 = 0$

Interlude: Progress Check 1.4

A company that manufactures and sells rubber chickens has determined that the profit earned from selling x **thousand** chickens is given by

$$P(x) = -0.8x^2 + 16.2x - 10 \text{ (thousand of \$)},$$

where $0 \leq x \leq 22$.

1. What level of production (to the nearest chicken) generates maximum profit? Support your result graphically.

2. Find all levels of sales that will generate a profit of at least $60,000, and give your solution in terms of numbers of chickens. Support your result graphically.

3. Find all levels of sales where the company breaks even, and give your solution(s) as a number of chickens. Support your result graphically.

4. If it costs the manufacturer $52,000 to produce 10,000 chickens, then how much revenue is generated from the sale of 10,000 chickens?

5. Let C denote the cost (in thousands of dollars) to produce and sell x rubber chickens and assume that C and x are linearly related. Write an equation that gives C in terms of x (use the information in the previous question).

6. Let R denote the revenue (in thousands of dollars) from producing and selling x thousand rubber chickens. Write an equation that gives R in terms of x. Simplify the equation as much as possible.

7. Sketch a complete graph of the equation from **6.** and use that graph to determine the number chickens that must be produced and sold in order to generate maximum possible revenue, and find the corresponding revenue. Give your results to the nearest chicken and nearest dollar, respectively.

Answers to Odd-Numbered Problems

Problem Set 1.1

1. The fourth table only.

3. **(a)** See table below **(b)** Plot similar to that shown in Example 1. Make sure you labeled and showed a scale on both axes.
(c) 14 million **(d)** 0.2 million people/year (200,000 people/year)

t	P
0	14
5	15
10	16
15	17
20	18
25	19

5. $P = 23t + 4{,}650$ **7.** **(a)** $T = \$4{,}200$ **(b)** 16 credits
(c) In your table for $c = 4$ to $c = 16$, $T = 3800, 4000, 4200, 4400, 4600, 4800, 5000, 5200, 5400, 5600, 5800, 6000$, and 6200 consecutively.
(d) 16 credits **(e)** Initial tuition costs of \$3,000.
(f) Variable costs of \$200 per credit. It is also the average rate of change in tuition cost per credit taken.

9. **(a)** $F = 2C + 30$
(b) See table: **(c)** 10°

C	Actual	Estimate
−10	14	10
0	32	30
10	50	50
25	77	80

11. **(a)** $y = \dfrac{3}{500}x + 400$ **(b)** 550 units; 640 units **(c)** \$100,000
(d) This rate of change means an additional 3 units are sold for each additional \$500 spent on marketing

13. 8 yrs.: \$40,000, 12 yrs.: \$10,000

15. **(a)** values in women's table: 100, 105, 110, 120, 130, 160, 175
values in men's table:106, 112, 118, 130, 142, 178, 196
(b) $y = 5x + 100$ **(c)** 145 lbs. **(d)** $y = 6x + 106$; 160 lbs.

17. 18. defective components

19. **(a)** $C = \dfrac{4}{25x} + 44$
(b) \$88
(c) 3,000 miles

21. fixed costs: \$1,000
Variable costs: \$5 per case

Problem Set 1.2

1. True **3.** False **5.** False. The line $4x - y = -1$ has a bigger slope.

7. $y = 3x + 0.2$ **9.** $y = 2x + \dfrac{1}{2}$

11. **(a)** $y = 5x + 6$ **(b)** $y = -3x + 24$ **(c)** $y = -2x + 4$ **(d)** $y = \dfrac{7}{5}x - 7$ **(e)** $y = \dfrac{3}{4}x + \dfrac{27}{2}$ **(f)** $y = -\dfrac{8}{3}x - \dfrac{32}{3}$.

13. **(a)** $x = -4$ **(b)** $y = 9$ **(c)** $x = 0$ **(d)** $y = 5$.

15. $A \leftrightarrow y_1, B \leftrightarrow y_3, C \leftrightarrow y_2, D \leftrightarrow y_5, E \leftrightarrow y_4$ **17.** **(a)** $y = -\frac{1}{3}x + \frac{13}{3}$ **(b)** $y = 3x - 9$

19. **(a)** after 16 years **(b)** a Hotbox oven is worth \$40, while and Equator-bake oven is worth \$0.

21. **(a)** $p = 8n - 69{,}200$ **(b)** Slope: \$8 profit per visitor; p-intercept: if there are no visitors, loss of \$69,200

Problem Set 1.3

1. **(b)** $y = 15.338x - 30,593.581$ **(c)** or the second quarter of years 2000 through 2007, e-commerce was increasing at an average rate of about 15.338 billion dollars per year.

3. **(a)** See calculator **(b)** $r \approx -½$ **(c)** $y = -0.184x + 30.402, r = -0.539$

5. **(b)** Estimates will vary, e.g. $H = 37t - 37$ **(c)** $H = 37.262t - 2.586, r = 0.9995$,
The correlation between the data set and the regression line is very good. However, rounding does not create a perfect correlation!
(d) Slope gives the average number of home-runs per year, which in this situation is about 37.
(e) 37 each: Yes. Henry cannot have any home-runs when he is retired.

7. **(b)** Yes; $0 \le t \le 4$ **(9)** Answers will vary.

Problem Set 1.4

1. **(a)** no **(b)** yes **(c)** no **(d)** no

3. Symmetry with respect to the x-axis **5.** No symmetry **7.** No symmetry to the x-axis

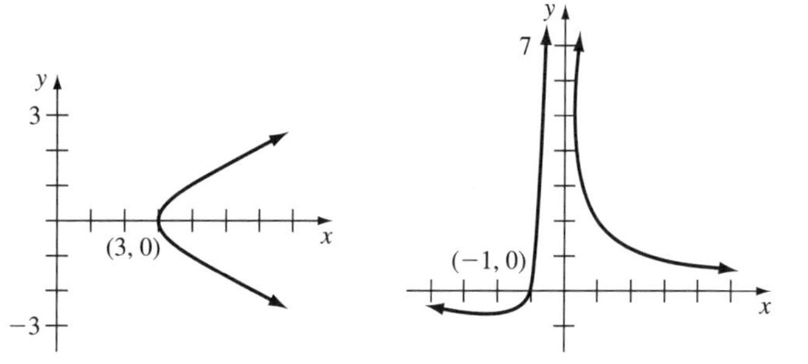

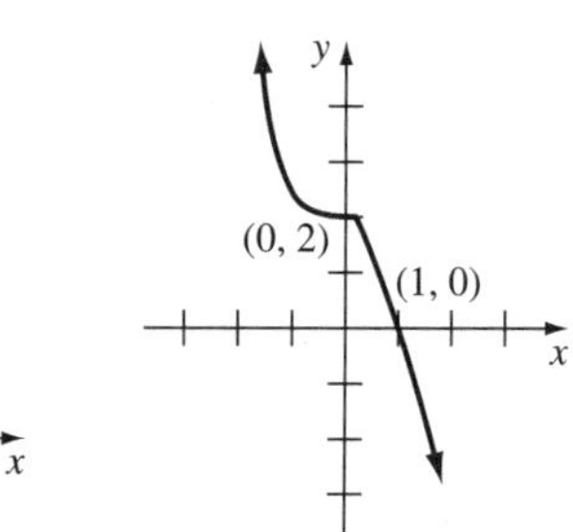

9. **(a)**

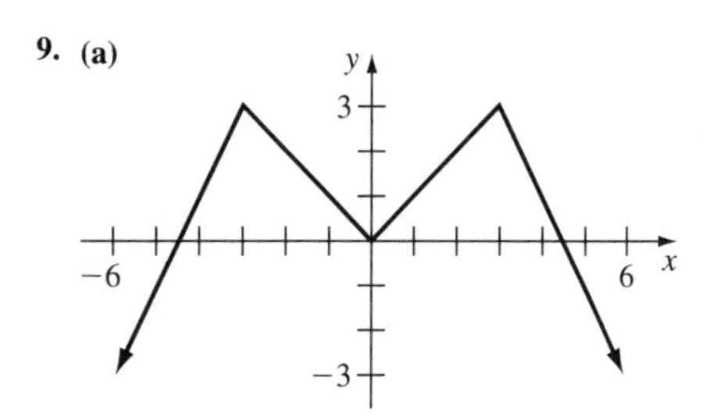

(b)

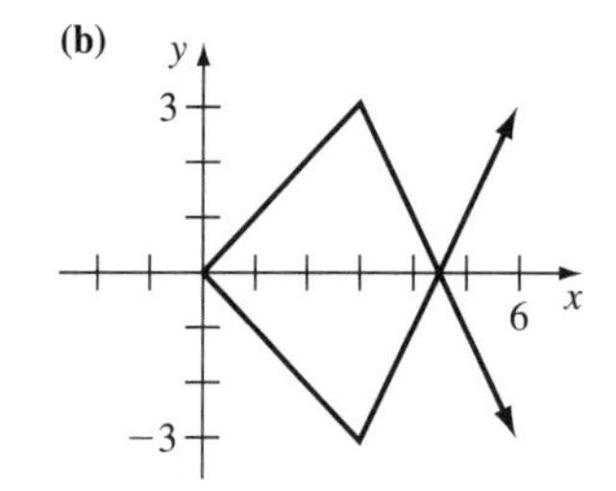

(c)

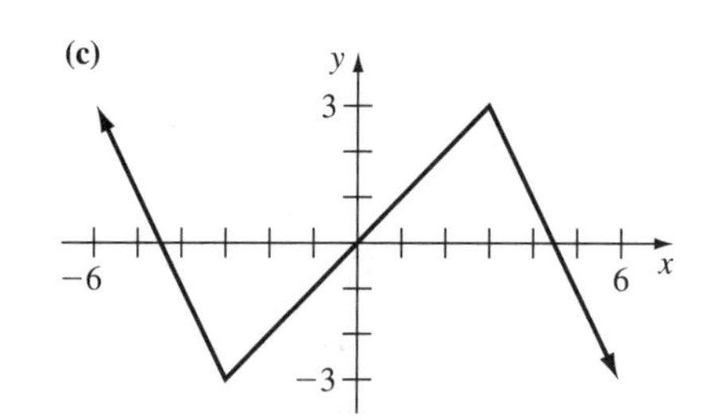

11. **(a)** $V = x(14 - 2x)^2$ **(b)** $\frac{7}{3}\text{cm} \times \frac{28}{3}\text{cm} \times \frac{28}{3}\text{cm}$

13. **(a)** $S = 2x^2 + \frac{512}{x}$ **(b)** 5.04in × 5.04in × 5.04in

15. 7500 calculators

17. **(a)** $x \approx -2.917, x \approx -1.028$ **(b)** $[-3.712, 1.212]$
(c) $x \approx -1.697, x \approx -1.856$ **(d)** $[-2.372, 3.372]$

Functions and Their Graphs

CHAPTER CONTENTS

Overview

The concept of a function is central to mathematics. As we shall see, functions are used to model special relationships between quantities and an understanding of how functions work and different ways to represent functions will be invaluable for solving a variety of problems. In particular, the ability to analyze a function graphically is a significant benefit when the function is modeling a relationship between two quantities that is nonlinear, so that rate of change is not constant.

OBJECTIVES

1. Define the Concept of Function
2. Describe Functions Numerically and Algebraically
3. Use Functional Notation

2.1 Functions from a Numerical and Algebraic View

Consider the following statements.

"Income tax is a function of how much you earn."

"My grade on an exam is a function of how much time I study."

"TV ad price is a function of ratings."

We could easily replace the phrase "is a function of" with "depends on" in each statement. This kind of relationship between quantities, where one quantity depends on another quantity (or quantities), is a fundamental idea in mathematics. We use the term *function* to refer to any such relationship. In particular, we will investigate the idea of one quantity being a function of another quantity.

What Is a Function?

As just stated, in the most general case we think of a function as a relationship between quantities, where one quantity determines the other. It is useful, when we say "quantity B is a function of quantity A," to think of particular values of quantity A as *inputs* and corresponding values of quantity B as *outputs*, with some rule (or process) prescribing how any particular value of quantity A determines precisely one corresponding value of quantity B. More formally, we define a **real-valued function of a real variable** in three parts, as follows:

Function

A function consists of:

- A collection of real number inputs, called the **domain** of the function
- A collection of corresponding real number outputs, called the **range** of the function
- A rule of correspondence that describes how each particular input generates exactly one corresponding output

It is important to note here that each input for a given function generates only one corresponding output, but distinct inputs may actually generate the same output.

Table 1 is adapted from the 2011 Form 1040 Federal Income Tax booklet. It describes federal income tax owed as a function of taxable income for a filer filing as *single*. It shows, in theory, how every taxable income level generates a corresponding tax liability.[1]

TABLE 1
If your taxable income was

Over-	**But not over**	*Your tax is*	**Of the amount over-**
$0	$8,500	10%	$0
8,500	34,500	$850 + 15%	8,500
34,500	83,600	4,750 + 25%	34,500
83,600	174,400	17,025 + 28%	83,600
174,400	379,150	42,449 + 33%	174,400
379,150	———	110,016.50 + 35%	379,150

The domain of the "income tax function" is all possible taxable incomes, represented by the real number interval $(0, \infty)$. The range is the collection of all possible corresponding tax amounts, which the interval $(0, \infty)$ also represents.

A question of some merit is, "How do we describe functions?" In some cases, a verbal description of domain, range and rule of correspondence may indeed be possible (imagine doing this with the income tax function above), but it is typically not a practical approach. In the situation above, the table provided an adequate display of the rule of correspondence between taxable income and tax liability, and from the context we were able to deduce a domain and range. We now turn our attention to ways of describing functions.

Describing Functions Numerically

Table 2 has as its basis information from the periodical *The New York Times*. It lists the approximate U.S. oil imports from Mexico for 2001–2006.

TABLE 2

Year (t)	2001	2002	2003	2004	2005	2006
Imports (I) (million gallons/day)	1.35	1.5	1.55	1.6	1.5	1.5

This table describes oil imports, I, as a function of year, t. The domain is the set

$$D = \{2001, 2002, 2003, 2004, 2005, 2006\}$$

while the range is the set

$$R = \{1.35, 1.5, 1.55, 1.6\}$$

The table gives the rule of correspondence between t and I. Notice, however, that the table **does not** describe t as a function of I, as the value $I = 1.5$ corresponds to

[1] The income tax tables provided in the tax booklet are actually only constructed to give tax liabilities in $50 increments of taxable income up to $50,000, so a comparison between the tables and the calculation rules given in Table 1 may yield slightly different results. Also, in practice, dollar amounts are usually rounded to the nearest dollar.

$t = 2002$, $t = 2005$, and $t = 2006$. We say here that I is described *numerically* as a function of t, as the table gives specific values of I corresponding to specific values of t.

In the situation given above, since the description of I was as a function of t, we say that t is the **independent variable** and I is the corresponding **dependent variable**. The distinction between independent variables and dependent variables will become more important later.

Table 3 below gives the median age of the U.S. population for certain years, as recorded by the U.S. Bureau of the Census in *Statistical Abstract of the United States*, 1998. The table describes median age as a function of year.

TABLE 3

Year	1920	1930	1940	1950	1960	1970	1980	1990
Median Age	25.3	26.4	29.0	30.2	29.5	28.0	30.0	32.8

We will use the notation $M(y)$ (read "M of y") to denote the median age of the U.S. population in year y. Specifically, $M(1920)$ is the median age in 1920, $M(1930)$ is the median age in 1930, and so on. Consequently, we write $M(1920) = 25.3$, $M(1930) = 26.4$, etc. We call this notation *functional notation*, and it is a useful shorthand for describing the relationship between inputs and outputs for a given function. In particular, mathematicians commonly use it when describing functions algebraically. We now turn our attention to that method of describing functions.

Describing Functions Algebraically

When we use an explicit formula or an equation to describe how inputs and outputs are related, we say that we are describing a function *algebraically*. For instance, consider the familiar formula for the area, A, of a circle of radius r: $A = \pi r^2$. Considering r as the independent variable, this equation describes A as a function of r. For each particular radius, we have an explicit set of instructions for how to calculate the corresponding area of the circle with that radius, namely, "square the radius and multiply the result by π." In this context, a natural domain and a corresponding range arise for the function in question. The domain is all possible radii r (so, all positive real numbers), and the range is the set of all corresponding areas, which is also the set of all positive real numbers. Using interval notation, the interval $(0, \infty)$ gives both the domain and the range.

The usefulness of functional notation becomes more apparent when working with functions described algebraically. For instance, consider the equation $f(x) = 3x^2 + 4$. As before, when using this notation, x represents the input and the symbol $f(x)$ represents the corresponding output. The letter f is nothing more than a "nickname" for the function described. The equation, taken as a whole, gives the *rule of the function.*

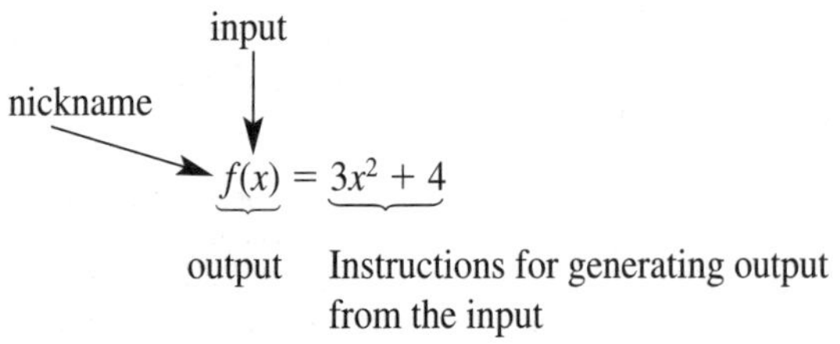

input

nickname

$$\underbrace{f(x)}_{\text{output}} = \underbrace{3x^2 + 4}_{\text{Instructions for generating output from the input}}$$

output Instructions for generating output from the input

Mathematicians call the process of determining an output associated with some input *evaluating the function* at that particular input.

EXAMPLE 1 Evaluating a Function

Let f be the function defined by $f(x) = 3x^2 + 4$. Evaluate f at the inputs -2, 0, and $\frac{3}{2}$.

Solution Using functional notation, we have

$$f(-2) = 3(-2)^2 + 4 = 16$$

$$f(0) = 3(0)^2 + 4 = 4$$

$$f\left(\frac{3}{2}\right) = 3\left(\frac{3}{2}\right)^2 + 4 = \frac{43}{4}$$

❖

We should note here that the equations $f(x) = 3x^2 + 4$ and $p = 3q^2 + 4$ define *exactly the same function* if we understand that p is the dependent variable and q is the independent variable. Choice of variables is largely a matter of personal preference. Similarly, the choice of the letter f to represent the function was also of little consequence. It is common practice to use various alphabet characters to name functions. For example, the equation $g(z) = 3z^2 + 4$ defines a function "named" g, but it produces outputs in precisely the same way as the function named f described by $f(x) = 3x^2 + 4$. The distinction is merely the choice of notation, as f and g produce outputs according to the same rule, "square the input, multiply by three, and then add four."

Although it is not necessarily apparent at this juncture why we would wish to do so, it is also possible to evaluate a function defined by a formula at inputs that are algebraic expressions rather than specific real numbers. Functional notation proves very helpful at illustrating this.

EXAMPLE 2 Evaluating a Function at an Algebraic Input

Let g be the function defined by $g(x) = 4 - x - 5x^2$. Evaluate the following.

(a) $g(5a)$ (b) $g(x + 2)$ (c) $g(t + h) - g(t)$

Solution In each case, we merely replace the independent variable x in $g(x) = 4 - x - 5x^2$ with the indicated input and simplify the resulting expression.

(a) $g(5a) = 4 - (5a) - 5(5a)^2 = 4 - 5a - 5(25a^2) = 4 - 5a - 125a^2$

(b) $$\begin{aligned} g(x + 2) = 4 - (x + 2) - 5(x + 2)^2 &= 4 - x - 2 - 5(x^2 + 4x + 4) \\ &= 4 - x - 2 - 5x^2 - 20x - 20 \\ &= -18 - 21x - 5x^2 \end{aligned}$$

(c) $$\begin{aligned} g(t + h) - g(t) &= [4 - (t + h) - 5(t + h)^2] - [4 - t - 5t^2] \\ &= [4 - t - h - 5t^2 - 10th - 5h^2] - [4 - t - 5t^2] \\ &= -h - 10th - 5h^2 \end{aligned}$$

❖

Remark: A common mistake when working with functional notation and expressions of the form $f(t + h)$ is to treat $f(t + h)$ and $f(t) + h$ as equal. In general, this is not the case. That is, in general, $f(t + h) \neq f(t) + h$.

An Agreement on Domains

When working with a function defined numerically, the issue of the domain of the function is not a difficult one, as the inputs are explicitly listed. In the case of a function defined algebraically, though, that is most often not the case. In some instances, the placement of the function within the context indicates an appropriate domain, as we saw with the equation defining the area of a circle as a function of radius. In other instances, there may not be sufficient context present to make such an assertion. We adopt a convention on domains of functions defined algebraically.

Domain Convention

Given a function defined algebraically, if we do not have specific information to the contrary, then we will assume the domain of the function to be all possible real number inputs for which the formula defining the rule of the function produces a corresponding real number output.

EXAMPLE 3 Determining Domains

Find the domains of each of the functions given below, according to the domain convention.

(a) $f(x) = x^2$ (b) $g(t) = \sqrt{5 - 4t}$ (c) $h(v) = \dfrac{v^2 + 1}{v^2 - 1}$

Solution

(a) $f(x) = x^2$ produces a real number output no matter what x is, so the domain of f is the set of all real numbers. This corresponds to the real number interval $(-\infty, \infty)$.

(b) $g(t) = \sqrt{5 - 4t}$ produces a real number output only if $5 - 4t \geq 0$. This means, then, that $g(t)$ is only defined when $t \leq \frac{5}{4}$. This corresponds to the interval $\left(-\infty, \frac{5}{4}\right]$.

(c) $h(v) = \dfrac{v^2 + 1}{v^2 - 1}$ produces a real number output only if $v^2 - 1 \neq 0$. So, any value of v will produce a real number output except $v = 1$ and $v = -1$. Consequently, the domain of h is all real numbers except 1 and -1. ❖

Do All Equations Define Functions?

As noted earlier (Exercise #9, Section 1.1), the Fahrenheit and Celsius temperature scales are linearly related by the formula $F = \frac{9}{5}C + 32$, where F represents Fahrenheit temperature and C represents Celsius temperature. This equation certainly defines F as a function of C, since every particular value of C generates precisely one corresponding value of F. Notice, though, that we can rewrite the equation as one that expresses C in terms of F:

$$F = \frac{9}{5}C + 32$$

$$F - 32 = \frac{9}{5}C$$

$$\frac{5}{9}(F - 32) = C$$

Notice that the last equation clearly defines C as a function of F, as every particular value of F will produce exactly one corresponding value of C. This suggests that some equations may, in fact, either *explicitly* or *implicitly* define more than one function.

EXAMPLE 4 An Equation Describing More Than One Function

Consider the linear equation $8p - 24q = 96$. We say that this equation implicitly defines p as a function of q and also implicitly defines q as a function of p because in each case the equation does not clearly give the rule of correspondence between inputs and outputs. However, it is not difficult to solve the equation for either of the variables. Solving for p yields

$$8p - 24q = 96$$

$$8p = 24q + 96$$

$$p = \frac{24q + 96}{8}$$

$$p = 3q + 12$$

The equation $p = 3q + 12$ explicitly defines p as a function of q. Similarly, solving the original equation for q yields

$$8p - 24q = 96$$

$$8p - 96 = 24q$$

$$\frac{8p - 96}{24} = q$$

$$q = \frac{1}{3}p - 4$$

The last equation explicitly defines q as a function of p. ❖

Notice that p and q have a linear relationship in the last example, so we say that the equation $8p - 24q = 96$ implicitly defines two distinct *linear functions*. Also, this last example suggests an approach that we may use in order to ascertain whether or not an equation in two variables defines one variable as a function of the other: If we can uniquely solve the equation for one of the variables, then the equation defines a function.

Equations Implicitly Defining Functions

Given an equation in two variables, say x and y,

- If we can explicitly and uniquely solve the equation for y, then the equation defines y as a function of x.
- If we can explicitly and uniquely solve the equation for x, then the equation defines x as a function of y.

The terms "explicitly" and "uniquely" are very important, as illustrated in the next example.

EXAMPLE 5 Determining if Equations Define Functions

For each equation, determine if the equation defines y as a function of x and if the equation defines x as a function of y.

(a) $x^2 - y = 16$ (b) $x = y^3 - 3y + 4$

Solution (a) Solving $x^2 - y = 16$ for y gives $y = x^2 - 16$, so the equation defines y as a function of x. Solving for x is more involved, but we see that

$$x^2 - y = 16$$
$$x^2 = y + 16$$
$$x = \sqrt{y + 16} \quad \text{or} \quad x = -\sqrt{y + 16}$$

Consequently, the equation does not define x as a function of y, as a single value of y produces two different corresponding values of x.

(b) The equation clearly defines x as a function of y. Notice, however, that solving the equation explicitly for y in terms of x is not possible using elementary solving techniques. So, at this moment, we are unable to ascertain whether or not the equation defines y as a function of x using this method (but an alternative approach will be discussed later). ❖

Rate of Change, Revisited

The defining characteristic of linearly related quantities (and consequently linear functions) is a constant average rate of change. When a function that is not linear expresses the relationship between two quantities, it seems reasonable to expect that when we look at average rate of change we will see different rates for different changes in the independent variable.

Suppose that y is a function of x; therefore we write $y = f(x)$ (we don't presume, however, that the function is described algebraically). We restate an earlier definition using functional notation.

Total Change and Average Rate of Change

Given $y = f(x)$, as x changes from $x = a$ to $x = b$, so that $\Delta x = b - a$

- $\Delta y = f(b) - f(a)$ (total change in y, or total change in $f(x)$)
- $\dfrac{\Delta y}{\Delta x} = \dfrac{f(b) - f(a)}{b - a}$ (average rate of change)

The expression $\dfrac{f(b) - f(a)}{b - a}$ is called a **difference quotient** for the function f. Expressions of this type are important in the study of calculus. There are several forms of difference quotients for functions, but regardless of the form a difference quotient takes, it is merely an average rate of change.

EXAMPLE 6 Calculating Average Rates of Change

Suppose that we throw a softball straight up in the air from ground level with an initial velocity of 80 feet per second. Basic physics tells us that the height of the ball above the ground is reasonably approximated by

$$h(t) = -16t^2 + 80t$$

where $h(t)$ is height in feet and t is seconds since the ball was thrown for $0 \leq t \leq 5$. Calculate and interpret the following quantities.

(a) $h(3)$ (b) $h(5) - h(2)$ (c) $\dfrac{h(5) - h(2)}{5 - 2}$ (d) $\dfrac{h(b) - h(a)}{b - a}$

Solution

(a) $h(3) = -16(3)^2 + 80(3) = 96$; three seconds after being thrown the softball is 96 feet above the ground.

(b) $h(5) - h(2) = [-16(5)^2 + 80(5)] - [-16(2)^2 + 80(2)] = 0 - 96 = -96$; the net change in height between $t = 2$ and $t = 5$ seconds is -96 feet, so the ball was 96 feet higher in the air at two seconds than at five seconds. In particular, the ball was 96 feet in the air at two seconds and was on the ground at five seconds.

(c) $\dfrac{h(5) - h(2)}{5 - 2} = \dfrac{-96}{3} = -32$; the *average velocity* of the softball between $t = 2$ and $t = 5$ seconds is -32 feet per second.

(d)
$$\begin{aligned}\frac{h(b) - h(a)}{b - a} &= \frac{(-16b^2 + 80b) - (-16a^2 + 80a)}{b - a} \\ &= \frac{16a^2 - 16b^2 - 80a + 80b}{b - a} \\ &= \frac{-16(b^2 - a^2) + 80(b - a)}{b - a} \\ &= \frac{-16(b - a)(b + a) + 80(b - a)}{b - a} \\ &= -16(b + \mathrm{a}) + 80\end{aligned}$$

This is the average velocity of the softball, in feet per second, from time $t = a$ to $t = b$. ❖

PROBLEM SET 2.1

1. Table 1 defines a function $y = f(x)$. Calculate $f(-2)$, $f(1)$, and $f(4)$.

TABLE 1

x	-2	-1	0	1	2	3	4
y	6	13	-10	-18	0	6	35

2. Determine the domain and range of the function f defined by Table 1.

3. Does Table 1 define x as a function of y? Explain.

4. Table 2 defines a function $p = g(w)$. Calculate $g(10)$, $g(15)$, and $g(20)$.

TABLE 2

w	10	15	20	25	30
p	-100	275	-400	625	-900

5. Does Table 2 define w as a function of p? Explain.

6. Determine the domain and range of the function g defined by Table 2.

7. Let $f(x) = 3 - 6x$. Calculate $f(-4)$, $f(0)$ and $f(5/2)$.

8. Let $g(t) = 3t^2 - 5t - 12$. Calculate $g(-2)$, $g(1)$ and $g(-3/4)$.

9. Let $M(k) = 2k - \dfrac{k}{1-k}$. Calculate $M(-1)$ and $M(2)$.

10. Find the domain of g if $g(t) = \dfrac{t^2 + 4}{t^2 - 3t - 10}$.

11. Find the domain of h if $h(p) = 1 - \dfrac{p^2 - 1}{p^2 + 1}$.

12. Find the domain of z if $z(y) = \sqrt{3y - 8}$.

13. The description of a function f is: "For a given input x, first add 4, square the result, and then divide that by three." Write the rule of f.

14. The description of a function T is: "For a given input s, cube the input, subtract that result from 10 and then find the square root of the result." Write the rule of T.

15. Let $r(x) = \sqrt{x^2 - 8x}$. Find all values of x such that $r(x) = \sqrt{20}$.

16. Let $f(y) = (2y - 5)(y + 9)$ for $-6 \le y \le 6$. Find all values of y in the domain of f such that $f(y) = 0$.

17. A rectangle is to have a perimeter of 24 inches. If one side of the rectangle has length l inches, then give a formula that defines the area of the rectangle as a function of l.

18. Let $A(d)$ denote the area of a circle of diameter d. Write the rule of $A(d)$.

19. Let $f(x) = 2x + 4$ and suppose $h \neq 0$. Simplify the following.

(a) $f(2a)$ **(b)** $f(p + 1)$

(c) $f(t + h) - f(t)$ **(d)** $\dfrac{f(t + h) - f(t)}{h}$

20. Let $g(r) = 1 - r^2$ and suppose $h \neq 0$. Simplify the following.

(a) $g(z^2)$ **(b)** $g(p - q)$

(c) $g(b + h) - g(b)$ **(d)** $\dfrac{g(b + h) - g(b)}{h}$

21. For each of the following, determine if the equation defines y as a function of x and if the equation defines x as a function of y.

(a) $3x - 10y = 16$ **(b)** $x^2 - y = 30$

(c) $3x + 6y^3 - 40 = 0$

22. For each of the following, determine if the equation defines y as a function of x and if the equation defines x as a function of y.

(a) $x^2 - y^2 = 16$ **(b)** $\dfrac{5x - y}{4} = 0$

(c) $x^2y - 9x^2 = 45$

23. Table 3 shows the amount of money spent annually on corrections in Cobb County, where $t = 0$ corresponds to 2003.

TABLE 3

Year, t	0	1	2	3	4	5	6	7
Spending, $s(t)$, in \$ thousands	64	68	72	79	79	88	91	98

(a) Calculate and interpret $s(1)$, $s(4)$, and $s(7)$.

(b) Calculate the average rate of change of corrections spending from 2004 to 2008. Label your response appropriately.

24. One day I was standing at the edge of the rooftop of a tall building and threw my graphing calculator straight up in the air. Table 4 shows the height above ground, h, of my calculator t seconds after I had thrown it.

TABLE 4

t	0	1	2	3	4
h	100	124	116	76	4

(a) How tall is the building I was standing atop?

(b) According to Table 4, is h a function of t? Is t a function of h? Explain.

(c) Calculate the average velocity of my calculator from $t = 1$ to $t = 3$ seconds and label with appropriate units.

25. Experiments with a certain type of bacteria suggest that the number of bacteria in a particular culture after t hours can be reasonably approximated by

$$N(t) = -\frac{1}{3}t^3 + 4t^2 + 20t + 2$$

thousand bacteria for $0 \le t \le 15$.

(a) Calculate and interpret $N(2)$.

(b) Calculate $\dfrac{N(15) - N(12)}{15 - 12}$. What does this quantity represent?

26. A plant has determined that the average cost of producing x riding lawnmowers (in dollars per mower) in a day is given by

$$A(x) = 0.3x^2 + 21x - 251 + \frac{2500}{x}$$

for $1 \le x \le 25$.

(a) Calculate the average cost of producing 15 lawnmowers.

(b) Calculate and interpret $\dfrac{A(20) - A(10)}{20 - 10}$. Include appropriate units with your response.

27. The size, $S(t)$, of a malignant tumor is given by $S(t) = 2^t$ cubic millimeters, where t is the number of months since the tumor was discovered.

(a) Calculate and interpret $S(4) - S(0)$. Include appropriate units with your response.

(b) Calculate and interpret $\dfrac{S(3) - S(0)}{3 - 0}$. Include appropriate units with your response.

28. The weekly profit, $P(x)$, for a certain product is given by

$$P(x) = 80 + 20x - 0.5x^2$$

for $0 \le x \le 20$, where x indicates the amount of money spent on advertising, and both x and $P(x)$ are measured in thousands of dollars.

(a) Calculate the profit when \$8000 is spent on advertising.

(b) Calculate and interpret $\dfrac{P(18) - P(10)}{18 - 10}$. Include appropriate units with your response.

29. Let $f(x) = 5x - 8$. Simplify each of the following (assume $h \ne 0$ and $a \ne b$).

(a) $\dfrac{f(t + h) - f(t)}{h}$

(b) $\dfrac{f(b) - f(a)}{b - a}$

(c) $\dfrac{f(t + h) - f(t - h)}{2h}$

30. Let $g(t) = 2t^2 - t + 7$. Simplify each of the following (assume $h \ne 0$ and $a \ne b$).

(a) $\dfrac{g(x + h) - g(x)}{h}$

(b) $\dfrac{g(b) - g(a)}{b - a}$

(c) $\dfrac{g(x + h) - g(x - h)}{2h}$

31. Let $k(p) = -4p^2 + 6p + 1$. Simplify each of the following (assume $h \ne 0$ and $a \ne b$).

(a) $\dfrac{k(z + h) - k(z)}{h}$

(b) $\dfrac{k(b) - k(a)}{b - a}$

(c) $\dfrac{k(z + h) - k(z - h)}{2h}$

32. Let $f(v) = \dfrac{1}{v+2}$. Simplify each of the following (assume $h \neq 0$ and $a \neq b$).

(a) $\dfrac{f(t+h) - f(t)}{h}$

(b) $\dfrac{f(b) - f(a)}{b - a}$

(c) $\dfrac{f(t+h) - f(t-h)}{2h}$

33. Let $g(u) = \dfrac{2u}{u+4}$. Simplify each of the following (assume $h \neq 0$ and $a \neq b$).

(a) $\dfrac{g(q+h) - g(q)}{h}$

(b) $\dfrac{g(b) - g(a)}{b - a}$

(c) $\dfrac{g(q+h) - g(q-h)}{2h}$

OBJECTIVES

1. Describe Functions Graphically
2. Analyze Function Behavior Using Graphs

2.2 Functions From a Graphical View

As seen earlier, a graphical demonstration of the relationship between two quantities can be very useful (see Example 8, Section 1.4). Since functions describe such relationships, it seems natural that we should be able to look at functions from a graphical perspective, as well as numerical and algebraic perspectives. We now turn our attention to functions defined by graphs.

Functions Defined Graphically

The graph shown in Figure 1 shows the temperature of a small town in Minnesota over a period of time one day in August.

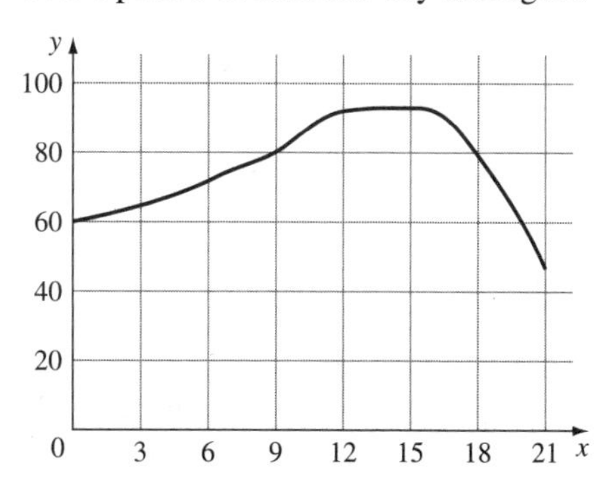

Figure 1

In this graph, x represents the number of hours since 6 A.M. and y represents the temperature, in degrees Fahrenheit. Since there are no two points on the graph with the same x-coordinate, we say here that this graph **defines** y **as function of** x. The horizontal axis of the graph represents inputs, while the vertical axis represents outputs. The rule of correspondence is given by points on the graph, which we interpret as (input, output).

If we let T represent the function defined by the graph in Figure 1, then we can write $y = T(x)$. Functional notation retains its meaning in a graphical setting. For instance, here we estimate from the graph that $T(0) = 60$ and $T(12) = 92$. The domain of T is the interval $[0, 21]$ and we estimate the range of T as the interval $[45, 95]$.

When considering a graph that may define a function, we adopt the convention that inputs are represented on the horizontal axis and outputs are represented on the vertical axis. As seen above, it can be relatively easy to ascertain the domain and range of a function defined graphically. The drawback to functions defined graphically, also as seen above, is that we may need to estimate in order to evaluate the function.

EXAMPLE 1 Analyzing a Function Defined Graphically

Figure 2 shows the temperature, $D(t)$, in degrees Fahrenheit, of a potato put into a hot oven as a function of time, t, in minutes, since the potato was put into the oven.

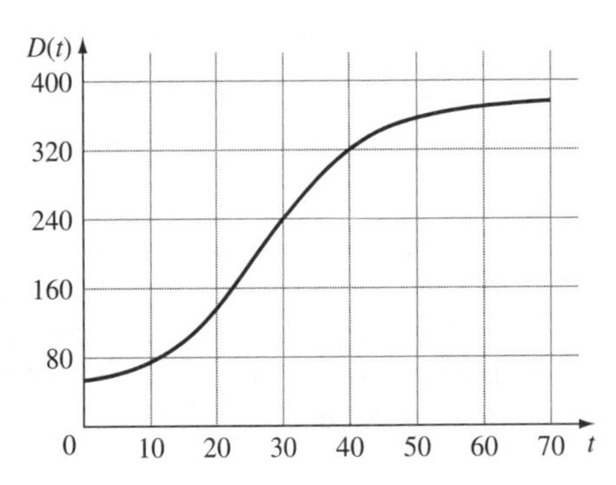

Figure 2

(a) Estimate the domain and range of the function D. Interpret the results.
(b) Estimate $D(35)$ and interpret that quantity.
(c) When does the potato reach a temperature of $240°F$?
(d) Calculate the average rate of change of D with respect to t over the first hour the potato is in the oven. Include appropriate units.

Solution (a) The domain of D is the interval $[0, 70]$. We estimate the range of D to be the interval $[60, 375]$. This means that the graph shows the temperature of the potato for 70 minutes after it was put into the hot oven, and the temperature of the potato ranges from a low of about $60°F$ to a high of about $375°F$.
(b) From the graph, we estimate that $D(35) = 280$. So, 35 minutes after being put into the oven, the temperature of the potato is about $280°F$.
(c) There is only one value of t for which $D(t) = 240$, namely $t = 30$. So after half an hour, the potato has reached a temperature of $240°F$.
(d) $\dfrac{D(60) - D(0)}{60 - 0} \approx \dfrac{370 - 60}{60} = 5.16666\ldots$, so we estimate that the temperature of the potato was increasing at an average rate of about 5.17 degrees per minute the first hour the potato was in the oven. ❖

A function that is defined graphically may be defined by a finite set of points, as was the case in some of the scatterplots seen in Section 1.3. For instance, the graph in Figure 3 defines y as a function of x.

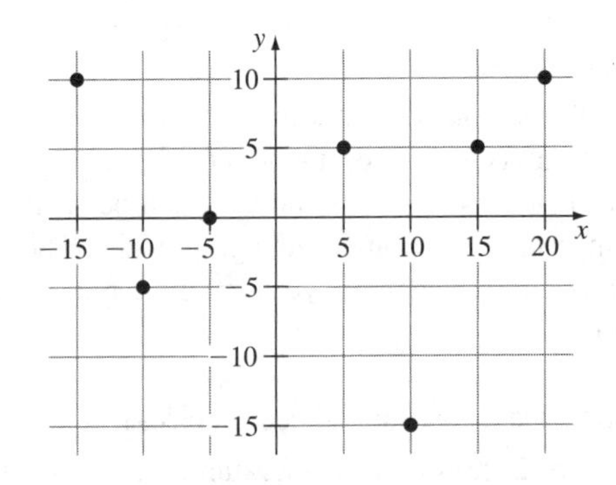

Figure 3

The domain of the function defined by this graph is the set

$$D = \{-15, -10, -5, 5, 10, 15, 20\}$$

while the range is the set

$$R = \{-15, -5, 0, 5, 10\}$$

Not All Graphs Define Functions

In the previous section we saw that some two-variable equations did not define either variable as a function of the other. It is natural, then, to wonder if there are graphs that do not define functions. It is evident that any graph in the xy-plane containing two distinct points with the same x-coordinate cannot define y as a function of x, as a single input would correspond to more than one output. See Figure 4 for examples of such graphs. The dashed lines are not part of the graphs.

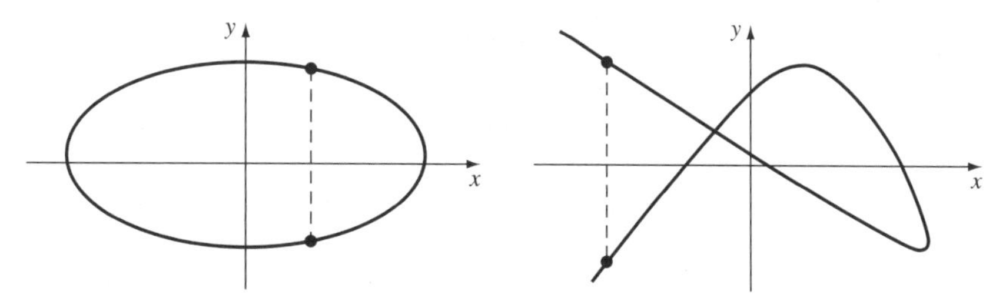

Figure 4

Notice that in each of the graphs shown in Figure 4, there are numerous places where a vertical line can be drawn so as to intersect the graph at more than one point. This suggests a visual test for ascertaining whether or not a graph in the plane defines a function.

Vertical Line Test

Given a graph in the xy-plane, the graph defines y as a function of x if, and only if, no vertical line intersects the graph at more than one point.

Graphs of Functions Defined Algebraically

Suppose that f is an algebraically-defined function. *The graph of f* is the set of all points in the plane of the form $(x, f(x))$, where x ranges over the entire domain of f. An accurate and complete graph of a function can be a very useful tool for analyzing the behavior of the function, particularly in cases where the formula defining the function is complicated. Graphing calculators are of great value in this endeavor, and we assume that most graphing of algebraically-defined functions will be done with the aid of a grapher.

EXAMPLE 2 **Graphing a Function Defined Algebraically**

Let f be the function defined by $f(x) = 0.5x^2 + 2x - 3$. Figure 5 shows the graph of f produced by a graphing calculator using window settings $-10 \le x \le 5$ and $-10 \le y \le 10$.

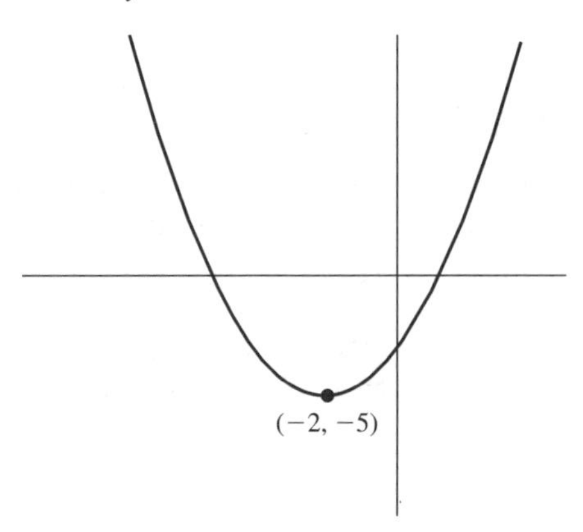

Figure 5

By inspecting the formula defining f, we see that the domain of f is the set of all real numbers. As we shall see later, f is called a *quadratic function*, and as such has a completely predictable graph shape (a parabola). So, the graph shown in Figure 5 is a *complete graph*, in that it shows all relevant information about the shape of the graph of f. Consequently, we see from the graph that the range of f is the interval $[-5, \infty)$. ❖

A certain amount of caution needs to be exercised when using graphing calculators to graph functions. We shall spend a fair amount of time analyzing specific classes of algebraically-defined functions in order to garner some insight as to what to expect the graphs of these function to look like, as the accuracy and usefulness of a graph produced by a calculator is highly dependent on the viewing window chosen for display. In addition, important features of a graph may not be visible. For instance, consider the function r defined by $r(x) = \dfrac{x^2 - 9}{x - 3}$. By inspection, we see that the domain of r is the set of all real numbers, excluding 3. Consequently, the graph of r cannot have a point when $x = 3$. Figure 6 shows a sketch of the graph of r.

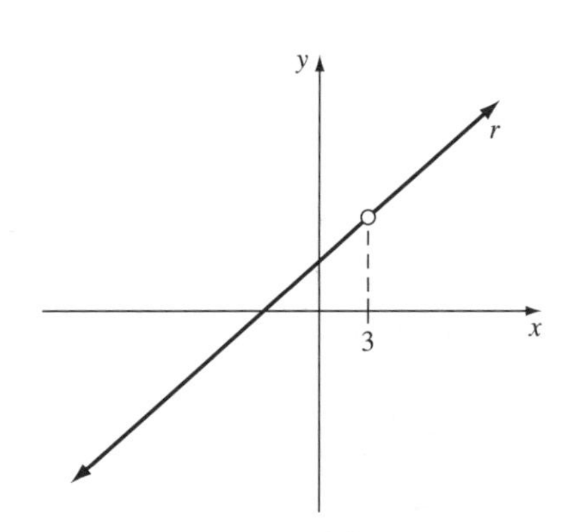

Figure 6

By comparison, a graphing calculator might show what *appears* to be a straight line; because of the resolution available on the calculator viewscreen, the gap in the graph of r will not necessarily appear for every choice of window settings.

Graphs as Analytical Tools

Figure 7 shows the graph of some function f. Gridlines each mark one unit.

We make several observations about f:

- the domain of f is $(-\infty, 8]$
- the range of f is $(-\infty, 5]$
- $f(x)$ is *increasing* on the intervals $(-\infty, -4]$ and $[-1, 2]$
- $f(x)$ is *decreasing* on the intervals $[-4, -1]$ and $[5, 8]$
- $f(x)$ is *constant* on the interval $[2, 5]$
- the approximate x-intercepts of the graph of f are -5.9, -2.4, and 0.8
- the approximate y-intercept of the graph of f is -2.7

All of these features are readily recognized by inspection of the graph of f.

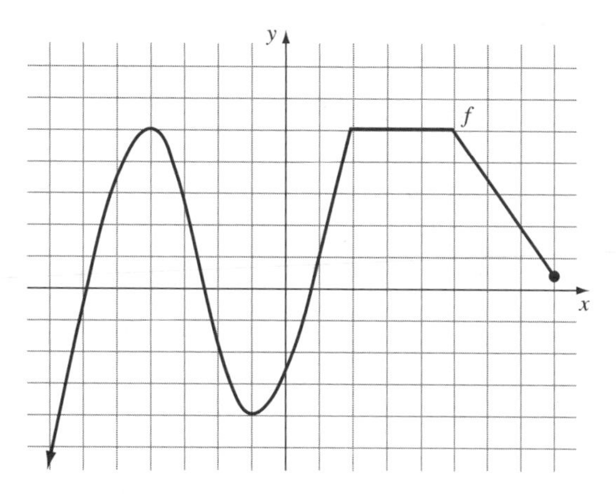

Figure 7

We use the term *monotonicity* to refer to increasing, decreasing, and constant behavior of a function. We define the terms *increasing*, *decreasing*, and *constant* more formally as follows:

Increasing, Decreasing, Constant Behavior (Monotonicity)

Given a function f,

- $f(x)$ is **increasing on an interval I** if $f(a) < f(b)$ whenever a and b are in I and $a < b$
- $f(x)$ is **decreasing on an interval I** if $f(a) > f(b)$ whenever a and b are in I and $a < b$
- $f(x)$ is **constant on an interval I** if $f(a) = f(b)$ whenever a and b are in I

A function's increasing, decreasing, and constant behavior is demonstrated graphically by the direction the graph takes as we trace the graph from left to right:

- if the function is increasing (increasing outputs) on an interval, then the graph of the function rises from left to right on that interval
- if the function is decreasing (decreasing outputs) on an interval, then the graph of the function falls from left to right on that interval
- if the function is constant on an interval, then the graph of the function is a horizontal line on that interval

We also restate definitions of intercepts using functional notation:

Intercepts

Given a function f, graphed in the xy-plane,

- the y-intercept of the graph of f is $f(0)$, provided $f(0)$ is defined
- the x-intercepts of the graph of f are those values of x such that $f(x) = 0$, provided such values exist

When working with algebraically-defined functions, we often use features of graphing calculators to answer questions about monotonicity and intercepts.

EXAMPLE 3 **Analyzing a Function Using Its Graph**

Based on data from the Board of Trustees of the Social Security Administration, the assets of the Social Security "trust fund" may be approximated by

$$f(t) = -0.0129t^4 + 0.3087t^3 + 2.176t^2 + 62.8466t + 506.2955$$

for $0 \leq t \leq 35$, where $f(t)$ is measured in millions of dollars and t is in years, with $t = 0$ corresponding to 1995.

(a) Determine the intervals where $f(t)$ is increasing, and the intervals where $f(t)$ is decreasing. Interpret the results.
(b) Determine the intercepts of the graph of f and interpret the results.

Solution It will prove to be beneficial to have a graph of f. See Figure 8.

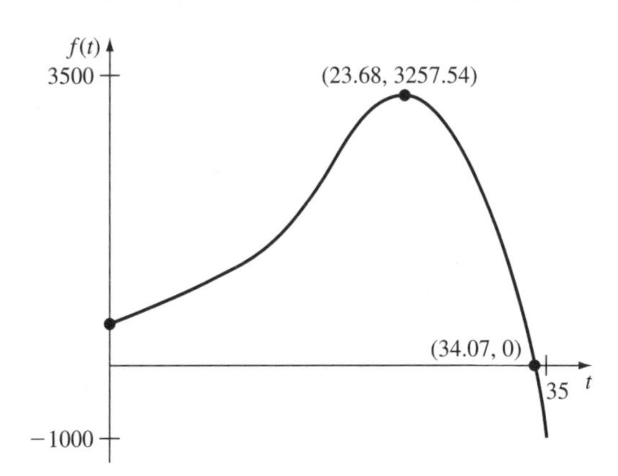

Figure 8

The listed coordinates in Figure 8 have been rounded to two decimal places. All were obtained using built-in routines on a graphing calculator.

(a) We see that $f(t)$ is increasing for $0 \leq t \leq 23.68$ and $f(t)$ is decreasing for $23.68 \leq t \leq 35$. So, the assets in the Social Security Trust fund are expected to increase from 1995 until sometime during 2018, at which point in time the asset level will begin to decrease.
(b) The y-intercept is

$$f(0) = 506.2955$$

So, as of the beginning of 1995, the "trust fund" contained about \$506,295,500 in assets.

The t-intercept is approximately 34.07, so the model predicts that shortly after the start of 2029 the "trust fund" will be depleted. ❖

Graphs are also invaluable tools for comparing two (or more) functions.

EXAMPLE 4 **Graphical Analysis of Two Functions**

Figure 9 shows the graphs of two functions f and g. Gridlines each mark one unit.

(a) Determine the interval(s) where $g(x) > f(x)$.
(b) Determine the interval(s) where both $f(x)$ and $g(x)$ are decreasing.
(c) Determine all values of x such that $f(x) \geq g(x)$ and $g(x) \leq -2$.

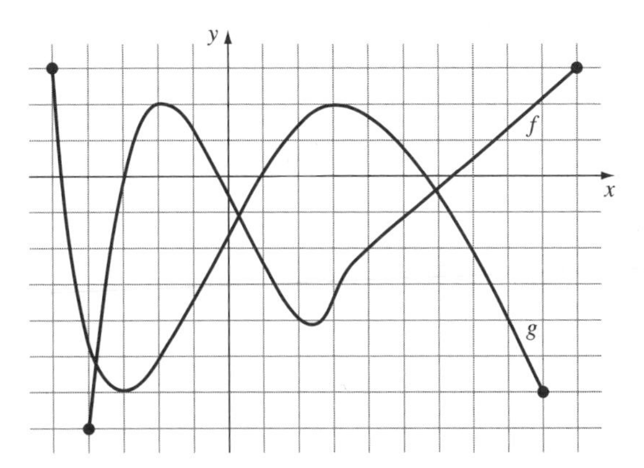

Figure 9

Solution We first note that the domain of f is the interval $[-4, 10]$ and the domain of g is the interval $[-5, 9]$ so that any direct comparisons of f and g will be restricted to the inputs that the functions have in common, namely the numbers in the interval $[-4, 9]$. We also estimate that $f(x) = g(x)$ when $x = -3.8$, $x = 0.3$, and $x = 5.9$. We give our best approximations when necessary in the following.

(a) The graph of g is strictly above the graph of f on the intervals $[-4, -3.8)$ and $(0.3, 5.9)$.

(b) We see that $f(x)$ is decreasing on the interval $[-2, 2.4]$ while $g(x)$ is decreasing on the intervals $[-5, -3]$ and $[3, 9]$. Since there is no overlap between the interval of decrease for f and the intervals of decrease for g, there are no intervals where both $f(x)$ and $g(x)$ are decreasing. Graphically, this means that there are no intervals where both graphs are falling from left to right.

(c) We see that $f(x) \geq g(x)$ on the intervals $[-3.8, 0.3]$ and $[5.9, 9]$. Also, $g(x) \leq -2$ on the intervals $[-4.5, -0.3]$ and $[6.9, 10]$. The overlap of the first pair of intervals and the second pair of intervals is $[-3.8, -0.3]$ and $[6.9, 9]$. Graphically, this corresponds to the intervals where the graph of f is either strictly above or meeting the graph of g and the graph of g is no higher than $y = -2$ on the grid. ❖

Some Standard Graphs

We conclude this section by listing some relatively simple functions. We refer to the graphs of these functions as *standard graphs*. Oftentimes we work with functions defined algebraically whose formulas are variations of the formulas of the functions on this list. As we will see later, the resulting graphs can often be obtained readily based on knowledge of these standard graphs.

The functions we consider are:

- the *identity function*, $f(x) = x$
- the *cubing function*, $f(x) = x^3$
- the *cube root function*, $f(x) = \sqrt[3]{x}$
- the *squaring function*, $f(x) = x^2$
- the *square root function*, $f(x) = \sqrt{x}$
- the *absolute value function*, $f(x) = |x|$

We leave it as an exercise to obtain these standard graphs.

$f(x) = x$

$f(x) = x^2$

$f(x) = x^3$

$f(x) = \sqrt{x}$

$f(x) = \sqrt[3]{x}$

$f(x) = |x|$

PROBLEM SET 2.2

For problems 1–4, use the given graphs to determine the indicated function values. Assume that gridlines each mark one unit. Give your best estimate when necessary.

1. $f(3), f(0), f(-2)$

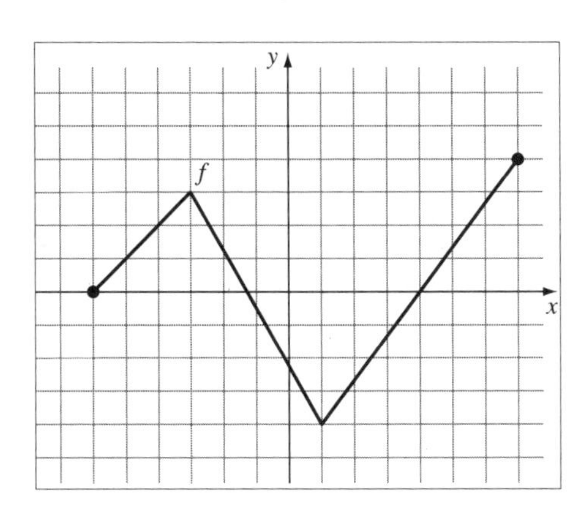

2. $g(-1), g(2), g(4)$

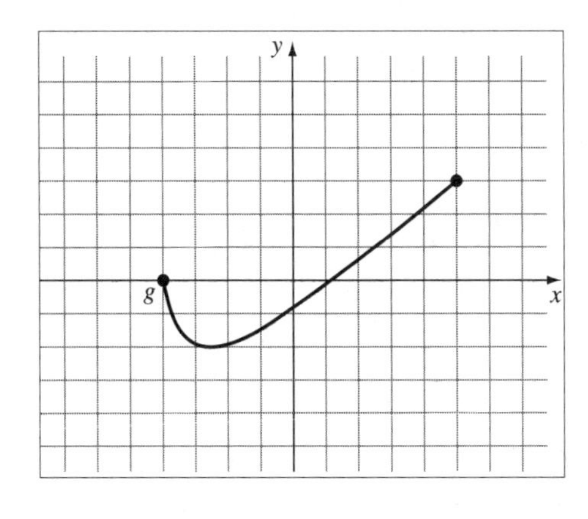

3. $h(1), h(-5), h(-2)$

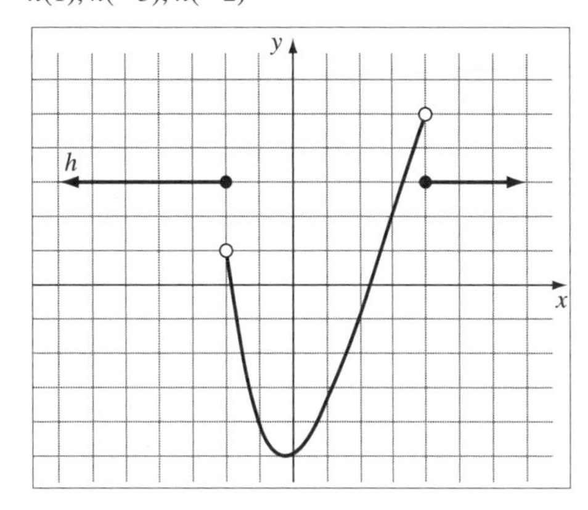

4. $k(-2), k(0), k(1)$

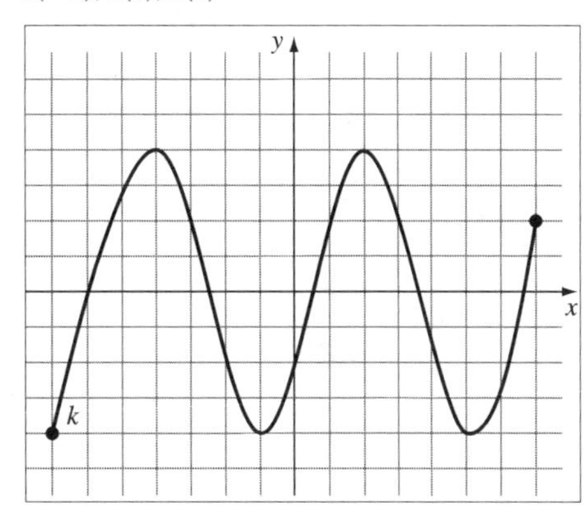

5. Find the domain and range of the functions defined by the graphs in problems 1–4.

For problems 6–9, use your calculator to sketch a graph of the function. Indicate the window settings that you use.

6. $f(x) = 3 - x - 2x^2$

7. $g(x) = \dfrac{x^2 - 9}{x + 3}$

8. $h(x) = \sqrt{64 - x^2}$

9. $k(x) = x^3 - 3x^2 + 4x - 1$

For problems 10–15, sketch a graph and use it to determine the domain and range of the function. Use interval notation.

10. $f(x) = \sqrt{x^2 - 4}$
11. $h(x) = x^2 - 9$
12. $g(x) = x^2 - 6x + 3$ for $-1 \le x \le 5$
13. $k(x) = 2^x$
14. $m(x) = \dfrac{x}{x^2 + 1}$
15. $s(x) = \dfrac{1}{3}x^3 + x^2 - 15x - 3$ for $-8 \le x \le 6$

For problems 16 and 17, refer to the given graphs of functions f and g.

16. Find each of the following. Assume that gridlines each mark one unit.

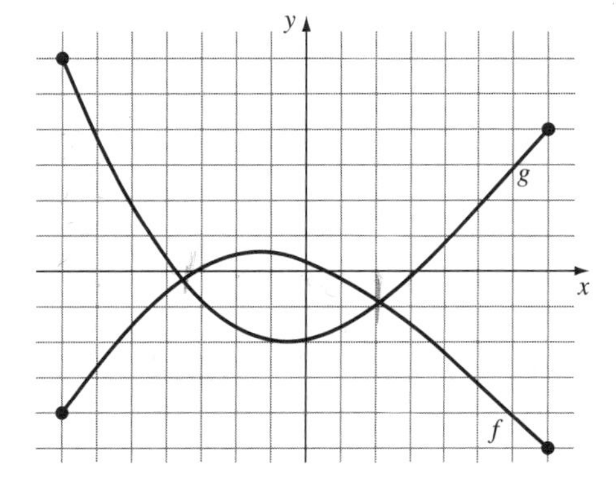

(a) $g(3)$
(b) all x such that $g(x) = 2$
(c) all x such that $f(x) = 0$
(d) the intervals where $f(x) > g(x)$
(e) the intervals where $g(x)$ is decreasing
(f) the intervals where $f(x)$ is increasing and $g(x)$ is decreasing

17. Find each of the following. Assume that gridlines each mark one unit.

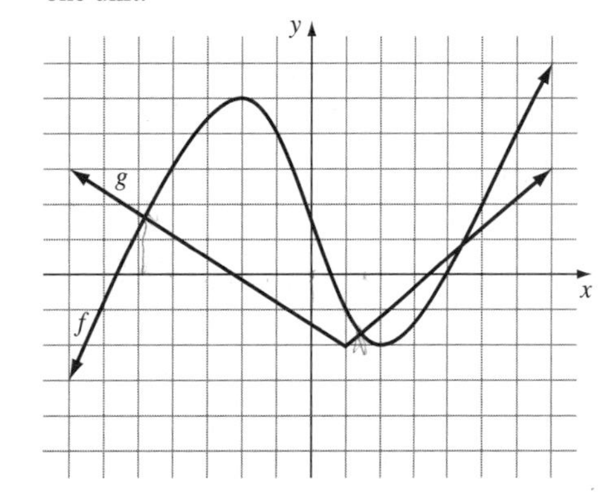

(a) $f(1)$
(b) $g(0)$
(c) the intervals where $f(x) < g(x)$
(d) the intervals where $f(x)$ is increasing
(e) all x such that $f(x) = g(x)$
(f) the intervals where $f(x) > 2$

For problems 18–23, find intervals (use interval notation) where the given function is increasing, intervals where it is decreasing, and all intercepts of the graph of the function. If rounding is necessary, then round to three decimal places.

18. $m(r) = -2r^2 + 8r - 18$
19. $f(t) = -0.2t^3 + 0.1t^2 + t - 1$
20. $g(x) = 0.5x^5 - 2x^4 - 0.6x + 3$ for $-10 \le x \le 10$
21. $h(p) = 2p^4 - 8p^3 + 2p^2 + 3p$ for $-5 \le x \le 5$
22. $k(x) = 2x^3 - x + 3$
23. $f(x) = -x^4 + 0.5x^3 - 2x^2 + x - 2$

For problems 24–29, determine if the given graphs define y as a function of x.

24.

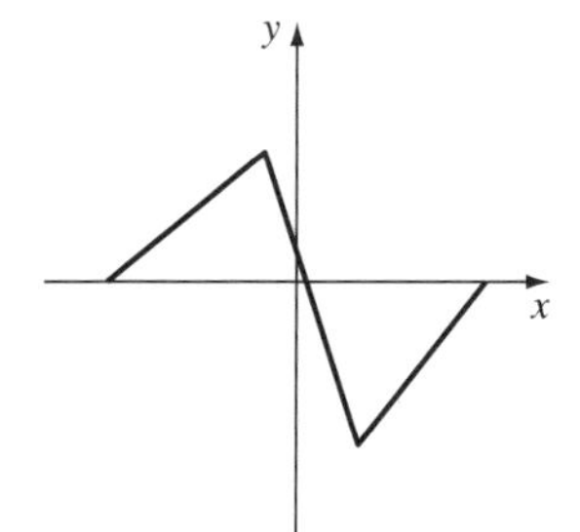

25.

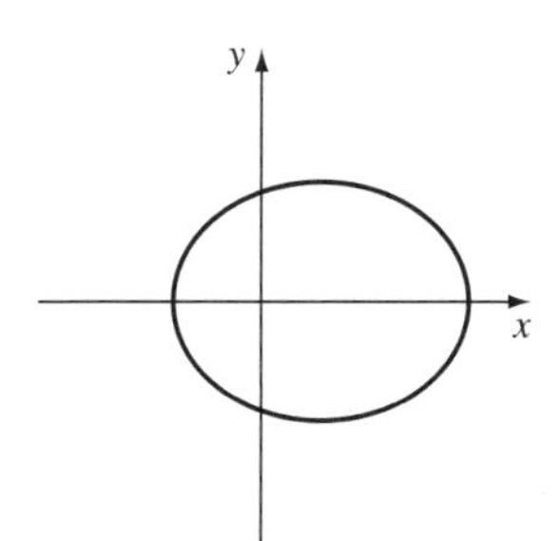

26.

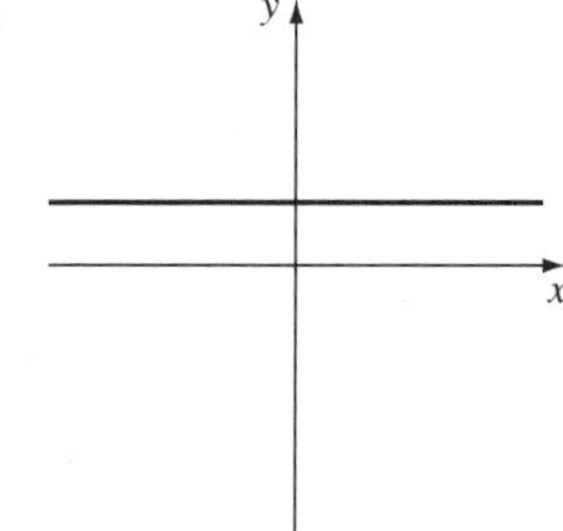

27.

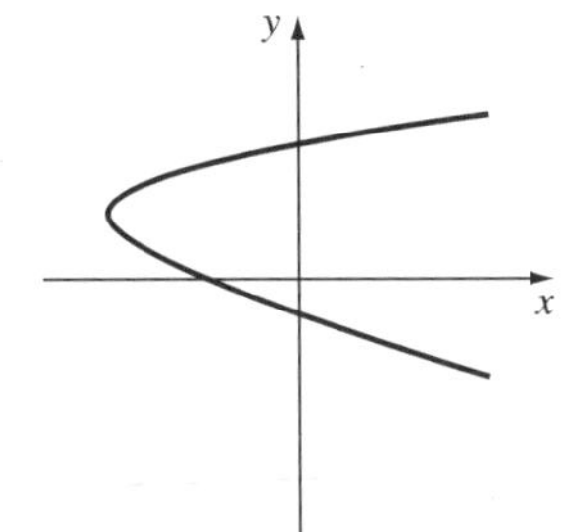

28.

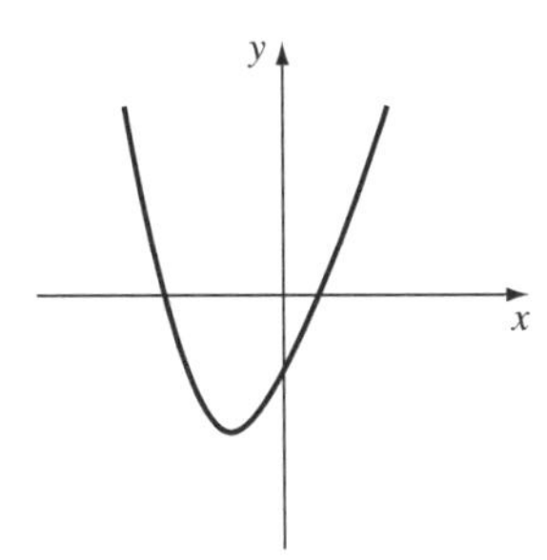

29.

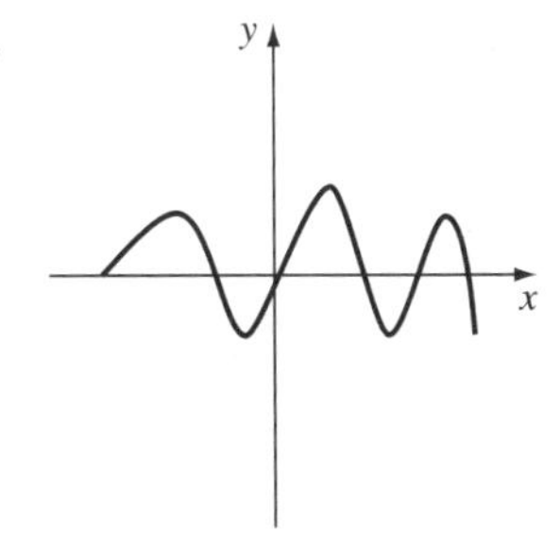

For problems 30–35, use graphical techniques to solve the given equations and inequalities. Round answers to three decimal places, where necessary. Use interval notation to express your answers.

30. $3x^2 - 2x + 1 \geq 0$

31. $x^3 - 2x + 5 = 2$

32. $x^4 - 2x^2 \leq 3$

33. $x^5 - 4x^2 < 2x - 4$

34. $x^4 + x^2 + 2 = 1$

35. $x^3 - 5x^2 > 2 - x$

For problems 36 and 37, sketch the graph of a function $f(x)$ that has **all** of the listed properties. (There may be several correct graphs.)

36. **(a)** domain is $[-6, 8]$
(b) range is $[-3, 5]$
(c) $f(-1) = 4, f(2) = 0$
(d) f is increasing on $[0, 3]$ only

37. **(a)** domain is $[-8, 8]$
(b) range is $[-10, 4]$
(c) f is decreasing on $[-8, -4]$ and constant on $[1, 3]$
(d) the average rate of change of f from $x = -1$ to $x = 3$ is 3

38. Suppose that I take a fresh turkey out of the refrigerator in the morning to prepare for Thanksgiving dinner. It is stuffed, then put into the oven. Once it is fully cooked, it sits for a short while before being carved. Sketch a plausible graph that shows the temperature of the turkey as a function of time on Thanksgiving Day.

39. It warmed up throughout the morning, and then suddenly got much cooler around noon, when a storm came through. After the storm, it warmed up before cooling off after sunset. Sketch a possible graph of this day's temperature as a function of time.

40. Right after a certain drug is administered to a patient with a rapid heart rate, the heart rate plunges dramatically and then slowly rises again as the drug wears off. Sketch a possible graph of the heart rate against time from the moment the drug is administered.

41. A bug starts out ten feet from a light, flies closer to the light, then farther away, then closer than before, then farther away. Finally the bug hits the bulb and flies off. Sketch a possible graph of the distance of the bug from the light as a function of time for a span of two minutes.

OBJECTIVE

1. Solve Applied Problems

2.3 More Modeling and Problem-Solving Using Graphs

If we can model the relationship between two quantities with a function, then the graph of that function is often the most useful tool for answering questions about how the quantities are related. We will draw upon previous material and use capabilities of graphing calculators to solve problems that would be extremely difficult if handled only algebraically.

EXAMPLE 1 Analyzing Cost, Revenue, and Profit

PUQ Publishing has entered into an agreement with all of the institutions in the Independent Canadian University System (ICUS) to be the exclusive provider of texts for the mandatory freshman course, *Mathematics of Hockey*. The publishing company knows that it will have start-up costs of $800,000 for the project and that the cost of publishing the texts will amount to $18,000 per 1000 texts produced. The agreement with ICUS also states that textbooks will each be sold at a price, $\$p$, as given by

$$p = 75 - 0.5x$$

where x measures the number of books ordered, in thousands. The publisher can provide at most 120,000 textbooks.

(a) Define a function C that gives the cost, in thousands of dollars, of publishing x thousand textbooks.
(b) Define a function R that gives the expected revenue, in thousands of dollars, from selling x thousand textbooks.
(c) Define a function P that gives the expected profit, in thousands of dollars, from selling x thousand textbooks.
(d) To the nearest book, what size of textbook order will result in the largest possible revenue for PUQ Publishing? What is the maximum possible revenue?
(e) How many textbooks must be ordered by ICUS in order to guarantee a profit for PUQ Publishing?

Solution Note that for all functions defined in this example, x represents the input variable and $0 \leq x \leq 120$.

(a) $C(x) = 800 + 18x$
(b) In general, *revenue* = *(number of units sold)* × *(price per unit)*, so here we have

$$R(x) = xp = x(75 - 0.5x)$$
$$R(x) = 75x - 0.5x^2$$

Notice that since x is in thousands of textbooks and p is in dollars per book, $R(x)$ is in thousands of dollars.

(c) Profit is revenue minus cost, so

$$P(x) = R(x) - C(x)$$
$$P(x) = (75x - 0.5x^2) - (800 + 18x)$$
$$P(x) = -0.5x^2 + 57x - 800$$

(d) We will determine the number of books that maximizes revenue by analyzing the graph of R. In order to gain some insight into the behavior of R, we have evaluated the function at a variety of inputs and used the results to find an appropriate viewing window on a graphing calculator to see the graph of R. Figure 1 shows a sketch of this graph.

x	$R(x)$
20	1300
40	2200
60	2700
80	2800
100	2500
120	1800

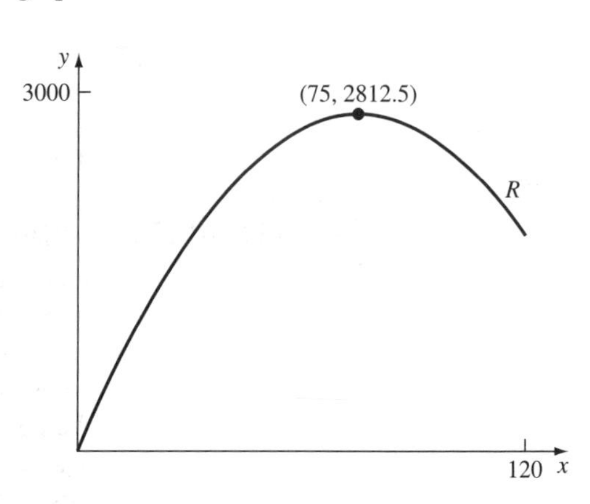

Figure 1

Using a maximum-finding routine on a graphing calculator, we find that the maximum output for R occurs when $x = 75$. This means that an order of 75,000 textbooks will result in the maximum possible revenue, which is \$2,812,500.

(e) The publishing company will experience a profit provided revenue exceeds costs. In other words, whenever $R(x) > C(x)$, the company is profitable. Figure 2 shows graphs of C and R.

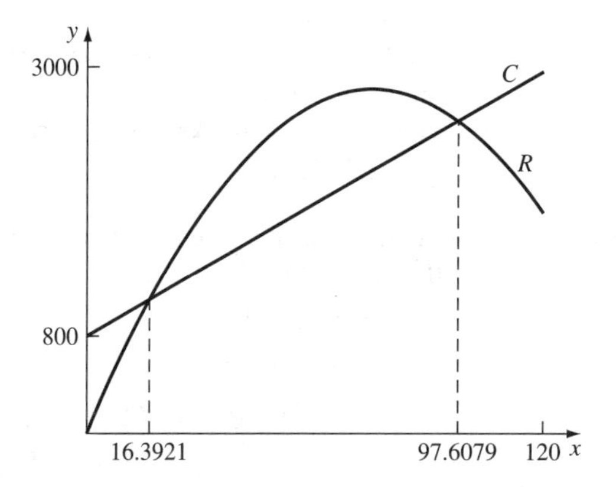

Figure 2

Using an intersection routine on a graphing calculator, we find that $R(x) = C(x)$ when $x \approx 16.3921$ and when $x \approx 97.6079$. We have estimated these values to the nearest one thousandth since x is given in thousands of books. From the graphs, we see that $R(x) > C(x)$ for values of x in between these estimates. We conclude that PUQ Publishing will be profitable provided the size of the book order is between 13,393 books and 97,607 books. Note that the endpoints of the range of order size are adjusted to the nearest textbook that gaurantees revenue larger than cost. ❖

It should be noted that determining when $R(x) > C(x)$ in Example 1 could also be viewed algebraically as solving the inequality

$$75x - 0.5x^2 > 800 + 18x$$

Inequalities of this sort, which involve expressions that are not linear, can be very difficult to solve algebraically, so we typically opt instead for a graphical solution. Also, it is possible at times to use techniques from calculus to answer questions like that of the maximum possible revenue, but we will generally elect to use graphs in those situations.

In the next example, we construct a function that models a physical setting.

EXAMPLE 2 **Enclosing a Flower Bed**

A flowerbed will be dug alongside the wall of a building. The perimeter of the bed will be formed by four rigid pieces of fence, as shown in Figure 3. Each piece of fence is 10 feet long. Let x denote the width of the flowerbed (in feet) along the building wall.

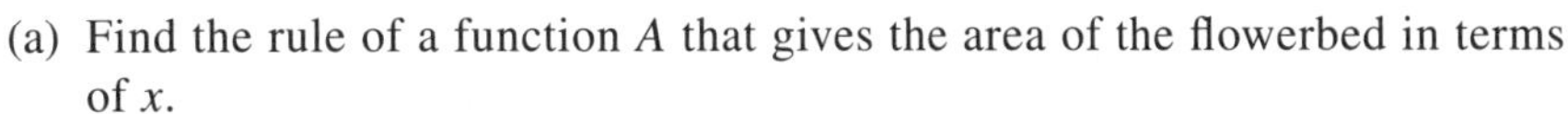

(a) Find the rule of a function A that gives the area of the flowerbed in terms of x.

(b) Determine the value of x that results in the flowerbed of maximum area and find that area.

(c) Determine the value(s) of x that result in a flowerbed with an area of 200 square feet.

Figure 3

Solution We first note that $0 \le x \le 20$. Also, the flowerbed can be viewed as being comprised of a rectangle adjacent to an isosceles triangle of height h, as shown in Figure 3.

(a) The area of the flowerbed is

$$\text{(area of rectangular part)} + \text{(area of triangular part)}$$

So,

$$A(x) = 10x + \frac{1}{2}xh$$

Now, by applying the Pythagorean Theorem to one of the right triangles illustrated in Figure 3, we have

$$h^2 + \left(\frac{x}{2}\right)^2 = 10^2$$

$$h^2 = 100 - \frac{x^2}{4}$$

$$h = \sqrt{100 - 0.25x^2}$$

Consequently,

$$A(x) = 10x + 0.5x\sqrt{100 - 0.25x^2}$$

(b) As in Example 1, we form a table of selected values for $A(x)$ (rounded to two decimal places) to help determine an appropriate viewing window and use a graphing calculator to obtain a sketch of the graph of A. See Figure 4.

x	$A(x)$
3	44.83
6	88.62
9	130.19
12	168
15	199.61
18	219.23

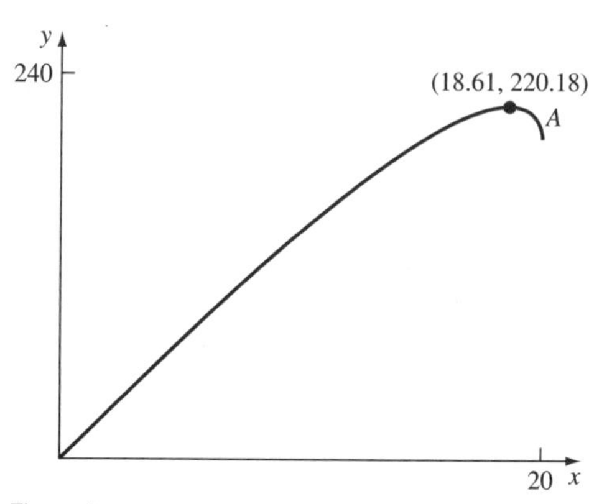

Figure 4

Using a maximum-finding routine on a graphing calculator, we find that the maximum output for A occurs when $x \approx 18.61$. This yields a corresponding area of about 220.18 square feet.

(c) We will graph the horizontal line with equation $y = 200$ and find all points of intersection of that line with the graph of A. This gives us a graphical solution to the equation $A(x) = 200$. See Figure 5.

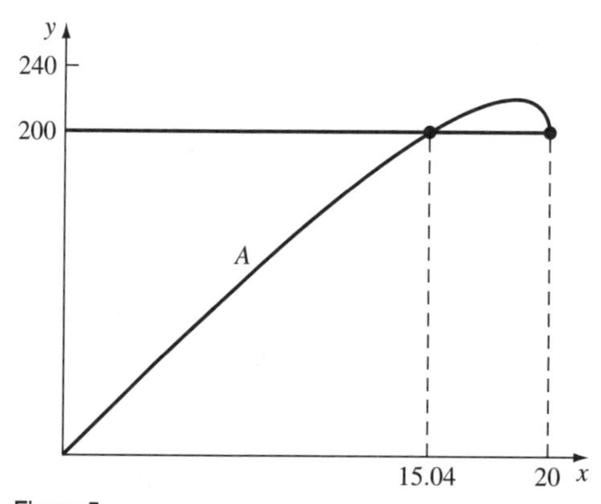

Figure 5

Rounded to two decimal places, we find that the flowerbed will have an area of 200 square feet when $x \approx 15.04$ feet and when $x = 20$ feet. Notice, though, that if $x = 20$, then the flowerbed will be rectangular. ❖

PROBLEM SET 2.3

1. The profit, in thousands of dollars, realized by a corporation is given by

$$P(a) = -2a^3 + 35a^2 - 100a + 200,$$

where a is the amount of money spent on advertising, in thousands of dollars and $0 \le a \le 12$. Find the advertising level that minimizes profit and the advertising level that maximizes profit. Give each answer to the nearest dollar.

2. Research shows that x hundred units of a commodity can be sold at a price of $\$p$ per unit according to the *price-demand equation* $p = 100 - 0.5x$ for $0 \le x \le 160$. In addition, the cost of producing and selling x hundred units of the commodity is given, in hundreds of dollars, by $C(x) = 900 + 26x$.
   - **(a)** Give the rule of the revenue function, R, for the commodity.
   - **(b)** Give the rule of the profit function, P, for the commodity.
   - **(c)** Find the levels of production (and sales) where cost and revenue are equal.
   - **(d)** Determine the range of sales that guarantees revenue of at least \$400,000. Give your answer to the nearest unit of commodity.
   - **(e)** Find all levels of sales that guarantee a profit of exactly \$90,000. Give your answer(s) to the nearest unit of commodity.

3. Using the profit function from the previous problem, determine the level of sales of the commodity that maximizes profit.

4. A manufacturer wants to design an open box that has a square base and a surface area of 108 square inches. What dimensions will produce a box with the maximum possible volume?

5. A textbook is being prepared in such a way that each page of the text should contain 24 square inches of print. The margins at the top and bottom of the page are 1.5 inches and the margin on each side is 1 inch. Find the dimensions of the page so as to minimize the amount of paper used.

6. A package in the shape of a right circular cylinder can be sent by parcel post if the length plus the girth of the package does not exceed 108 inches. Find the radius and length of such a package of maximum possible volume that can still be sent by parcel post.

7. A closed box with a square base is to have a volume of 343 cubic inches. The material for the sides and the top costs \$0.02 per square inch, and the material for the base costs \$0.04 per square inch. Determine the dimensions of the box that minimizes the cost of the materials.

8. Chrysler Corporation reports that annual revenue for the years 1988–1996 can be modeled by $R(t) = -146.38t^3 + 6014.19t^2 - 75{,}778.84t + 333{,}246.23$ millions of dollars per year, where $t = 8$ corresponds to 1988.
   - **(a)** During which year was revenue the greatest?
   - **(b)** During which year was revenue the least?
   - **(c)** During which year(s) was revenue closest to \$58,000,000,000?

9. A rancher has 200 feet of fencing available to enclose two adjacent rectangular corrals. Each corral is to be the same size. What dimensions for each corral will maximize the area enclosed by the two corrals?

10. The owner of an apple orchard has determined that the annual yield per apple tree is fairly constant at 412 pounds when no more than 45 trees are planted per acre. For each additional tree planted on an acre, the annual yield per tree drops by about 4 pounds due to overcrowding.
    - **(a)** Write the rule of a function A that describes the annual yield of an acre of trees in terms of n, the number of trees planted per acre.
    - **(b)** Using your model from **(a)**, sketch a graph and determine the number of trees that should be planted per acre in order to maximize the annual yield per acre.
    - **(c)** Determine how many trees per acre guarantees an annual yield per acre of at least 18,000 pounds. Illustrate your result using a graph.

11. Four feet of wire is to be cut into two pieces and used to form a square and a circle.
    - **(a)** How much wire should be used for the square and how much for the circle in order to enclose the least total area?
    - **(b)** How much wire should be used for the square and how much for the circle in order to enclose the most total area?

12. The owner of a retail lumber store wants to construct a fence to enclose a rectangular outdoor storage area adjacent to the store, using all of one wall of the store as part of one side of the storage area. The wall of the store is 100 feet long. Find the dimensions of the enclosure of maximum area if 240 feet of fencing is to be used.

13. A Norman window has the shape of a rectangle surmounted by a semicircle. If the window is to have a perimeter of 30 feet, then what dimensions of the rectangular part of the window allow for the maximum amount of sunlight through the window (round lengths to two decimal places)?

14. A grain silo has the shape of a right circular cylinder surmounted by a hemisphere. If a silo is to be constructed to have a capacity of 2000 cubic feet, then what height and radius of the silo will require the least amount of construction material?

Interlude: Progress Check 2.3, Part 1

1. A gas tank has ends that are hemispheres and a cylindrical midsection. Suppose that the midsection is 8 feet long and let r be the radius of the hemispherical ends, in feet.

(a) Let $V(r)$ denote the volume of the tank. Give the rule of $V(r)$. Recall that the volume of a sphere of radius r is $\frac{4}{3}\pi r^3$. Show any diagrams you use to find your result.

(b) Determine the appropriate radius, accurate to two decimal places, to obtain a volume of 560 cubic feet. Label your result appropriately and show all graphs that you used to obtain that result.

2. When a popular style of running shoe is priced at $80, Runner's Emporium sells an average of 96 pairs per week. Based on comparative research, management believes that for each decrease of $2.50 in price, four additional pairs of shoes could be sold weekly.

(a) Let n denote the number of $2.50 decreases in price and let q denote the number of pairs of shoes sold. Write an equation that gives n in terms of q.

(b) Write the rule of a function R that gives the expected weekly revenue from sales of this particular style of shoe as a function of p, the price per pair (in dollars).

(c) Determine the selling price per pair that will result in maximum possible revenue, as well as the maximum possible revenue. Show a graph that supports your claims.

(d) Determine the lowest price (to the nearest cent) that will result in weekly revenue of at least $7000 from sales of the shoes. Show a graph that supports your claims.

Interlude: Progress Check 2.3, Part 2

For these problems, construct a function that models the situation. Draw diagrams and declare what all variables represent. Show any graphs that you use to draw your conclusions.

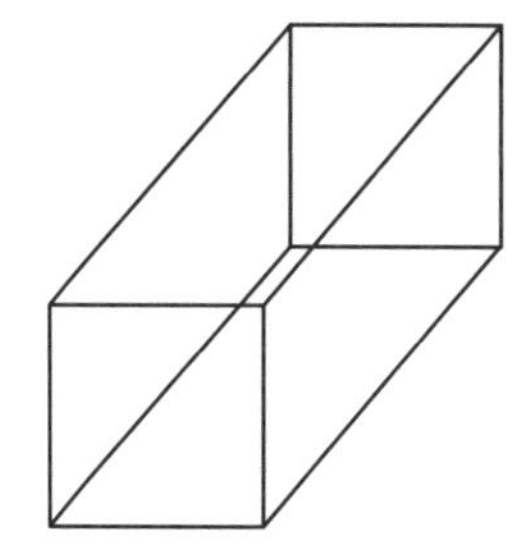

Figure 6

1. According to the Federal Express website (www.fedex.com), any package weighing less than 151 pounds that exceeds 165 inches in length plus girth (girth is distance around front of package) is to be considered an "extra-large" package. Consider a box with a square front (see Figure 6). Determine the largest-volume package with a square front that Federal Express will accept and not declare to be an "extra-large package (assume that the weight is less than 151 pounds).

2. A participant in an orienteering event must get to a specific tree in the woods as fast as possible. If she travels 300 meters east along a trail, she will then be 600 meters directly south of the tree. She can run at a rate of 120 meters per minute along the trail, but she can only move through the woods at a rate of 50 meters per minute. Your task is to determine the path she should take in order to get to the tree as quickly as possible. Carefully state your conclusion.

OBJECTIVES

1. Define Arithmetic Combinations of Functions
2. Define Composition of Two Functions
3. Construct and Graph Piecewise-defined Functions
4. Describe Transformations of Functions

2.4 New Functions from Old

Example 1 of the previous section constructed a profit function from cost and revenue functions. In this section we consider other ways of combining functions to produce new functions.

Arithmetic Combinations

The standard arithmetic operations of addition, subtraction, multiplication and division provide a natural model for ways of combining the rules defining functions to produce new functions. For instance, given two functions f and g, we can form the corresponding **sum function**, which we denote by $f + g$, whose rule is given by the formula

$$(f + g)(x) = f(x) + g(x)$$

for all x common to the domains of f and g. In other words, the rule of $f + g$ is "given any input, x, the output is the sum of the outputs generated by f and g." We analogously define **difference**, **product** and **quotient** functions.

Sum, Difference, Product and Quotient Functions

Given two functions f and g, the corresponding

- Sum function, $f + g$, is defined by $(f + g)(x) = f(x) + g(x)$
- Difference function, $f - g$, is defined by $(f - g)(x) = f(x) - g(x)$
- Product function, $f \cdot g$, is defined by $(f \cdot g)(x) = f(x) \cdot g(x)$
- Quotient function, $\frac{f}{g}$, is defined by $\left(\frac{f}{g}\right)(x) = \frac{f(x)}{g(x)}$, for $g(x) \neq 0$

Note that the domains of the sum, difference, product and quotient functions are all inputs common to both f and g, except that we must exclude from the domain of the quotient function any inputs for g that generate an output of zero.

EXAMPLE 1 Forming Sum, Difference, Product and Quotient Functions

Let f and g be the functions defined by $f(x) = x^2$ and $g(x) = \sqrt{x - 5}$. Find the rules of $f + g, f - g, f \cdot g$, and $\frac{f}{g}$. Determine the domain of each.

Solution The domain of f is the interval $(-\infty, \infty)$, while the domain of g is the interval $[5, \infty)$. So, for the sum, difference and product functions, the domains are the inputs common to f and g, namely, the interval $[5, \infty)$. In the case of the quotient function, we must exclude the input 5 as $g(5) = 0$, so the domain of $\frac{f}{g}$ is the interval $(5, \infty)$.

The rules of the sum, difference, product and quotient functions are, respectively,

$$(f+g)(x) = f(x) + g(x) = x^2 + \sqrt{x-5}$$
$$(f-g)(x) = f(x) - g(x) = x^2 - \sqrt{x-5}$$
$$(f \cdot g)(x) = f(x) \cdot g(x) = x^2\sqrt{x-5}$$
$$\left(\frac{f}{g}\right)(x) = \frac{f(x)}{g(x)} = \frac{x^2}{\sqrt{x-5}}$$

❖

It is rather challenging to describe the sum, difference, product and quotient functions when *f* and *g* are not both defined algebraically. However, we can still resort to the definitions of these functions in order to evaluate them at specific inputs.

EXAMPLE 2 Evaluating Sum, Difference, Product and Quotient Functions

Figure 1 shows the graphs of two functions, h and k. Gridlines each mark one unit.

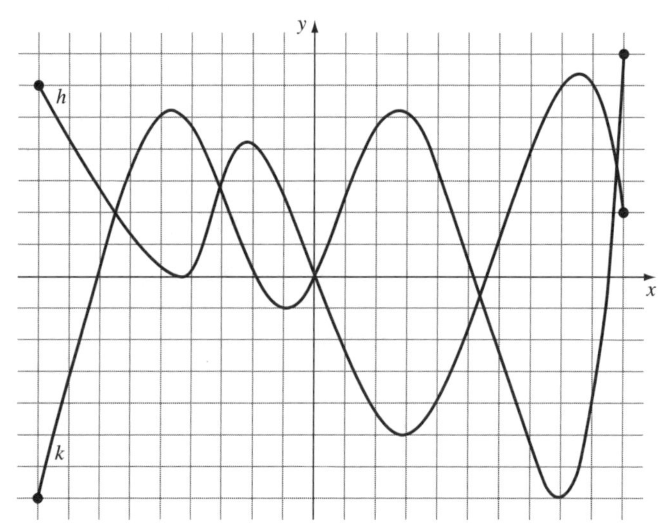

Figure 1

(a) Calculate $(h+k)(2)$

(b) Calculate $(h-k)(-1)$

(c) Calculate $(h \cdot k)(-5)$

(d) Calculate $\left(\frac{h}{k}\right)(-4)$

(e) Calculate $(h-h)(-8)$

(f) Calculate $(k+k)(6)$

(g) Calculate $(h \cdot h)(3)$

(h) Calculate $(k-h)(-1)$

(i) Calculate $\left(\frac{k}{h}\right)(-4)$

Solution As usual when working with functions defined graphically, we estimate all function values.

(a) $(h+k)(2) = h(2) + k(2) = -4.3 + 4.6 = 0.3$

(b) $(h-k)(-1) = h(-1) - k(-1) = 2.4 - (-1) = 3.4$

(c) $(h \cdot k)(-5) = h(-5) \cdot k(-5) = (0.2)(5) = 1$

(d) $\left(\frac{h}{k}\right)(-4) = \frac{h(-4)}{k(-4)} - \frac{0.2}{4.5} = \frac{2}{45}$

(e) $(h - h)(-8) = h(-8) - h(-8) = 0$
(f) $(k + k)(6) = k(6) + k(6) = -2.9 + (-2.9) = -5.8$
(g) $(h \cdot h)(3) = h(3) \cdot h(3) = (-5)(-5) = 25$
(h) $(k - h)(-1) = k(-1) - h(-1) = -1 - 2.4 = -3.4$
(i) $\left(\frac{k}{h}\right)(-4) = \frac{k(-4)}{h(-4)} = \frac{4.5}{0.2} = \frac{45}{2}$ ❖

Composing Functions

Suppose that a cylindrical tank with a radius of 2 feet is filling with water at a constant rate such that the height of the water in the tank is increasing at a rate of 6 inches per minute. Consequently, the height of water in the tank is a function of the amount of time the tank has been filling. How do we describe the volume of water in the tank as a function of the amount of time the tank has been filling?

If we let h denote the height of the water in the tank, in feet, after t minutes of filling, then

$$h = 0.5t$$

Similarly, if V is the volume of water (in cubic feet) in the tank given a water height of h feet, then

$$V = \pi(2)^2 h$$
$$V = 4\pi h$$

We have one equation defining h as a function of t and one equation defining V as a function of h, so we will write $h = g(t)$ and $V = f(h)$. Clearly, in order to calculate V at some particular time t, we must first evaluate $h = g(t)$ and then use that result as an input for f to calculate $V = f(h)$. Using functional notation, we will have calculated

$$V = f(g(t))$$

For instance, ten minutes into the filling process, we see that

$$h = g(10) = 0.5(10) = 5 \text{ feet}$$

so

$$V = f(5) = 4\pi(5) = 20\pi \text{ cubic feet}$$

In other words, when $t = 10$,

$$V = f(g(10)) = f(5) = 20\pi$$

We say in this situation that we formed a **composition of** g **and** f (or we **composed** g **and** f) in order to write V as a function of t. In general, given two functions f and g, the corresponding **composite function of** g **and** f, denoted by $f \circ g$, is the function whose rule is given by

$$(f \circ g)(x) = f(g(x))$$

The order in which the function names appear in a composition is important. Given two functions f and g, the composition notation $f \circ g$ indicates that g acts first, followed by f, whereas the composition notation $g \circ f$ indicates that f acts first, followed by g.

We must take care when considering the domain of the composite function $f \circ g$. Inputs must come from the domain of g, but they must be such that the corresponding outputs generated by g are in the domain of f.

Domains of Compositions

Given two functions f and g, the domain of $f \circ g$ is all numbers x in the domain of g such that $g(x)$ is in the domain of f.

EXAMPLE 3 **Calculating Compositions**

Let $f(x) = x^2 + 4$ and $g(x) = 3x - 7$. Calculate each of the following.

(a) $(f \circ g)(2)$ (b) $(g \circ f)(2)$ (c) $(f \circ f)(-3)$

(d) $(f \circ g)(x)$ (e) $(g \circ f)(x)$ (f) $(g \circ g)(x)$

Solution

(a) $(f \circ g)(2) = f(g(2)) = f(3(2) - 7) = f(-1) = (-1)^2 + 4 = 5$

(b) $(g \circ f)(2) = g(f(2)) = g(2^2 + 4) = g(8) = 3(8) - 7 = 17$

(c) $(f \circ f)(-3) = f(f(-3)) = f((-3)^2 + 4) = f(13) = 13^2 + 4 = 173$

(d) $(f \circ g)(x) = f(g(x)) = f(3x - 7) = (3x - 7)^2 + 4$
$= 9x^2 - 42x + 49 + 4 = 9x^2 - 42x + 53$

(e) $(g \circ f)(x) = g(f(x)) = g(x^2 + 4) = 3(x^2 + 4) - 7 = 3x^2 + 5$

(f) $(g \circ g)(x) = g(g(x)) = g(3x - 7) = 3(3x - 7) - 7 = 9x - 28$ 

Remark: As is illustrated in parts (a) and (b), as well as in parts (d) and (e) above, in general, $(f \circ g)(x) \neq (g \circ f)(x)$.

Piecewise-Defined Functions

Sometimes a function "changes gears" based on inputs, in that different rules for calculating outputs become employed, depending on the specific input. The "income tax function" from Section 2.1 is one such example, as it used one of six possible rules to calculate tax liability, depending on taxable income. If we denote this function by T and let i denote taxable income, then we write the rule of T as

$$T(i) = \begin{cases} 0.10i & \text{if } 0 \le i \le 8375 \\ 837.50 + 0.15(i - 8375) & \text{if } 8375 < i \le 34{,}000 \\ 4681.25 + 0.25(i - 34{,}000) & \text{if } 34{,}000 < i \le 82{,}400 \\ 16{,}781.25 + 0.28(i - 82{,}400) & \text{if } 78{,}850 < i \le 171{,}850 \\ 41{,}827.25 + 0.33(i - 171{,}850) & \text{if } 171{,}850 < i \le 373{,}650 \\ 106{,}421.25 + 0.35(i - 373{,}650) & \text{if } i > 373{,}650 \end{cases}$$

We say here that T is a **piecewise-defined** function.

EXAMPLE 4 **Evaluating and Graphing a Piecewise-Defined Function**

Suppose that f is the function defined by the rule

$$f(x) = \begin{cases} x^2 & \text{if } x \le 1 \\ 2x + 1 & \text{if } x > 1 \end{cases}$$

(a) Evaluate $f(-3)$, $f(1)$, and $f(5)$.

(b) Sketch a graph of f.

Solution (a) Since $-3 \leq 1, f(-3) = (-3)^2 = 9$
Since $1 \leq 1, f(1) = 1^2 = 1$
Since $5 > 1$, $f(5) = 2(5) + 1 = 11$

(b) We must take special care when graphing piecewise-defined functions to pay close attention to the places where the rule of the function switches formulas. In this example, the graph of f will be the graph of the squaring function over the interval $(-\infty, 1]$ and a line with slope 2 over the interval $(1, \infty)$. See Figure 2. We use an open circle to indicate that the point (1, 3) is not on the graph of f.

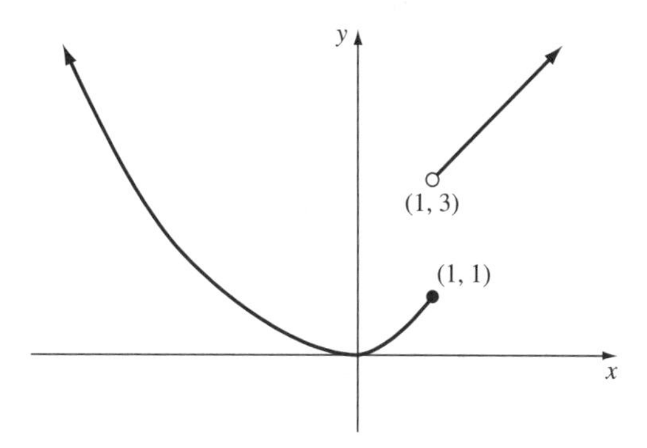

Figure 2

EXAMPLE 5 Constructing a Piecewise-Defined Function

The city of Gooberville operates a large parking ramp. The basis for parking charges is the amount of time a vehicle is parked. If a vehicle is parked for one hour or less, then the charge is \$3.00. If the vehicle is parked for a length of time between one and eight hours, then the fee is \$3.00 plus \$0.05 for each minute over one hour of parking time. If the vehicle is parked for a length of time from 8 to 16 hours, then a flat fee of \$24 is charged. Write the rule of the function C that gives the cost (in dollars) of parking a vehicle for t minutes, with $0 \leq t \leq 960$. Sketch a graph of C.

Solution There will be three formulas incorporated in the rule defining C since there are three stated schemes for assigning parking fees. Specifically,

$$C(t) = \begin{cases} 3 & \text{if } 0 \leq t \leq 60 \\ 3 + 0.05(t - 60) & \text{if } 60 < t \leq 480 \\ 24 & \text{if } 480 < t \leq 960 \end{cases}$$

Figure 3 shows a sketch of the graph of C.

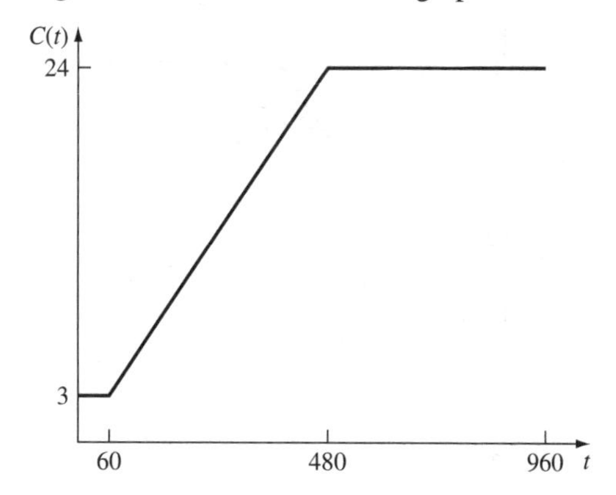

Figure 3

Transformations of Functions

Due to cuts in state aid, the city of Gooberville has experienced budget revenue shortfalls and is investigating ways to increase revenue, lest services like street maintenance, fire and rescue, and so on, get cut. In particular, the city commission is looking into the possibility of raising fees for the use of city parking ramps. One proposal under consideration is to simply add a \$2.00 surcharge to the existing fee structure (see Example 5). If $S(t)$ denotes the parking fee for t minutes under this proposal, then we see that $S(t) = C(t) + 2$ for $0 \leq t \leq 960$. Figure 4 shows a graph of the function S.

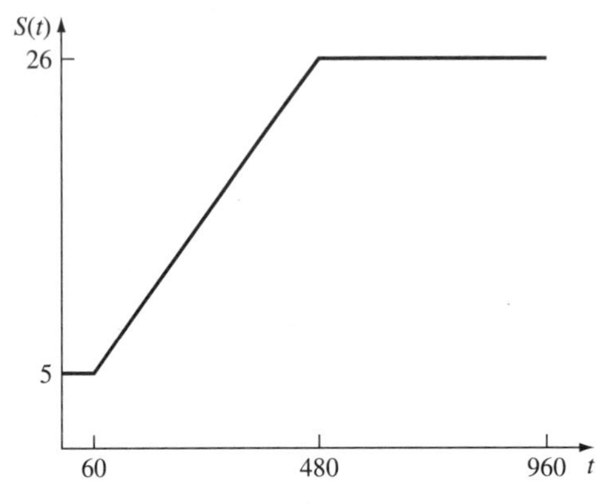

Figure 4

Notice that the graph of S is merely the graph of C from Example 5, but shifted up two units.

Another proposal regarding modification to the parking ramp fee structure is to keep the same three formulas for calculating fees, but to move the "switching" times back thirty minutes. This means that the flat \$3.00 fee would apply to vehicles parked thirty minutes or less, and so on. If we let $M(t)$ denote the cost of t minutes of parking under this proposal, we see, for instance, that

$$M(30) = C(60)$$
$$M(60) = C(90)$$
$$M(90) = C(120)$$

and so on. In general, $M(t) = C(t + 30)$ for $0 \leq t \leq 930$. Figure 5 shows a sketch of the graph of M. Comparing it to the graph of C, we see that the graph of M is simply the graph of C shifted thirty units to the left.

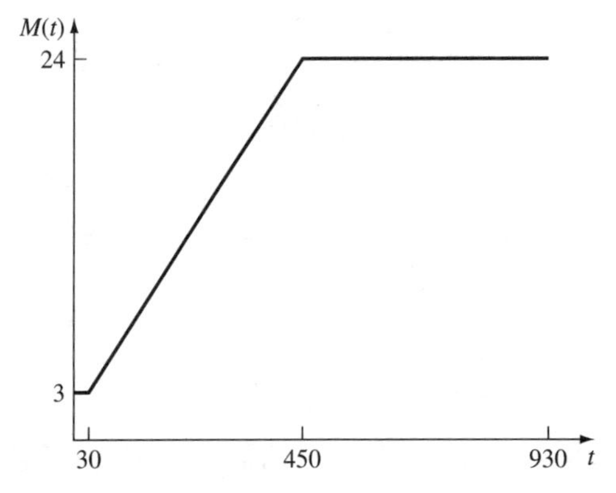

Figure 5

We say that the functions S and M are **transformations** of the function C. Notice how the graphs of S and M have the same shape as the graph of C, but shifted vertically and horizontally, respectively. We present a sequence of exercises involving transformations of some basic functions to motivate some general results about specific types of transformations and how the graphs of the resulting functions relate to the graphs of the basic functions.

Exercise 1 Let $f(x) = x^3$, $g(x) = f(x) + 4 = x^3 + 4$, and $h(x) = f(x) - 2 = x^3 - 2$. Complete the table below and sketch graphs of f, g and h on the same set of axes. Then describe in your own words how the graphs of g and h relate to the graph of f.

x	$f(x)$	$g(x)$	$h(x)$
-10			
-8			
-6			
-4			
-2			
0			
2			
4			
6			
8			
10			

Graphs of f, g and h

Relationships:

Exercise 2 Let $f(x) = |x|$, $g(x) = f(x - 3) = |x - 3|$, and $h(x) = f(x + 5) = |x + 5|$. Complete the table below and sketch graphs of f, g and h on the same set of axes. Then, describe in your own words how the graphs of g and h relate to the graph of f.

x	$f(x)$	$g(x)$	$h(x)$
-10			
-8			
-6			
-4			
-2			
0			
2			
4			
6			
8			
10			

Graphs of f, g and h

Relationships:

Exercises 1 and 2 illustrate **vertical** and **horizontal translations**, respectively. We generalize the observations made in the exercises as follows:

Vertical and Horizontal Translations

Suppose that f is some function and k is a positive constant.

- The graph of $g(x) = f(x) + k$ is the graph of f shifted upward k units.
- The graph of $g(x) = f(x) - k$ is the graph of f shifted downward k units.
- The graph of $g(x) = f(x + k)$ is the graph of f shifted horizontally k units to the left.
- The graph of $g(x) = f(x - k)$ is the graph of f shifted horizontally k units to the right.

Exercise 3 Let $f(x) = x^2$, $g(x) = \frac{1}{2}f(x) = \frac{1}{2}x^2$ and $h(x) = 3f(x) = 3x^2$. Complete the table below and sketch graphs of f, g and h on the same set of axes. Then describe in your own words how the graphs of g and h relate to the graph of f.

x	$f(x)$	$g(x)$	$h(x)$
-10			
-8			
-6			
-4			
-2			
0			
2			
4			
6			
8			
10			

Graphs of f, g and h

Relationships:

Exercise 3 illustrates **vertical scaling**. The graph of g is the graph of f compressed towards the x-axis by a scaling factor of one-half. The graph of h is the graph of f stretched away from the x-axis by a factor of three. We generalize as follows:

Vertical Scaling

Suppose that f is some function and k is a positive constant.

- If $0 < k < 1$, then the graph of $g(x) = k \cdot f(x)$ is the graph of f compressed towards the x-axis by a factor of k.
- If $k > 1$, then the graph of $g(x) = k \cdot f(x)$ is the graph of f stretched away from the x-axis by a factor of k.

Exercise 4 Let $f(x) = \sqrt{x}$, $g(x) = f(-x) = \sqrt{-x}$ and $h(x) = -f(x) = -\sqrt{x}$. Complete the table below and sketch graphs of *f*, *g* and *h* on the same set of axes. Then describe in your own words how the graphs of *g* and *h* relate to the graph of *f*.

x	$f(x)$	$g(x)$	$h(x)$
-10			
-8			
-6			
-4			
-2			
0			
2			
4			
6			
8			
10			

Graphs of *f*, *g* and *h*

Relationships:

Exercise 4 illustrates **reflections**. The graph of *g* is the graph of *f* reflected across the *y*-axis. The graph of *h* is the graph of *f* reflected across the *x*-axis. We generalize as follows:

Reflections

Suppose that *f* is some function.

- The graph of $g(x) = f(-x)$ is the graph of *f* reflected across the *y*-axis.
- The graph of $g(x) = -f(x)$ is the graph of *f* reflected across the *x*-axis.

We conclude with two examples dealing with combinations of basic transformations. In each case, the order in which we apply transformations is important.

EXAMPLE 6 Combining Transformations

Suppose that $g(x) = -2|x + 1| - 3$. Then we can consider *g* as resulting from a sequence of transformations of $f(x) = |x|$. Following how the input *x* generates its corresponding output $g(x)$, we see that the rule of *g* is derived from the rule of *f* by the following steps:

$$f(x) = |x| \xrightarrow{\text{Step 1}} |x + 1| \xrightarrow{\text{Step 2}} -2|x + 1| \xrightarrow{\text{Step 3}} -2|x + 1| - 3$$

The sequence of graphs in Figure 6 shows how we can obtain the graph of *g* from the graph of *f*. Notice that the sequence is

- Step 1: shift one unit to the left
- Step 2: scale by a factor of 2 and reflect across the *x*-axis
- Step 3: shift downward 3 units

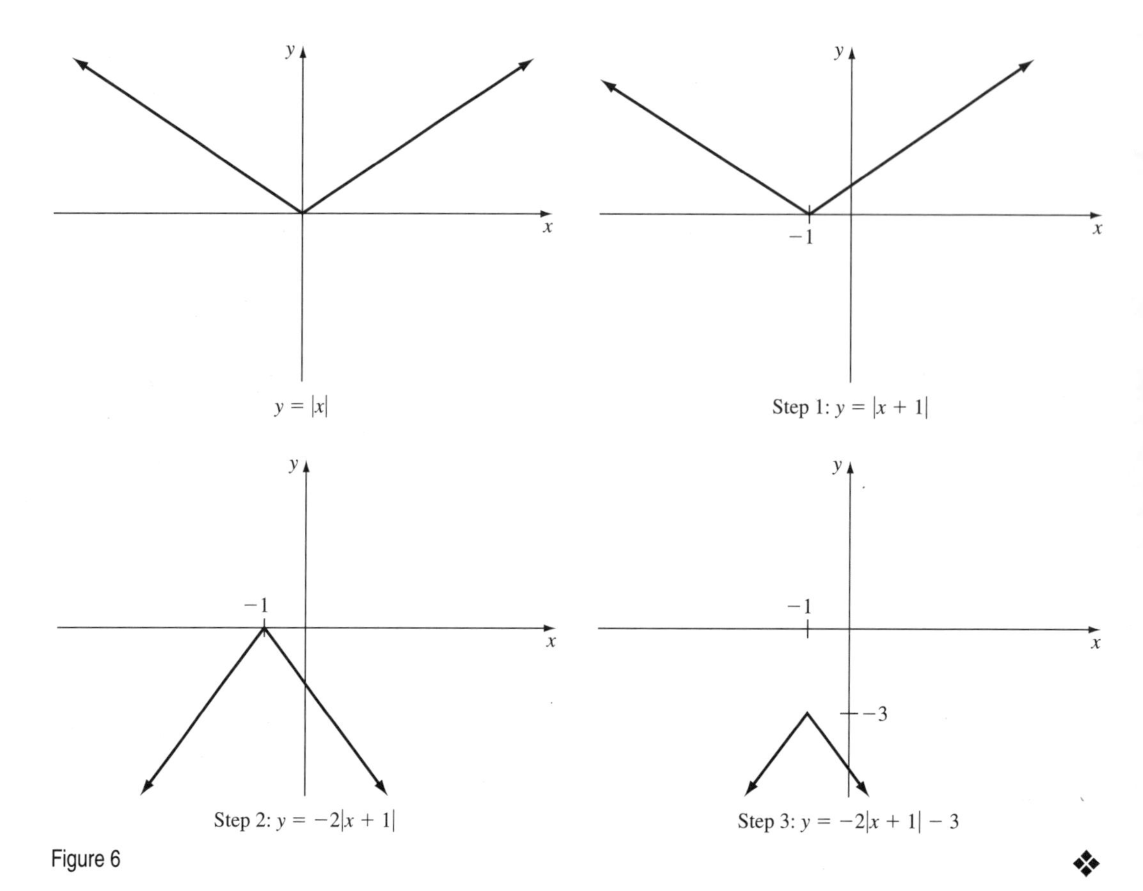

Figure 6

❖

EXAMPLE 7 Combining Transformations

We can obtain the graph of a function g from the graph of $f(x) = x^2 - 3x + 2$ by the following sequence of transformations: Stretch the graph of f away from the x-axis by a factor of 3, shift it horizontally 4 units to the right, shift it vertically upwards by 1 unit, and reflect across the x-axis. Find the rule of g.

Solution We start with the formula defining $f(x)$ and apply the stated transformations in the prescribed order to modify the rule of f to obtain the rule of g:

$$x^2 - 3x + 2 \to 3(x^2 - 3x + 2) \to 3((x-4)^2 - 3(x-4) + 2)$$
$$\to 3((x-4)^2 - 3(x-4) + 2) + 1$$
$$\to -[3((x-4)^2 - 3(x-4) + 2) + 1]$$

So,

$$g(x) = -[3((x-4)^2 - 3(x-4) + 2) + 1]$$
$$g(x) = -3(x-4)^2 + 9(x-4) - 6 - 1$$
$$g(x) = -3(x-4)^2 + 9(x-4) - 7$$

❖

PROBLEM SET 2.4

Use Table 1 for problems 1 and 2.

TABLE 1

x	1	2	3	4	5
$f(x)$	3	4	1	5	2
$g(x)$	5	3	1	4	2

1. Complete the following table.

x	1	2	3	4	5
$(f+g)(x)$					
$(f-g)(x)$					
$(f\circ g)(x)$					
$(g\circ f)(x)$					

2. Complete the following table.

x	1	2	3	4	5
$(f\cdot g)(x)$					
$(f/g)(x)$					
$(f\circ f)(x)$					
$(g\circ g)(x)$					

3. Let $f(x) = x - x^2$ and $g(x) = 5x + 2$. Calculate each of the following.

(a) $(f+g)(3)$ **(b)** $(f-g)(4)$
(c) $(f\circ g)(-2)$ **(d)** $(g\circ f)(1)$

4. Let $g(t) = \dfrac{1}{t}$ and $h(t) = 8 - 3t$. Calculate each of the following.

(a) $(h\cdot g)(2)$ **(b)** $(g/h)(6)$
(c) $(h\cdot g)(-1/2)$ **(d)** $(g\cdot h)(4/3)$

5. Let $p(x) = x^2 + 3x - 4$ and $q(x) = x - 1$. Find the rules of each of the following.

(a) $p+q$ **(b)** $q-p$
(c) $p\cdot q$ **(d)** p/q

Use the graphs in Figure 7 for problems 6 and 7.

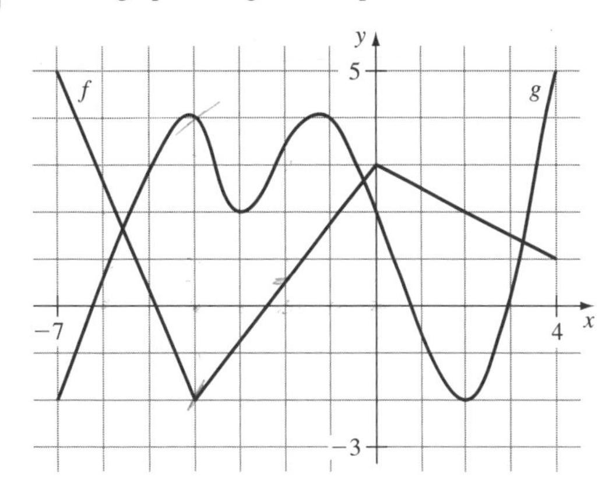

Figure 7

6. Calculate each of the following.

(a) $(f+g)(-2)$ **(b)** $(f-g)(4)$
(c) $(f\circ g)(-3)$ **(d)** $(g\circ f)(2)$
(e) $(f\circ f)(-4)$

7. Calculate each of the following.

(a) $(f\cdot g)(-6)$ **(b)** $(f/g)(0)$
(c) $(f\circ g)(-4)$ **(d)** $(g\circ f)(-4)$
(e) $(g\circ g)(-1)$

8. Let $f(x) = x^2$ and $g(x) = 3x - 2$. Find the rules and domains of each of the following functions.

(a) $f\circ g$ **(b)** $g\circ f$ **(c)** $f\circ f$ **(d)** $g\circ g$

9. Let $g(t) = \sqrt{t}$ and $h(t) = \dfrac{t^2}{1+t}$. Find the rules and domains of each of the following functions.

(a) $h\circ g$ **(b)** $g\circ h$ **(c)** $h\circ h$ **(d)** $g\circ g$

10. Suppose that $h(x) = \sqrt{x^2+4}$. Find the rules of two functions f and g so that $h(x) = (f\circ g)(x)$.

11. Suppose that $h(x) = x^2 + 2x + 1$. Find the rules of two functions f and g so that $h(x) = (f\circ g)(x)$.

12. Suppose that $h(x) = \dfrac{1+2x}{1+x}$. Find the rules of two functions f and g so that $h(x) = (f\circ g)(x)$.

13. Suppose that $h(x) = \dfrac{6}{\sqrt{5-x}}$. Find the rules of two functions f and g so that $h(x) = (f\circ g)(x)$.

14. A circular puddle of water is growing in size in such a way that after t seconds the radius r is given by $r = \dfrac{t}{t+1}$ feet.

(a) Express the area, A, of the puddle as a function of time, t.
(b) Calculate the area of the puddle after 4 seconds.
(c) When will the area of the puddle be 2 square feet?

15. A 26-foot ladder is leaning against a wall. The bottom of the ladder is four feet from the base of the wall when the bottom of the ladder starts sliding away from the wall at a rate of 3 inches per second. Let h be the height above the ground, in feet, of the top of the ladder. Write h as a function of the time, t, in seconds, during which the bottom of the ladder has been sliding away from the wall. Identify the domain and range of this function.

16. The weekly cost of producing x units in a manufacturing process is $C(x) = 60x + 750$ dollars. The number of units produced in t hours is $x(t) = 50t$. Find how much time must elapse until the cost increases to \$15,000.

17. Oftentimes it is useful to convert from one unit of measurement to another (inches to feet, for instance).

(a) Give the rule of a function F that converts x inches to feet.
(b) Give the rule of a function Y that converts x feet to yards.
(c) Give the rule of the function $Y \circ F$. What is a practical interpretation of this function?

18. Sketch graphs of the following functions.

(a) $f(x) = \begin{cases} 3x+1 & \text{if } x \le 2 \\ x^2 & \text{if } x > 2 \end{cases}$

(b) $g(x) = \begin{cases} 2x & \text{if } x \le 0 \\ 5 & \text{if } 0 < x < 3 \\ 4 - x^2 & \text{if } x \ge 3 \end{cases}$

19. Sketch graphs of the following functions.

(a) $f(x) = \begin{cases} x^2+1 & \text{if } x \ne 0 \\ 4 & \text{if } x = 0 \end{cases}$

(b) $g(x) = \begin{cases} x^3 & \text{if } x < -1 \\ |x| & \text{if } -1 \le x \le 4 \\ \sqrt{x} & \text{if } x > 4 \end{cases}$

20. A salesperson has a base monthly salary of \$1,500. If she sells at least \$1,000 worth of merchandise in a month, she earns a commission of 8% on the sales. If she sells more than \$2500 in merchandise, she also receives a \$1000 bonus. Let E be the function that gives her monthly salary based on s dollars of merchandise sold.

(a) Find the rule of E.
(b) Sketch a graph of E.
(c) Calculate $E(950)$, $E(1200)$ and $E(3500)$.
(d) What level of sales does she need to maintain each month in order to have a monthly salary of at least \$3,000?

21. Figure 8 shows the cost of a hospital stay under three different health insurance plans (A, B and C). The cost under each plan is a function of the number of days spent in the hospital.

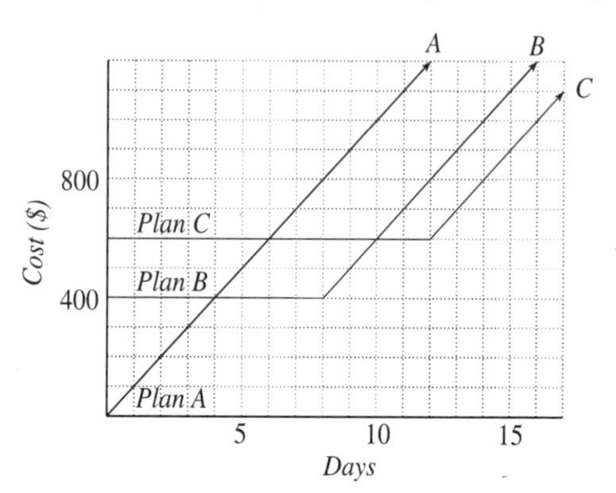

Figure 8

(a) Let $A(t)$ denote the cost of a stay of t days under Plan A. Give the rule of $A(t)$.
(b) Let $B(t)$ denote the cost of a stay of t days under Plan B. Give the rule of $B(t)$.
(c) Let $C(t)$ denote the cost of a stay of t days under Plan C. Give the rule of $C(t)$.

22. The Gooberville Museum of Elvis Memorabilia charges an admission fee of \$30 for tour groups of 10 people or less. However, if a group has more than 10 people, then there is an additional charge of \$1.50 for each person in excess of 10 in the size of the group. Let A be the function that gives the admission fee for a group of x people. Find the rule of A.

23. Let $f(x) = x^2 + 1$. Describe a sequence of transformations that will transform the graph of f into the graph of $g(x) = 2(x-3)^2 + 8$.

24. Let $f(x) = \sqrt{5-x}$. Describe a sequence of transformations that will transform the graph of f into the graph of $g(x) = 8 - 3\sqrt{6-x}$.

25. Let $f(x) = \sqrt{1-x^3}$. Describe a sequence of transformations that will transform the graph of f into the graph of $g(x) = \frac{1}{4}\sqrt{1+x^3} + 2$.

In problems 26–30, find the rule of the function g whose graph is obtained from the graph of the function f by the indicated sequence of transformations in the given order.

26. $f(x) = \sqrt{x}$; shift graph 4 units horizontally to the left, then vertically down 6 units.

27. $f(x) = 1 - x^2$; stretch graph away from the x-axis by a factor of 2, shift vertically 3 units up, then reflect across the y-axis.

28. $f(x) = |x - 1|$; reflect graph across the x-axis, compress the graph towards the x-axis by a factor of 2/3, then shift horizontally 2 units to the right.

29. $f(x) = x^3 + 1$; shift horizontally 3 units to the left, stretch the graph away from the x-axis by a factor of 9, then reflect across the x-axis.

30. $f(x) = \frac{1}{x}$; shift graph vertically down 2 units, horizontally 1 unit to the right, and then compress towards the x-axis by a factor of 3/8.

Use the graph in Figure 9 for problems 31–35.

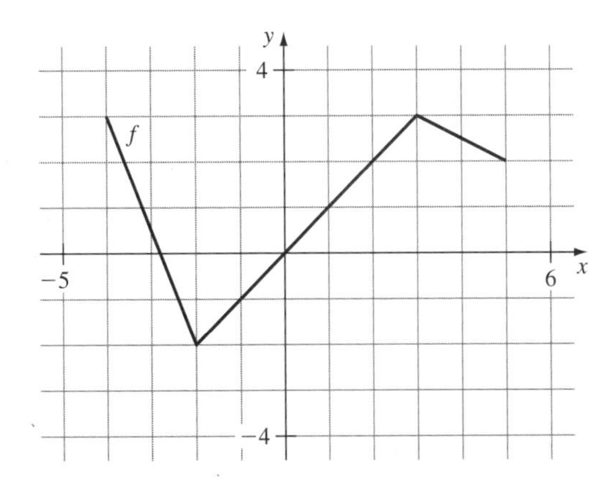

Figure 9

31. Sketch the graph of $g(x) = f(-x) + 3$.

32. Sketch the graph of $h(x) = 2f(x) - 1$.

33. Sketch the graph of $k(x) = -\frac{1}{2}f(x + 3)$.

34. Sketch the graph of $n(x) = -3f(x - 1) + 2$.

35. Sketch the graph of $r(x) = 2 - f(1 - x)$.

36. A roofing company gives you an initial estimate for fixing your leaky roof. The estimate is that materials will cost \$800. In addition, the company will charge you for labor at a rate of \$50 per hour. Let $C(x)$ denote the cost of fixing your roof, where x is the number of hours of labor.

(a) Find a formula for $C(x)$ and sketch the graph of C.

(b) Let $D(x) = C(x) - 100$. Give a practical interpretation of $D(x)$ in terms of your roof repair costs.

37. A dance club charges a cover charge of \$6.00 to get in the door and then serves beverages for \$2.50 each. Let $d(x)$ be the cost of a night at the club, where x is the number of beverages purchased.

(a) Find a formula for $d(x)$.

(b) Suppose that management decides to raise the cover charge by \$2.00. If $c(x)$ is the cost of a night at the club, where x is the number of beverages purchased, then express c as a transformation of d.

(c) Suppose that management raises the cover charge to \$12 and keeps the beverage charge the same, but with a minimum two-drink purchase requirement. Let $n(x)$ be the cost of a night at the club, where x is the number of beverages purchased. Express n as a transformation of d (with appropriate restrictions on x).

Interlude: Progress Check 2.4, Part 2

1. The daily profit, in thousands of dollars, for a particular manufacturer is described by $f(t) = \begin{cases} 16 - 1.25t & \text{if } 0 \le t \le 8 \\ 3t - 18t & \text{if } 8 < t \le 12 \end{cases}$ for one particular year. Here, t measures months, with $t = 0$ corresponding to January 1st.

(a) Sketch a graph of f. Label axes and provide appropriate scale.

(b) Estimate the day(s) during the year when the daily profit was $18,000. Give the date(s).

2. The graphs in Figure 10 are graphs of two functions f and g. Gridlines each mark one unit. Let $h(x) = (f \circ g)(x)$. Fill in the entries in Table 2 giving your best estimates. On the provided grid, carefully sketch the graph of h.

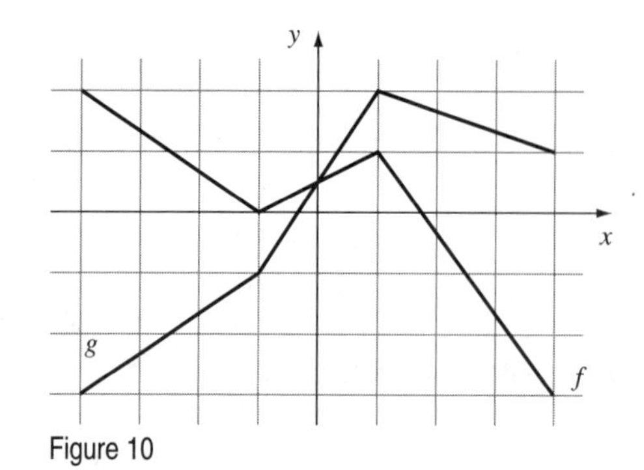

Figure 10

TABLE 2

x	$f(x)$	$g(x)$	$(f \circ g)(x)$
-4			
-3			
-2			
-1			
0			
1			
2			
3			
4			

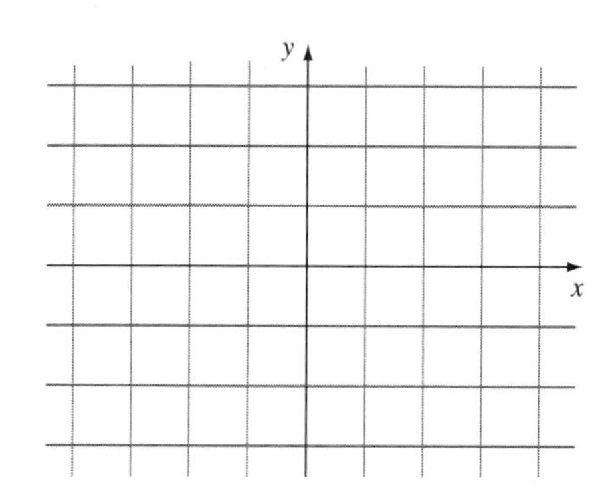

3. Let $f(x) = \dfrac{\sqrt{x-4}}{10}$ and $g(x) = 2x^2 - 8$. Simplify $(g \circ f)(x)$ and determine the domain of that composition.

Interlude: Progress Check 2.4, Part 3

Suppose that f is the function defined graphically in Figure 11. Assume that gridlines each mark one unit.

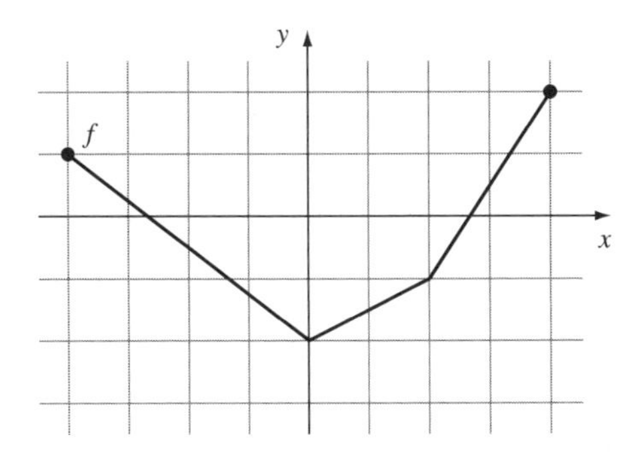

Figure 11

1. Sketch the graph of $g(x) = -1 - 2f(x)$ on the provided grid.

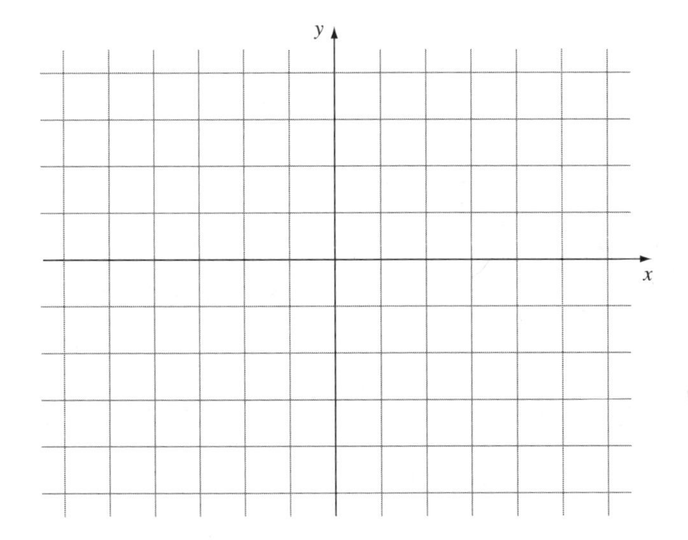

2. Sketch the graph of $h(x) = f(-x) - 3$ on the provided grid.

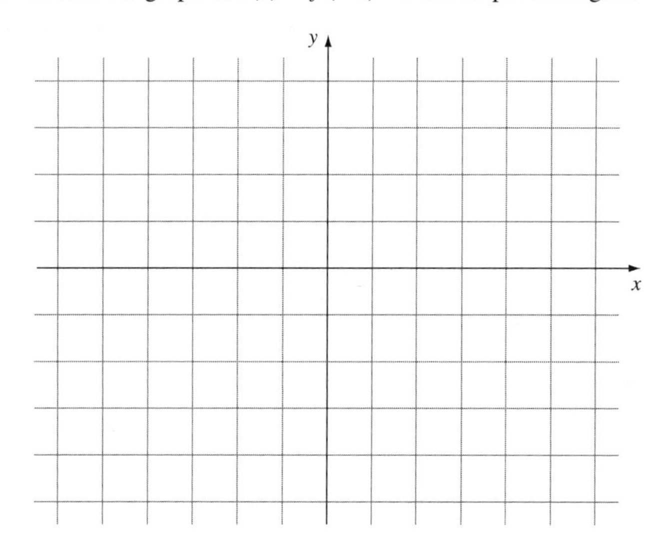

3. Sketch the graph of $k(x) = \frac{1}{2}f(x - 2) + 3$ on the provided grid.

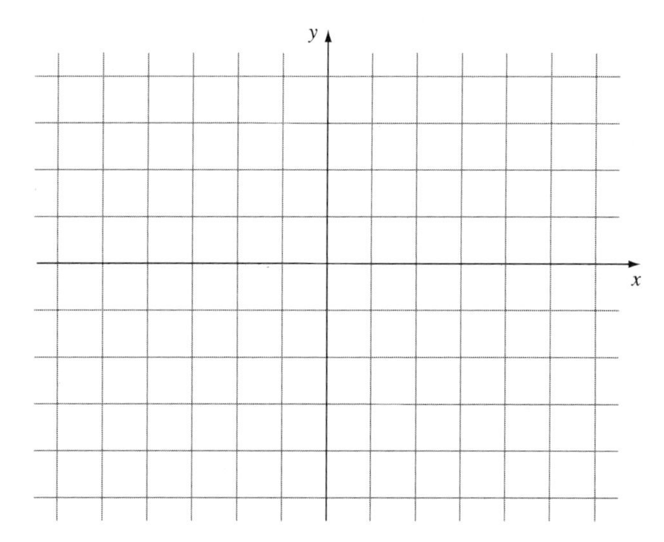

OBJECTIVES

1. Discuss Proportionality
2. Revisit Average Rate of Change and Difference Quotients

2.5 More on Describing Change

We have examined the ideas of total change and average rate of change as ways to describe a dynamic relationship between two quantities. As we progress, we will discuss other ways of describing how one quantity changes relative to changes in a related quantity. Now we turn to a discussion of a way of describing how one quantity changes with respect to another that involves the notion of *proportionality*. For instance, consider the following statements.

- Force is proportional to acceleration.
- The area of a circle is proportional to the square of its radius.
- The force of attraction between two heavenly bodies is inversely proportional to the square of the distance between them.

All of the "proportional" statements describe one quantity as a function of another. We begin by considering what we call **direct proportionality**.

Direct Proportionality

The definition of direct proportionality is as follows:

Direct Proportionality

A quantity y is directly proportional to a quantity x if

$$y = kx$$

for some fixed constant k. Here, k is called the **constant of proportionality**.

As an example of direct proportionality, suppose that a person works for a fixed hourly wage, say w dollars per hour. If S is the salary earned for t hours of work, then we see that

$$S = w \cdot t$$

so S is directly proportional to t. The constant of proportionality is w, the hourly wage.

EXAMPLE 1 Direct Proportionality

For a certain gas contained in a container of fixed volume, the pressure P (measured in newtons per square meter) is directly proportional to the temperature T (in kelvins). Suppose that at a temperature of 75°K the pressure is 30 newtons per square meter. Determine the pressure when the temperature is 100°K.

Solution Since P is directly proportional to T, we have

$$P = kT$$

for some constant k. But we know that $P = 30$ when $T = 75$, so that

$$30 = k \cdot 75$$

$$k = \frac{30}{75} = \frac{2}{5}$$

Consequently, $P = \frac{2}{5}T$. Then, we calculate P when $T = 100$:

$$P = \frac{2}{5} \cdot 100$$
$$P = 40$$

The pressure is 40 newtons per square meter when the temperature is 100°K. ❖

Note here that if a quantity y is directly proportional to a quantity x, then x and y are linearly related. Many relationships involving direct proportionality are not linear though, as demonstrated in the next example.

EXAMPLE 2 Proportionality to a Power of a Variable

The volume V of a sphere is directly proportional to the cube of its radius, r. If a sphere with a radius of 3 inches has a volume of 36π cubic inches, find a formula for V in terms of r.

Solution We have that V is directly proportional to r^3, so

$$V = kr^3$$

for some constant k. Since we know that $V = 36\pi$ when $r = 3$, we have

$$36\pi = k \cdot 3^3$$
$$k = \frac{36\pi}{27} = \frac{4}{3}\pi$$

Consequently, $V = \frac{4}{3}\pi r^3$ ❖

Inverse Proportionality

Suppose that a freight train makes a daily run of 400 miles. Let v denote the average velocity of the train on its shipping run. If t is the number of hours required to make the trip, then we know

$$v = \frac{\text{distance}}{\text{time}} = \frac{400}{t}$$

We say in this case that v is **inversely proportional** to t. In general:

Inverse Proportionality

A quantity y is inversely proportional to a quantity x if

$$y = \frac{k}{x}$$

for some nonzero constant k. Again, k is called the constant of proportionality.

EXAMPLE 3 Inverse Proportionality

The weight W that a 2-inch by 4-inch piece of lumber can safely support is inversely proportional to it length l. A 10-foot piece of this lumber can safely support 420 pounds. Determine how much weight an 8-foot piece could safely support.

Solution Since W is inversely proportional to l, we have

$$W = \frac{k}{l}$$

for some nonzero constant k. Since $W = 420$ when $l = 10$,

$$420 = \frac{k}{10}$$
$$k = 4200$$

Consequently, $W = \frac{4200}{l}$. Then, we calculate W when $l = 8$:

$$W = \frac{4200}{8}$$
$$W = 525$$

So, an 8-foot piece of the lumber can safely support 525 pounds. ❖

Mathematicians describe some relationships involving proportionality in terms of three or more quantities that involve combinations of direct and inverse proportionality.

Joint Proportionality and Combined Proportionality

If some quantity y is directly proportional to the product of two or more quantities, say quantities $x_1, x_2, \ldots, x_n$, then we say that y is **jointly proportional** to those quantities. If the description of how y is proportional to two or more quantities also involves inverse proportionality, we say that the proportionality is **combined proportionality**.

EXAMPLE 4 Joint Proportionality

The volume V of a right circular cone varies jointly with the height of the cone and the square of the radius of the base of the cone. A cone with a height of 4 feet and a base radius of 3 feet has a volume of 12π cubic feet. Find the volume of a right circular cone with a height of 6 feet and a base radius of 2 feet.

Solution Let h and r denote the height and radius, respectively, in feet. Since V is jointly proportional to h and r^2, we have

$$V = khr^2$$

for some constant k. Now, $V = 12\pi$ when $h = 4$ and $r = 3$, so

$$12\pi = k \cdot 4 \cdot 3^2$$
$$k = \frac{12\pi}{36} = \frac{\pi}{3}$$

Consequently, $V = \frac{\pi}{3}hr^2$. Calculating V when $h = 6$ and $r = 2$, we have

$$V = \frac{\pi}{3} \cdot 6 \cdot 2^2$$
$$V = 8\pi$$

❖

So, a right circular cone of height 6 feet and base radius 2 feet has a volume of 8π cubic feet.

EXAMPLE 5 Combined Proportionality

The stress in the material of a pipe subject to internal pressure is jointly proportional to the internal pressure and the internal diameter of the pipe and inversely proportional to the thickness of the pipe. Suppose the material stress is 75 pounds per square inch when the internal diameter of the pipe is 3 inches, the thickness is 0.5 inches and the internal pressure is 20 pounds per square inch. Find a formula for material stress in terms of internal pressure, internal diameter and thickness.

Solution Let S be the material stress, P the internal pressure, d the internal diameter and t the thickness of the pipe (all in the appropriate units). Then S is proportional to the product of P, d and the reciprocal of t. That is,

$$S = k\frac{Pd}{t}$$

for some constant k. We have that $S = 75$ when $P = 20$, $d = 3$ and $t = 0.5$. So, we have

$$75 = k\frac{20 \cdot 3}{0.5}$$
$$k = \frac{75 \cdot 0.5}{20 \cdot 3} = 0.625$$

Consequently, $S = 0.625\frac{Pd}{t}$.

❖

Average Rate of Change, Revisited

Suppose that $y = f(x)$. Recall that the average rate of change of $f(x)$ as x changes from $x = a$ to $x = b$ is given by the difference quotient

$$\frac{f(b) - f(a)}{b - a}$$

We now consider a geometric interpretation of average rate of change when f is not describing a linear relationship.

Suppose that f is a function whose graph is depicted in Figure 1.

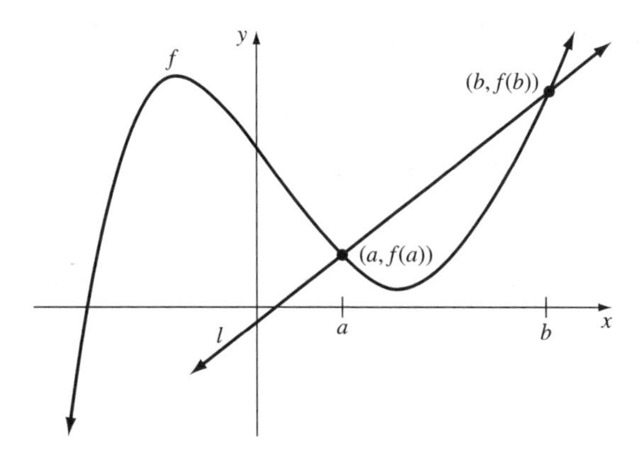

Figure 1

A line l has been drawn that passes through the points with coordinates $(a, f(a))$ and $(b, f(b))$. This line is called a **secant line** through the indicated points. The slope of l is given by

$$m = \frac{\Delta y}{\Delta x} = \frac{f(b) - f(a)}{b - a}$$

In other words, the slope of secant line l is the average rate of change of $f(x)$ from $x = a$ to $x = b$.

EXAMPLE 6 Comparing Average Rates of Change

Figure 2 gives the graph of a function g. Consider only the marked values of t.

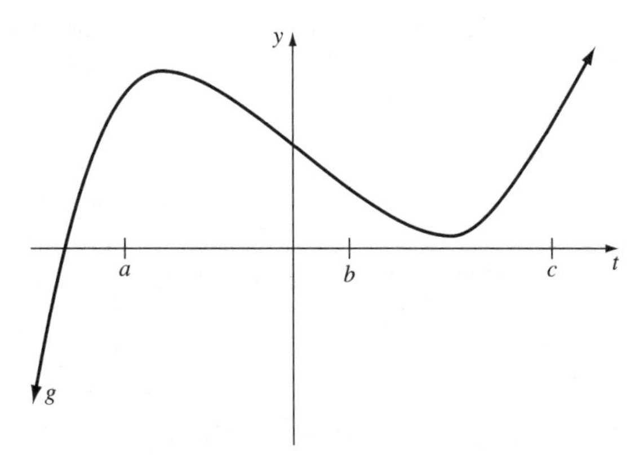

Figure 2

(a) For which pair of marked inputs is the average rate of change of $g(t)$ between those inputs greatest?
(b) For which pair of marked inputs is the average rate of change of $g(t)$ between those inputs least?
(c) For which pair of marked inputs is the average rate of change of $g(t)$ between those inputs closest to zero?

Solution We will sketch secant lines and compare slopes. See Figure 3.

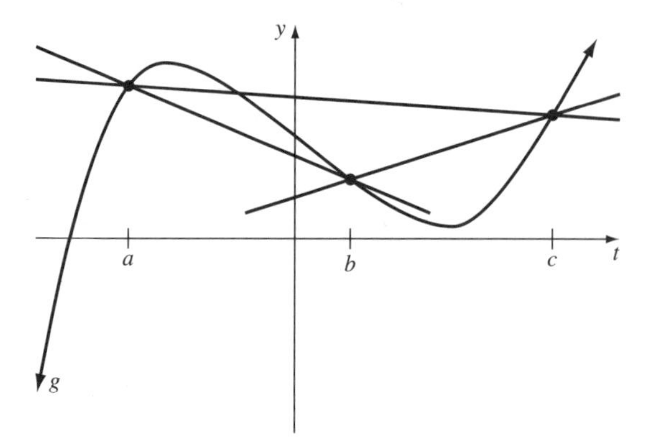

Figure 3

(a) The secant line between the points on the graph where $t = b$ and $t = c$ is the only line that has positive slope. So, the average rate of change between $t = b$ and $t = c$ is greatest.
(b) The secant line between the points on the graph where $t = a$ and $t = b$ is the line with the most negative slope. So, the average rate of change between $t = a$ and $t = b$ is least.
(c) The secant line between the points on the graph where $t = a$ and $t = c$ is the line with the slope closest to zero. So, the average rate of change between $t = a$ and $t = c$ is closest to zero. ❖

Suppose that f is a function and that t is some particular input from the domain of f. Let h denote a "small change" in input, so that h is nonzero and $t + h$ is also in the domain of f. Then the average rate of change of f between input t and input $t + h$ is

$$\frac{f(t+h)-f(t)}{(t+h)-t} = \frac{f(t+h)-f(t)}{h}$$

The last expression is another form of difference quotient describing average rate of change. So:

Difference Quotients, Average Rate of Change, Slope

Given a function f, the difference quotient

$$\frac{f(t+h)-f(t)}{h},\ h \neq 0$$

- gives the average rate of change of $f(x)$ from $x = t$ to $x = t + h$
- gives the slope of the secant line through the points on the graph of f corresponding to $x = t$ and $x = t + h$

EXAMPLE 7 Calculating Average Rates of Change

If you drop a bowling ball from a high elevation, it can be shown that the distance the ball has fallen, in feet, is reasonably approximated by $d(x) = 16x^2$, where x is the number of seconds since dropping the ball.

(a) Simplify the difference quotient $\dfrac{d(t+h)-d(t)}{h}$.

(b) Using the difference quotient from (a), let $t = 3$ and calculate the average velocity of the bowling ball from 2.9 seconds to 3 seconds, from 3 seconds to 3.1 seconds and from 3 seconds to 3.01 seconds.

Solution (a)

$$\frac{d(t+h)-d(t)}{h} = \frac{16(t+h)^2 - 16t^2}{h} = \frac{16t^2 + 32th + 16h^2 - 16t^2}{h}$$

$$= \frac{32th + 16h^2}{h} = 32t + 16h$$

(b) We fix $t = 3$ and choose appropriate values of h to calculate the desired average velocities. Specifically:

- For the average velocity from 2.9 seconds to 3 seconds, $h = -0.1$, so the average velocity is

$$32(3) + 16(-0.1) = 94.4 \text{ feet per second}$$

- For the average velocity from 3 seconds to 3.1 seconds, $h = 0.1$, so the average velocity is

$$32(3) + 16(0.1) = 97.6 \text{ feet per second}$$

- For the average velocity from 3 seconds to 3.01 seconds, $h = 0.01$, so the average velocity is

$$32(3) + 16(0.01) = 96.16 \text{ feet per second}$$

❖

PROBLEM SET 2.5

1. If y is directly proportional to x and $y = 6$ when $x = 4$, then calculate the value of y when $x = 12$.

2. If p is inversely proportional to q and $p = 38$ when $q = 9$, then calculate the value of p when $q = 5$.

3. If r is jointly proportional to s and t and $r = 324$ when $s = 8$ and $t = 0.5$, then calculate the value of r when $s = 12$ and $t = 0.8$.

4. If v is jointly proportional to the square root of g and the cube of h and $v = 0.25$ when $g = 36$ and $h = 2$, then calculate the value of v when $g = 20$ and $h = 4$.

5. Is the circumference of a circle directly proportional to the diameter of the circle? Explain.

6. While studying for final exams, the amount of money Hillary spends on fast food in a day is directly proportional to the number of hours she spends studying that day. Suppose that she spends \$8.70 on fast food if she studies for 6 hours. Give an equation that describes how the amount of money, m, she spends on food relates to h, the number of hours she studies, and interpret the constant of proportionality.

7. The weight of a body above the surface of Earth is inversely proportional to the square of the distance from the center of Earth. If a certain body weighs 95 pounds when it is 4000 miles from the center of Earth, how much will it weigh when it is 4030 miles from the center of Earth?

8. The horsepower that a shaft can safely transmit is jointly proportional to its speed (in revolutions per minute) and the cube of its diameter. A 3-inch diameter shaft made of a certain alloy can transmit 42 horsepower at 120 rpm.
   - **(a)** What horsepower can the shaft transmit at 100 rpm?
   - **(b)** What diameter must the shaft have in order to transmit 50 horsepower at 80 rpm?

9. The rate of vibration of a string under constant tension is inversely proportional to the length of the string. Suppose a string that is 60 inches long vibrates 196 times per second.
   - **(a)** What is the vibration rate of a string that is 56 inches long?
   - **(b)** What length string will generate a vibration rate of 224 times per second?

10. The volume V of a gas varies directly with its temperature T and inversely with its pressure P. A cylinder contains 80 liters of a gas at $300°K$ and a pressure of 120 atmospheres. If we lower the temperature to $290°K$ and compress the cylinder so that the gas occupies a volume of 60 liters, then what is the resulting pressure?

11. The thrust delivered by a ship's propeller is jointly proportional to the square of the propeller speed and the fourth power of the propeller diameter. Let T denote thrust, in pounds. Let r denote the propeller speed, in revolutions per minute, and let d denote the propeller diameter, in feet.
    - **(a)** Give a formula for T in terms of r and d.
    - **(b)** If we hold propeller speed constant but double the propeller's diameter, then what is the net effect on thrust?
    - **(c)** If we increase propeller speed by 50%, then by how much can we reduce propeller diameter to deliver the same thrust?

12. Let $f(x) = 2x - x^2$. Simplify $\dfrac{f(t+h) - f(t)}{h}$.

13. Let $g(x) = \dfrac{2}{x}$.
    - **(a)** Simplify $\dfrac{g(t+h) - g(t)}{h}$.
    - **(b)** Simplify $\dfrac{g(1+h) - g(1)}{h}$.
    - **(c)** Calculate the average rate of change of $g(x)$ from $x = 1$ to $x = 1.25$.

14. Suppose that f is a function such that $\dfrac{f(t+h) - f(t)}{h} = -2t^2 + 5th - 9h^2$. Calculate the average rate of change of $f(x)$ from $x = 2$ to $x = 2.4$.

15. A model rocket launches from a platform with an initial velocity of 32 feet per second. The height (in feet) of the rocket above ground t seconds after launch is given by $h(t) = -16t^2 + 32t + 8$. Find the rule of the function v that gives the average velocity of the rocket from the time of launch until time x seconds after launch.

16. Francis and Mark run along a straight stretch of road for 30 minutes. Both runners start from the same point.
    - **(a)** Sketch graphs showing the distance traveled by each runner if each has the same average speed for the 30 minutes of running, with Mark running at a constant speed and Francis running at a non-constant speed.
    - **(b)** Suppose that Mark runs at a constant speed, Francis runs at a non-constant speed, and Francis is never behind Mark, but after 30 minutes of running both have the same average speed. Sketch possible graphs showing the distance each runner has run as a function of time. Explain what is happening with Francis according to your graph.

17. A study of eating behavior found the cumulative intake (in grams) of one individual was related to time t (in minutes) according to $I(t) = 24 + 68t - 1.5t^2$ for $0 \le t \le 20$.

(a) Calculate $\frac{I(h) - I(0)}{h}$. What does this quantity represent?

(b) Find the time b such that the average rate of change of $I(t)$ from $t = 0$ to $t = b$ is as large as possible.

18. The graph in Figure 4 shows the number of orders placed at an online store for a certain dietary supplement over a period of several weeks. Calculate the average rates of change in orders with respect to time over the following time intervals and represent your answers graphically.

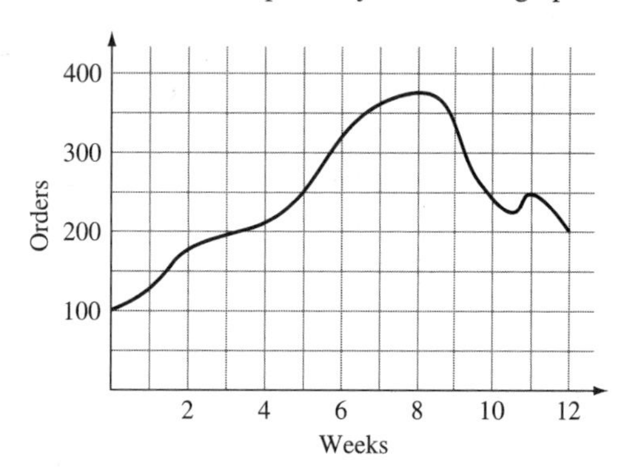

Figure 4

(a) week 2 to week 8
(b) week 0 to week 12
(c) week 8 to week 11
(d) week 6 to week 9

19. The velocity of air expelled during a cough can be modeled by $v(r) = 0.6r^2 - r^3$ centimeters per second, where r is the radius of the trachea, in centimeters.

(a) Simplify $\frac{v(r + h) - v(r)}{h}$.

(b) Evaluate the difference quotient from (a) for $r = 0.3$ and $h = 0.05$. What does this quantity represent?

20. Figure 5 shows the temperature, in degrees Fahrenheit, of a can of soda placed into a refrigerator as a function of the time that the can has been in the refrigerator. Find two intervals of time where, on average, the temperature of the soda was decreasing by 12° per hour. Explain how you arrived at those intervals.

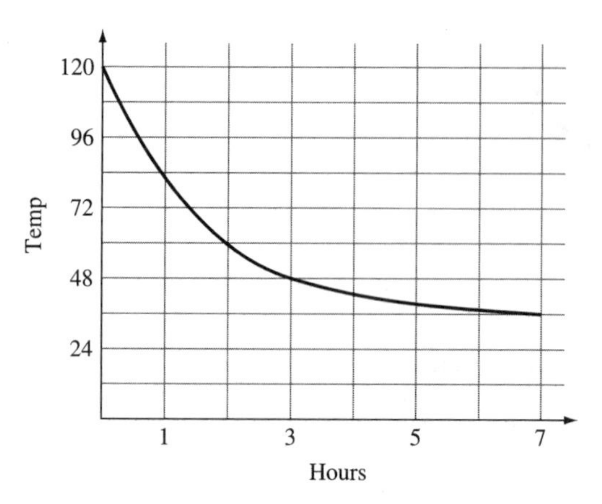

Figure 5

21. Figure 6 shows the graph of a function g.

(a) Between which pair of consecutive points is the average rate of change of g greatest? Closest to zero?

(b) Between which two pairs of consecutive points are the average rates of change of g closest to being equal?

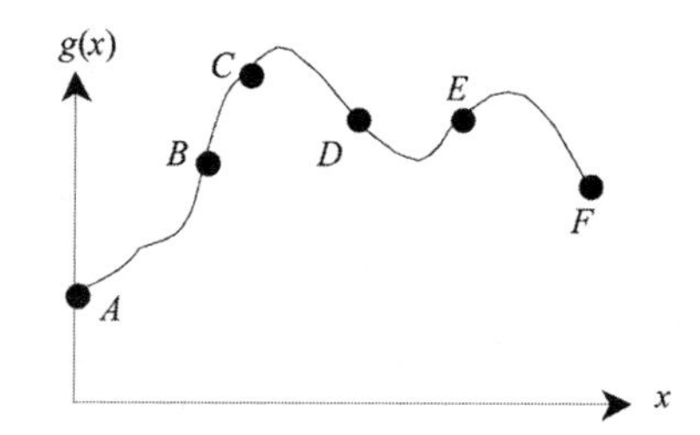

Figure 6

22. Figure 7 shows a company's weekly sales (in units of product).

(a) During which of the following time intervals was the average rate of change larger?

i. $0 \le t \le 4$ or $0 \le t \le 8$

ii. $0 \le t \le 8$ or $0 \le t \le 16$

(b) Estimate the average rate of change between $t = 0$ and $t = 16$. Interpret your answer in terms of sales. Include appropriate units in your response.

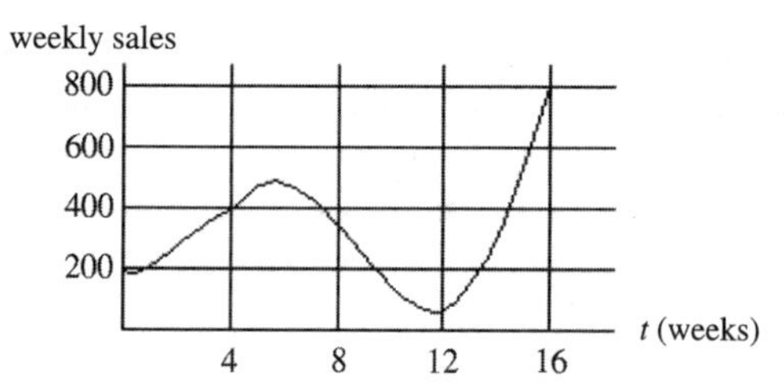

Figure 7

Interlude: Progress Check 2.5

1. Let $f(x) = 8\sqrt{x}$.

(a) Simplify the difference quotient $\dfrac{f(t+h) - f(t)}{h}$. Hint: Consider rationalizing the numerator.

(b) Evaluate your results from **(a)** when $t = 4$ to obtain $\dfrac{f(4+h) - f(4)}{h}$.

(c) Complete Table 2. Give all estimates accurate to four decimal places.

TABLE 2

h	-0.01	-0.001	-0.0001	0.0001	0.001	0.01
$\dfrac{f(4+h) - f(4)}{h}$						

(d) Do the values of $\dfrac{f(4+h) - f(4)}{h}$ appear to be approaching some particular fixed value as values of h become closer to zero? Explain.

2. The load that a horizontal beam can support varies jointly with the width of the beam, the square of its height, and inversely with its length. A beam that is 4 inches wide, 8 inches tall, and 12 feet long can support a load of one ton. Let S denote the load the beam can support in tons and let w, h, and l denote the width, height, and length, respectively, in feet.

(a) Find a formula for S in terms of w, h, and l. Don't approximate anything.

(b) Determine the load a similar beam that is 10 feet long can support, accurate to four decimal places and labeled appropriately.

3. Figure 8 shows the graph of a function p. Gridlines each mark one unit.

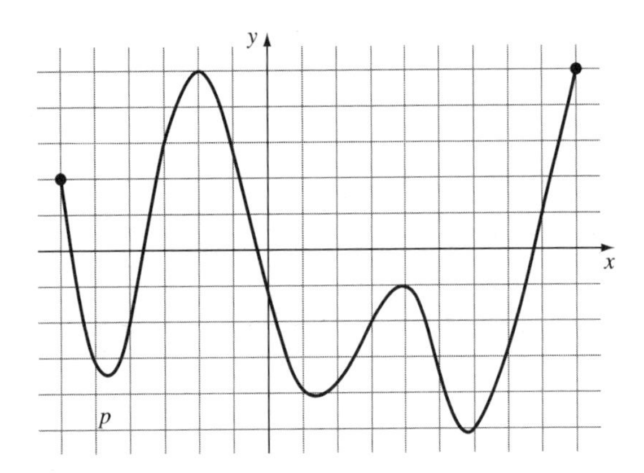

Figure 8

(a) Calculate the average rate of change of $p(x)$ from $x = -4$ to $x = 8$.

(b) Find all numbers t such that the average rate of change of $p(x)$ between $x = -4$ and $x = t$ is zero.

(c) Find the number b such that the average rate of change of $p(x)$ between $x = -3$ and $x = b$ is as small as possible (i.e, largest negative value). Explain your choice.

CHAPTER 2 READINESS, PART 1

Provide complete solutions in the space provided.

1. A function f is defined by the following rule: "For a given input x, the corresponding output $f(x)$ is generated by first doubling the input, adding three to that, squaring the result, and dividing by the square root of triple the input." Find the rule of f.

2. Let $f(x) = \sqrt{4 - x^2}$ and $g(x) = x^2 + 1$. Let $h(x) = (g \circ f)(x)$

   (a) Give the domain h. Use interval notation and show clearly how you arrived at your result.

   (b) Sketch a graph of h and state the range of h using interval notation.

3. Let $g(x) = -0.3x^3 + 1.4x^2 + 0.6x - 4.3$. Graph this function in the standard viewing window on your graphing calculator and make a corresponding sketch on the provided grid. Use the graph of g to answer the following. Give all estimates correct to three decimal places.

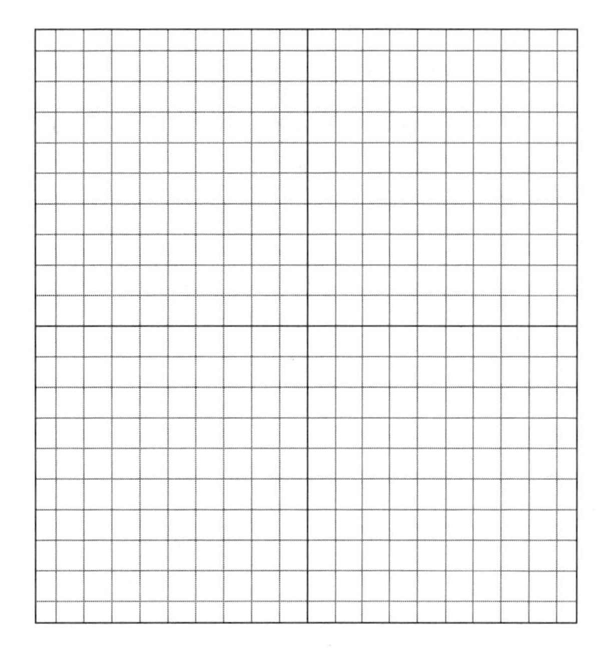

(a) Find the maximum output for $g(x)$ on the interval $[-2, 6]$.

(b) Solve the inequality $g(x) \geq 1$. Give the solution in interval notation.

(c) Determine the intervals where $g(x)$ is increasing. Use interval notation.

(d) Find all values of x in the interval $[-10, 10]$ such that $g(x) = -3$.

(e) Find all values of x in the interval $[-10, 10]$ such that $g(x) = 0$.

CHAPTER 2 READINESS, PART 3

We say that a function f is an **even function** if $f(-x) = f(x)$ for all x in the domain of f. We say that a function f is an **odd function** if $f(-x) = -f(x)$ for all x in the domain of f. A function may be neither an even function nor an odd function.

1. Determine if the following functions are even functions, odd functions, or neither.

(a) $f(x) = x^2$

(b) $g(x) = x^3$

(c) $h(x) = 3x - 5$

(d) $k(x) = 2x^3 - x$

(e) $m(x) = \dfrac{x}{x^2 + 1}$

2. Explain why the graph of an even function $y = f(x)$ must be symmetric with respect to the y-axis. Explain why the graph of an odd function must be symmetric with respect to the origin.

3. Let $f(x) = x^2$. Sketch graphs of each of the following.

(a) $g(x) = f(2x)$

(b) $h(x) = f(4x)$

(c) $k(x) = f\left(\frac{1}{2}x\right)$

(d) $m(x) = f\left(\frac{2}{3}x\right)$

4. Suppose f is some function. Based on experimental evidence, make a reasonable conjecture on how the graph of $g(x) = f(kx)$ is related to the graph of f for:

(a) $0 < k < 1$

(b) $k > 1$

Answers to Odd-Numbered Problems

Problem Set 2.1

1. $f(-2) = 6, f(1) = -18, f(4) = 35$ **3.** Domain: $\{-2, -1, 0, 1, 2, 3, 4\}$, Range: $\{-18, -10, 0, 6, 13, 35\}$

5. Yes. For each value of p, there is exactly one value for w. **7.** $f(-4) = 27$, $f(0) = 3$, $f\left(\frac{5}{2}\right) = -12$ **9.** $M(-1) = -\dfrac{3}{2}, M(2) = 6$

11. Domain: all real numbers. **13.** $f(x) = \dfrac{(x+4)^2}{3}$ **15.** $x = -2,\ x = 10$ **17.** $A(l) = l(12 - l)$

19. $f(2a) = 4a + 4,\ f(p + a) = 2p + 6,\ f(t + h) - f(t) = 2h,\ \dfrac{f(t + h) - f(t)}{h} = 2$

21. The equation defines: **(a)** y as a function of x and x as a function of y **(b)** y as a function of x **(c)** y as a function of x and x as a function of y

23. **(a)** $S(1) = 68$; in 2004, \$68,000 was spent on corrections. $S(4) = 79$; in 2007, \$79,000 was spent on corrections. $S(7) = 98$; in 2010, \$98,000 was spent on corrections. **(b)** \$5000/yr

25. **(a)** $N(2) = 55\frac{1}{3}$; There are approximately 55 bacteria after 2 hours. **(b)** -55. From the 12th to the 15th hour, the average rate of change in the number of bacteria is -55 bacteria/hour. (On average, the number of bacteria decreased by 55 bacteria/hour.)

27. **(a)** $S(4) - S(0) = 15$; four months after being discovered, the tumor has grown by 15 mm^3 **(b)** $2\frac{1}{3}$; during the first three months after discovery, the tumor was growing at an average rate of $2\frac{1}{3}$ mm^3/month

29. all parts: 5

31. **(a)** $-8z - 4h + 6$ **(b)** $6 - 4a - 4b$ **(b)** $6 - 8z$

33. **(a)** $\dfrac{8}{(q + h + 4)(q + 4)}$ **(b)** $\dfrac{8}{(a + 4)(b + 4)}$ **(c)** $\dfrac{8}{(q + h + 4)(q - h + 4)}$

Problem Set 2.2

1. $f(3) = -1.3, f(0) \approx -2.2, f(-2) \approx 1.2$ **3.** $h(1) = -3, h(-5) = 3, h(-2) = 3$
5. 1. domain $[-6, 7]$, range $[-4, 4]$; 2. domain $[-4, 5]$, range $[-2, 3]$; 3. Domain $(-\infty, \infty)$, range $[-5, 5)$; 4. domain $[-7, 7]$, range $[-4, 4]$

7.

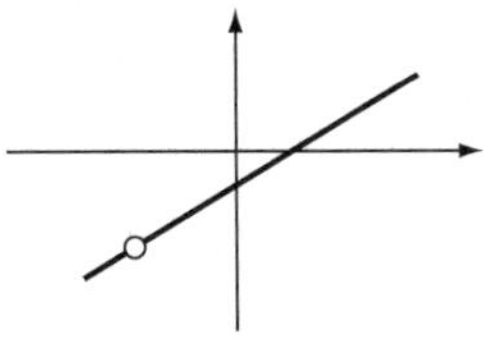

Standard viewing window

9.

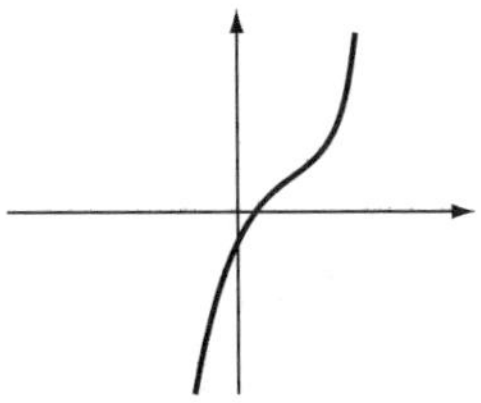

Standard viewing window

11. Domain $(-\infty, \infty)$, Range $[-9, \infty)$ **13.** Domain: $(-\infty, \infty)$, Range $(0, \infty)$ **15.** Domain: $[-8, 6]$, Range $[-30, 55\ 1/3]$
17. **(a)** -1; **(b)** -1.4; **(c)** $(-\infty, -4.9), (1.4, 4.4)$; **(d)** $(-\infty, -2], [2, \infty)$; **(e)** $x \approx -4.9, 1.4, 4.4$ **(f)** $(-4.7, -0.2), (5, \infty)$
19. f is increasing on $[-1.135, 1.468]$ and decreasing on $(-\infty, -1.135], [1.468, \infty)$, y-intercept: -1, t-intercepts: $t = -2.420, 1.204, 1.716$
21. h is increasing on $[-0.271, 0.5], [2.771, \infty)$ and decreasing on $(-\infty, -0.271], [0.5, 2.771]$, y-intercept: 0, p-intercepts: $-0.478, 0, 0.870, 3.608$
23. f is increasing on $(-\infty, 0.258]$ and decreasing on $[0.258, \infty)$, y-intercept: -2, no x-intercepts.
25. No **27.** No **29.** Yes **31.** $x \approx -1.893$ **33.** $(-\infty, -1.082), (0.827, 1.533)$ **35.** $[4.879, \infty)$

Problem Set 2.3

1. \$1667 spent on advertising minimizes profit; \$10,000 spent on advertising maximizes profit.
3. 7400 units **5.** 6 in × 9 in **7.** 6.115 in × 6.115 in × 9.173 in **9.** 25 ft × 30 1/3 ft
11. **(a)** 4 ft for the circle **(b)** 2.240 ft for the square, 1.760 ft for the circle
13. 8.40 ft × 4.20 ft

Problem Set 2.4

1. $(f+g)(x)$: 8, 7, 2, 9, 4; $(f-g)(x)$: -2, 1, 0, 1, 0; $(f+g)(x)$: 2, 1, 3, 5, 4; $(g+f)(x)$: 1, 4, 5, 2, 3
3. **(a)** 11 **(b)** -34 **(c)** -72 **(d)** 2
5. **(a)** $(p+q)(x) = x^2 + 4x - 5$ **(b)** $(p-q)(x) = -x^2 - 2x + 3$ **(c)** $(p \cdot q)(x) = (x^2 + 3x - 4)(x - 1)$ **(d)** $(\frac{p}{q})(x) = \dfrac{x^2 + 3x - 4}{x - 1}$
7. **(a)** 1.82 **(b)** 1.5 **(c)** **(d)** 3.5 **(e)** 5
9. **(a)** $(h \circ g)(t) = \dfrac{t}{1 + \sqrt{t}}$, domain $(0, \infty)$ **(b)** $(g \circ h)(t) = \sqrt{\dfrac{t^2}{1+t}}$, domain $(-1, \infty)$

   (c) $(h \circ h)(t) = \dfrac{t^4}{(1+t)(1+t+t^2)}$ domain all real numbers except -1 **(d)** $(g \circ g)(t) = \sqrt[4]{t}$ domain $[0, \infty)$
11. $f(x) = x + 1$, $g(x) = x^2 + 2x$
13. $f(x) = \dfrac{6}{\sqrt{x}}$, $g(x) = 5 - x$
15. $h(t) = \sqrt{676 - (4 + 0.25t)^2}$; domain $[0, 88]$, range $[0, \sqrt{660}]$
17. **(a)** $F = \dfrac{x}{12}$ **(b)** $Y = \dfrac{x}{3}$ **(c)** $Y \circ F = \dfrac{x}{36}$. This converts inches to yards.
19. **(a)**

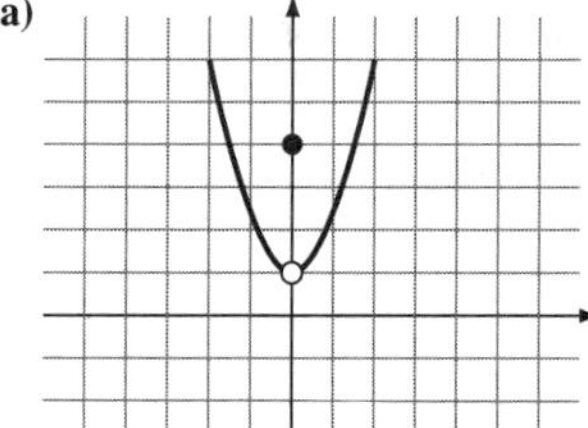

(b)

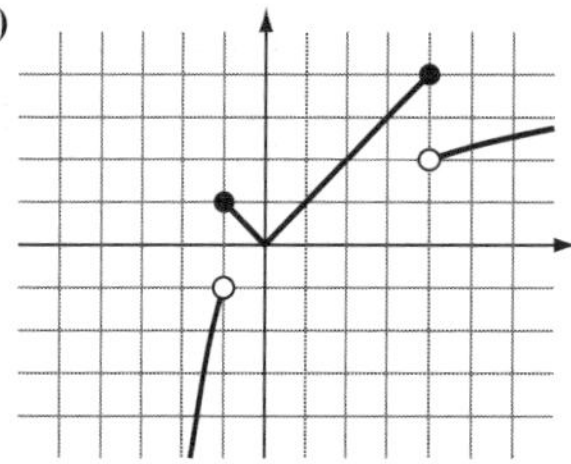

21. **(a)** $A(t) = 150t$ **(b)** $B(t) = \begin{cases} 200 \text{ if } 0 \le t \le 3 \\ 150t - 250 \text{ if } t > 3 \end{cases}$

    (c) $C(t) = \begin{cases} 400 \text{ if } 0 \le t \le 6 \\ 150t - 500 \text{ if } t > 6 \end{cases}$
23. Shift to the right 3, stretch away from the x-axis by a factor of 2, shift vertically up 7.
25. Reflect in the y-axis, compress towards the x-axis by a factor of $\frac{1}{4}$, shift vertically up 2.
27. $g(x) = 2(1 - x^2) + 3 = -2x^2 + 5$
29. $g(x) = -9\left[(x-1)^3 + 1\right]$
31.

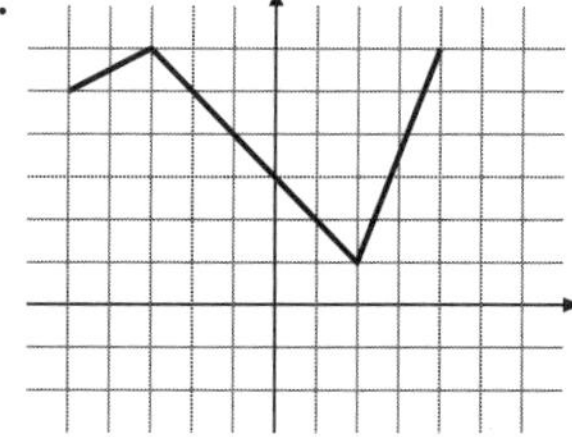

33.

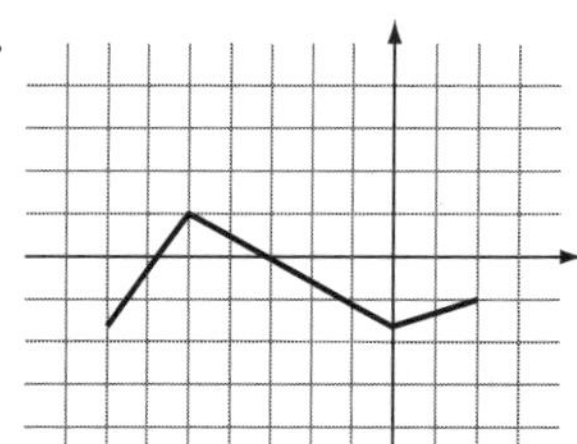

35.

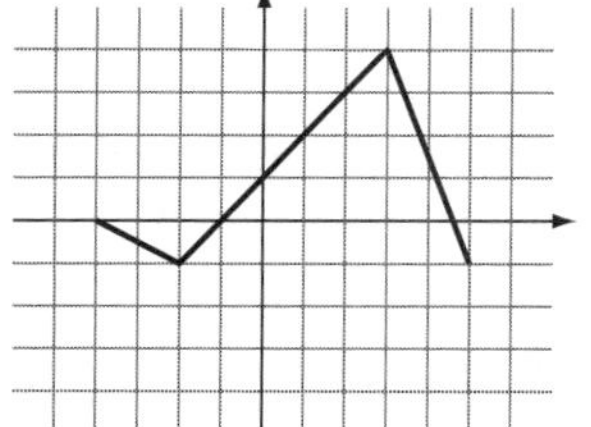

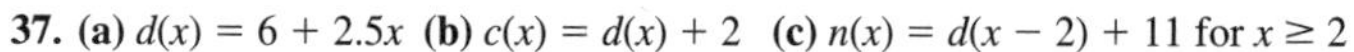

37. **(a)** $d(x) = 6 + 2.5x$ **(b)** $c(x) = d(x) + 2$ **(c)** $n(x) = d(x - 2) + 11$ for $x \ge 2$

Problem Set 2.5

1. 18
3. 777.6
5. Yes, since $C = kd$, with $k = \pi$.
7. 93.591 lbs
9. **(a)** 210 times per second **(b)** 52.5 in.
11. **(a)** $T = kr^2d^4$ **(b)** The thrust increases by a factor of 16. **(c)** The propeller diameter can be reduced by 18.35%.
13. **(a)** $\dfrac{-2}{t(t+h)}$ **(b)** $\dfrac{-2}{1+h}$ **(c)** -1.6
15. $-16x + 32$
17. **(a)** $68 - 1.5h$; This is the average intake in gms/min during the first h minutes of the study. **(b)** $b = 0$
19. **(a)** $1.2r + 0.6h - 3r^2 - 3rh - h^2$ **(b)** 0.0725; As trachea radius changes from 0.3 cm to 0.35 cm, the velocity of air increases at 0.0725cm/sec per centimeter increase in trachea radius, on average.
21. **(a)** B and C; D and E **(b)** C and D, E and F

Polynomial and Rational Functions

CHAPTER CONTENTS

The Mackinac Bridge, one of the world's longest suspension bridges, connects the upper and lower peninsulas of Michigan. How high above the roadway is the lowest point of the cable? The answer to this question is developed in Example 11 on p. 201.

In Chapter 2, we explored the general concepts associated with functions and their graphs. In this chapter, we consider *polynomial functions*—those functions whose rules are given by polynomials. In addition, we focus on showing the connection between the x intercepts, zeros, factors, and roots of polynomials both visually and algebraically. We also study *rational functions*—functions formed by quotients of polynomials. We will continue to emphasize graphs and applications, including mathematical modeling of real-world phenomena.

OBJECTIVES

1. Graph Functions of the Form $f(x) = a(x - h)^2 + k$
2. Solve Quadratic Inequalities Graphically
3. Determine Maximum and Minimum Values
4. Graph Horizontal Parabolas
5. Solve Applied Problems

3.1 Quadratic Functions

So far we have learned about first-degree, or linear, functions whose graphs are straight lines. We begin this chapter with a study of second-degree polynomial functions, referred to as *quadratic functions.* We will investigate the properties of such functions with particular emphasis on their graphs, which are called *parabolas.*

Definition Quadratic Function

A function of the form

$$f(x) = ax^2 + bx + c$$

where a, b, and c are constants and $a \neq 0$, is called a **quadratic function.**

Examples of quadratic functions are:

$$h(x) = x^2,$$

$$f(x) = 3x^2 - 1,$$

and

$$g(x) = -0.5x^2 - x + 3.$$

Graphing Functions of the Form $f(x) = a(x - h)^2 + k$

Figures 1a and 1b show the graphs of the quadratic functions

$$f(x) = x^2 \quad \text{and} \quad g(x) = -x^2$$

respectively.

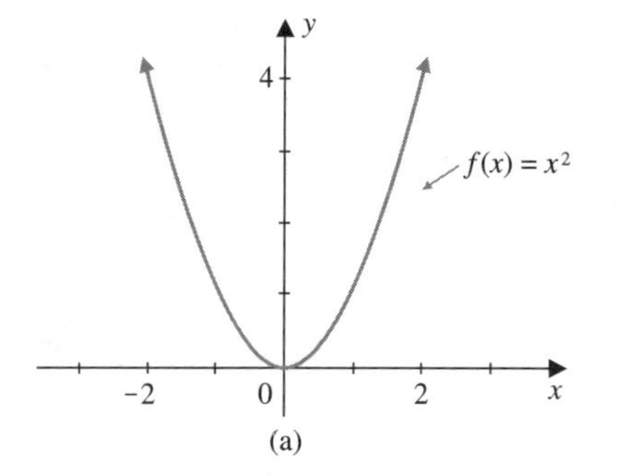

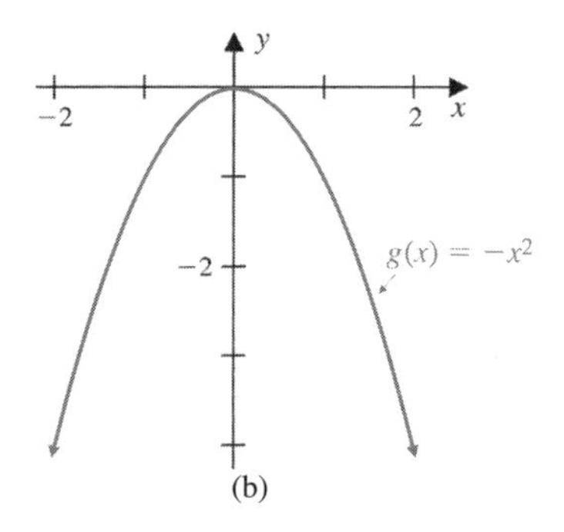

Figure 1

The graphs of these two functions are parabolas. In fact, the graph of any quadratic function is a *parabola,* with the following geometric characteristics, as displayed in Figure 2.

1. There is a vertical **axis of symmetry.**
2. The graph **opens upward** or **downward.**
3. The **vertex** of the parabola is either a *low point* on the graph (if it opens upward) or as *high point* (if it opens downward).

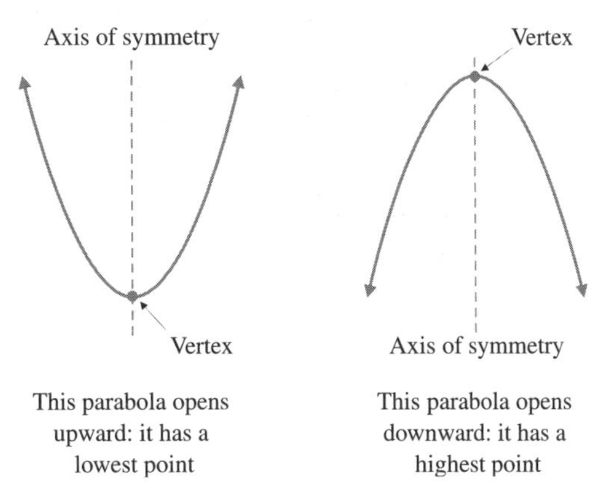

Figure 2

When a quadratic function

$$f(x) = ax^2 + bx + c$$

is rewritten in the special algebraic form

$$f(x) = a(x - h)^2 + k$$

we can use the transformation techniques introduced in Section 2.4 to graph it by using the standard graph of $y = x^2$. The diagram below reviews the effects of the values of a, h, and k on transformations, and Figure 3 illustrates a transformation from the graph of $y = x^2$ to the graph of f.

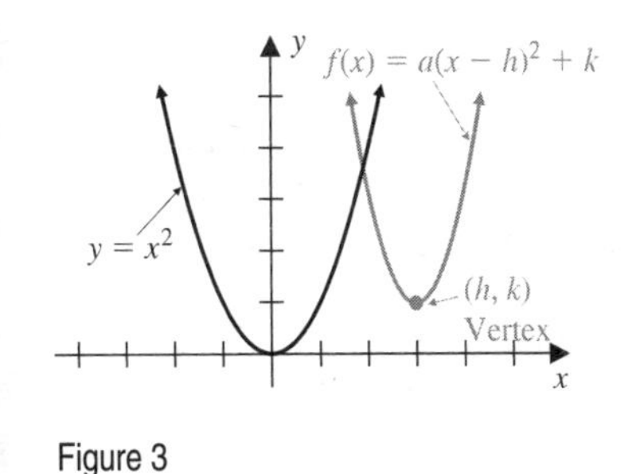

Figure 3

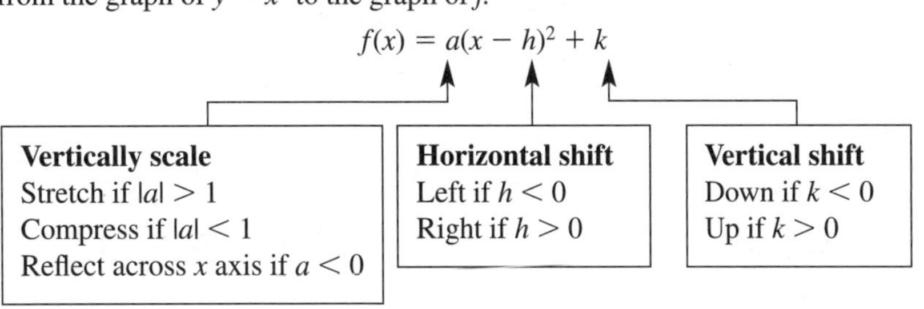

EXAMPLE 1 Sketching the Graph of a Parabola by Using Transformations

Use the graph of $y = x^2$, along with transformations, to graph the function

$$g(x) = -2(x - 1)^2 + 3.$$

Find the vertex and axis of symmetry of the parabola. Also determine the domain and range.

Solution The graph of g is obtained by transforming the graph of $y = x^2$ as follows:

(i) Vertically stretch the graph of $y = x^2$ by a factor of 2.

(ii) Reflect the resulting graph across the x axis.

(iii) Shift the latter graph to the right by 1 unit.

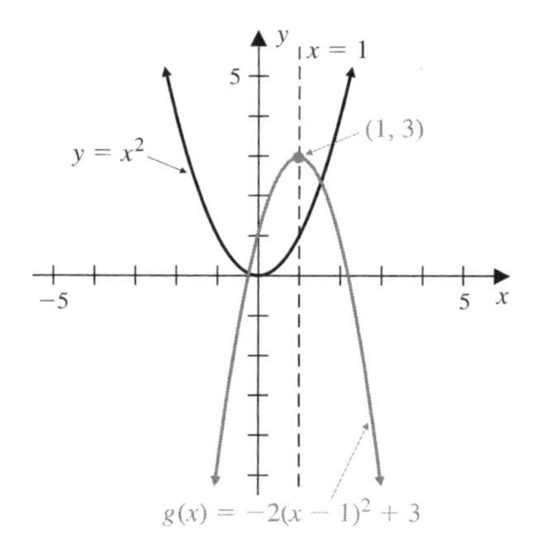

Figure 4

(iv) Finally, shift the graph obtained in step (iii) upward by 3 units to get the graph of $g(x) = -2(x - 1)^2 + 3$ (Figure 4).

Because of the 1-unit horizontal shift to the right and the 3-unit shift upward, the vertex (0, 0) of the graph of $y = x^2$ is shifted to the point (1, 3) to become the vertex of the graph of g. Also, the axis of symmetry of $y = x^2$ (the y axis) is shifted 1 unit to the right to become the line $x = 1$, which is the axis of symmetry of the graph of g. From the graph, we see that the domain of g is $\mathbb{R}$ and the range includes all numbers in the interval $(-\infty, 3]$. ❖

A quadratic function given in the *standard form*

$$f(x) = ax^2 + bx + c, a \neq 0$$

can be converted to the special algebraic form

$$f(x) = a(x - h)^2 + k$$

by using the technique of completing the square to produce a perfect square trinomial.

EXAMPLE 2 Converting a Quadratic Function to the Special Algebraic Form

Rewrite

$$f(x) = 3x^2 + 6x + 5$$

in the form

$$f(x) = a(x - h)^2 + k.$$

Find the vertex and axis of symmetry, and sketch the graph of f. Also determine the range of f.

Solution To rewrite the quadratic function f in the special algebraic form, we proceed as follows:

$$\begin{aligned} f(x) &= 3x^2 + 6x + 5 && \text{Given} \\ &= 3(x^2 + 2x) + 5 && \text{Group the } x \text{ terms and factor out 3} \\ &= 3\left(x^2 + 2x + \left(\frac{1}{2}\cdot 2\right)^2 - 1\right) + 5 && \text{Complete the square for } x^2 + 2x \text{ by adding and subtracting 1 inside the parentheses} \\ &= 3(x^2 + 2x + 1) - 3 + 5 && \\ &= 3(x+1)^2 + 2 && \text{Regroup, factor, and simplify} \\ &= 3[x - (-1)]^2 + 2 && \text{Special form} \end{aligned}$$

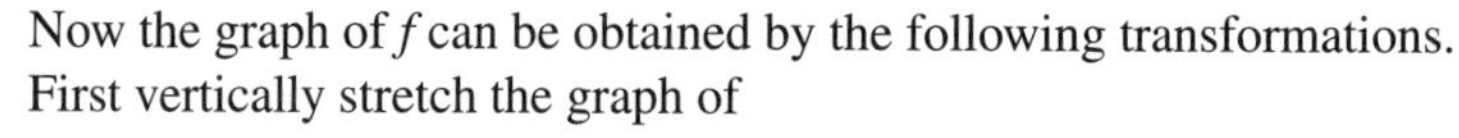

Now the graph of f can be obtained by the following transformations.
First vertically stretch the graph of

$$y = x^2$$

by a factor of 3 ($a = 3$).
Then shift the resulting curve 1 unit to the left ($h = -1$).
Finally, shift this latter curve 2 units upward ($k = 2$) (Figure 5).
The graph of f is a parabola that opens upward with vertex at $(-1, 2)$;
its axis of symmetry is the line $x = -1$.
The range includes all numbers in the interval $[2, \infty)$. ❖

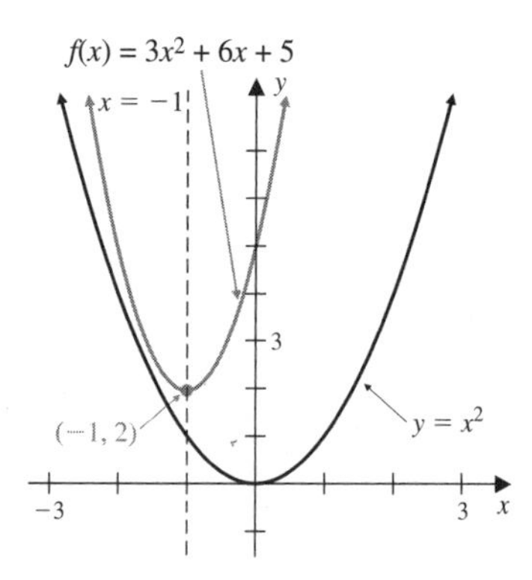

Figure 5

Table 1 summarizes the information about the graph of a quadratic function

$$f(x) = ax^2 + bx + c, a \neq 0$$

that can be read from its converted special algebraic form

$$f(x) = a(x - h)^2 + k.$$

TABLE 1
Characteristics of the Graph of $f(x) = ax^2 + bx + c = a(x - h)^2 + k$

If	Parabola Opens	Vertex	Axis of Symmetry	Illustration
$a > 0$	Upward	(h, k)	$x = h$	y, $x = h$, x, (h, k)
$a < 0$	Downward	(h, k)	$x = h$	y, (h, k), x, $x = h$

Up to now we have sketched graphs of quadratic functions defined by given equations. Sometimes we can reverse the process and use data from a parabolic graph to derive its equation.

EXAMPLE 3 Determining an Equation of a Parabola

Find an equation of the parabola, expressed in the form $f(x) = a(x - h)^2 + k$, that fits the data in Figure 6.

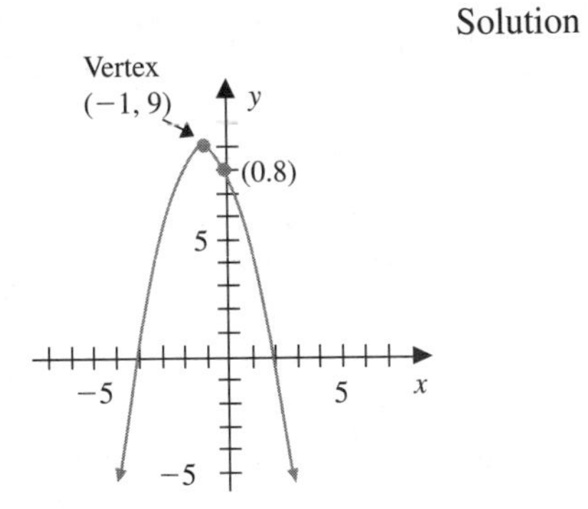

Figure 6

Solution Figure 6 shows the vertex at the point $(-1, 9)$, and the parabola opens downward. Using the special form $f(x) = a(x - h)^2 + k$ with $h = -1$ and $k = 9$, we get

$$f(x) = a(x + 1)^2 + 9$$

To find a, notice that (0, 8) is a point on the parabola, so its coordinates satisfy the equation $f(x) = a(x + 1)^2 + 9$. Thus

$$8 = a(0 + 1)^2 + 9 \quad \text{or} \quad 8 = a + 9$$

So $a = -1$.

Hence, an equation of the parabola is

$$f(x) = -(x + 1)^2 + 9.$$

By expanding the right side, we obtain another equation for the graph, namely, $f(x) = -x^2 - 2x + 8$. ❖

Solving Quadratic Inequalities Graphically

The real number solutions of the quadratic equation

$$ax^2 + bx + c = 0$$

correspond to the real zeros of f or x intercepts of the graph of

$$f(x) = ax^2 + bx + c.$$

For instance, if a quadratic equation has two distinct real solutions, then its associated function has two real-number zeros, or, equivalently, the graph has two x intercepts (Figure 7a). If the equation has one real solution, then there is one real zero and the graph has one x intercept (Figure 7b), and if the equation has only imaginary solutions, then the associated function has no real zeros and the graph has no x intercepts (Figure 7c).

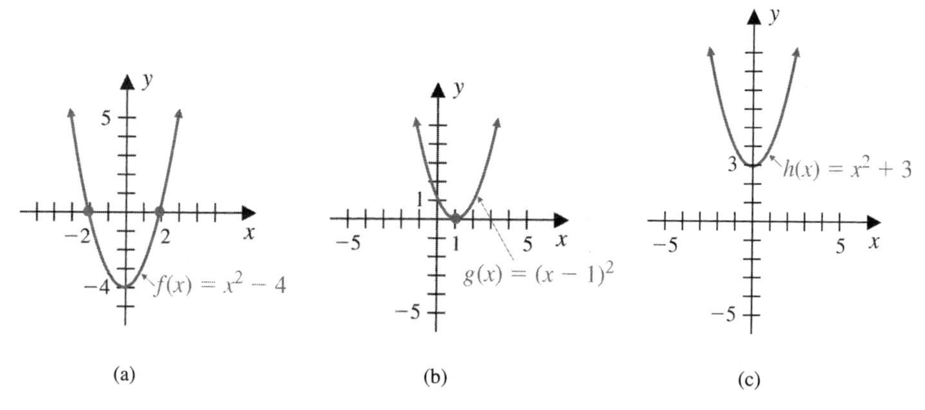

Figure 7

As we learned in Section 1.4, we can use the locations of the x intercepts along with the graph of a function to solve a quadratic inequality by determining where the graph is above or below the x axis, as the next example shows.

EXAMPLE 4 Solving Quadratic Inequalities Graphically

Use the graph of $f(x) = 3x^2 - 5x + 2$ to solve each inequality.

(a) $3x^2 - 5x + 2 \geq 0$ (b) $3x^2 - 5x + 2 < 0$.

Solution We begin by locating the x intercepts of the graph of f by finding the zeros of f; that is, we set $f(x) = 0$ and then solve the resulting equation. In this case, we have

$$3x^2 - 5x + 2 = (3x - 2)(x - 1) = 0$$

$$\text{Thus } x = \frac{2}{3} \text{ or } x = 1.$$

So the x intercepts are 2/3 and 1. Figure 8 shows the graph of f and the locations of the x intercepts.

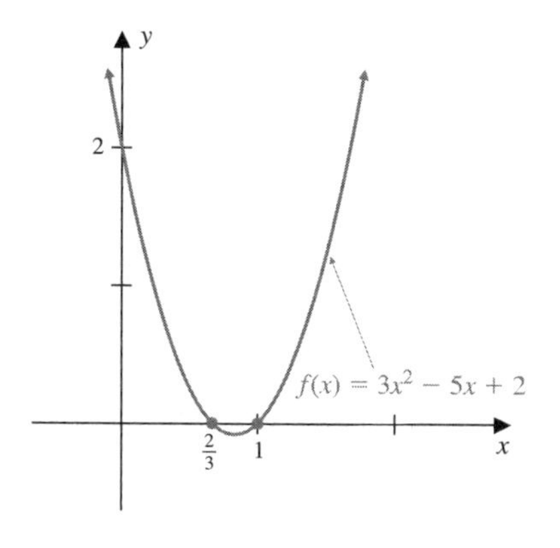

Figure 8

(a) By reading the graph in Figure 8, we see that $y = f(x) = 3x^2 - 5x + 2 \geq 0$ whenever (x, y) is a point on the graph of f that is above or on the x axis. So

$$3x^2 - 5x + 2 \geq 0 \text{ for all values of } x \text{ such that } x \leq \frac{2}{3} \text{ or } x \geq 1 .$$

Hence, the solution set of the inequality includes all values of x either in the interval $(-\infty, 2/3]$ or in the interval $[1, \infty)$.

(b) We observe in Figure 8 that $y = f(x) = 3x^2 - 5x + 2 < 0$ when the graph of f is below the x axis. Thus, the solution set of $3x^2 - 5x + 2 < 0$ includes all values of x such that $2/3 < x < 1$, that is, all numbers in the interval $(2/3, 1)$. ❖

Determining Maximum and Minimum Values

We know that the graph of a quadratic function is a parabola that opens either upward or downward. If the graph opens upward, the vertex—which occurs at the lowest point on the graph—is called the **minimum point** of the function. See Figure 5 on page 146, where the minimum point of $f(x) = 3x^2 + 6x + 5$ is $(-1, 2)$.

Analogously, if the graph opens downward, the vertex, or highest point on the graph, is called the **maximum point.** See Figure 6 on page 147, where the maximum point of $f(x) = -x^2 - 2x + 8$ is $(-1, 9)$.

The minimum or maximum point is called the **extreme point** of the graph, and its y coordinate is called the **extreme value** (either the **minimum** or the **maximum value**) of the function. Thus the minimum value of $f(x) = 3x^2 + 6x + 5$ is 2, and the maximum value of $f(x) = -x^2 - 2x + 8$ is 9.

By employing the technique of completing the square, we can rewrite

$$f(x) = ax^2 + bx + c$$

as

$$f(x) = a\left(x + \frac{b}{2a}\right)^2 + \left(c - \frac{b^2}{4a}\right) \quad \text{(See problem 18).}$$

Comparing the latter form with the special form

$$f(x) = a(x - h)^2 + k$$

we see that $h = \dfrac{-b}{2a}$, so the vertex or extreme point is

$$(h, k) = \left(-\frac{b}{2a}, f\left(-\frac{b}{2a}\right)\right).$$

This result provides us with an easy way of finding the extreme value of a quadratic function (Table 2).

TABLE 2
Extreme Value of $f(x) = ax^2 + bx + c$

If	Parabola Opens	Extreme Point or Vertex	Extreme Value	Illustration
$a > 0$	Upward	$\left(-\frac{b}{2a}, f\left(-\frac{b}{2a}\right)\right)$	$f\left(-\frac{b}{2a}\right)$ (Minimum)	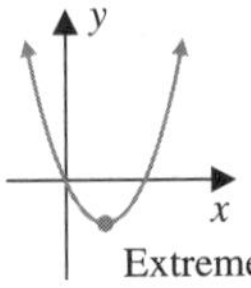 Extreme point $\left(-\frac{b}{2a}, f\left(-\frac{b}{2a}\right)\right)$ (minimum)
$a < 0$	Downward	$\left(-\frac{b}{2a}, f\left(-\frac{b}{2a}\right)\right)$	$f\left(-\frac{b}{2a}\right)$ (Maximum)	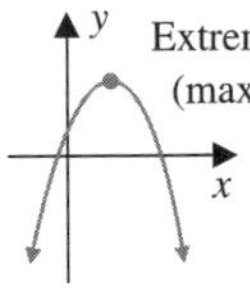 Extreme point $\left(-\frac{b}{2a}, f\left(-\frac{b}{2a}\right)\right)$ (maximum)

EXAMPLE 5 Finding the Extreme Value of a Quadratic Function

(a) Find the extreme value of

$$f(x) = -2x^2 + 12x - 16,$$

and determine whether the value is a maximum or minimum.

(b) Graph f, then interpret the graph by relating it to the result in part (a).

Solution (a) We use the results in Table 2 as follows.
The function

$$f(x) = -2x^2 + 12x - 16$$

is in the standard form

$$f(x) = ax^2 + bx + c,$$

where $a = -2$ and $b = 12$.
Since a is negative, there is a maximum value when

$$x = -\frac{b}{2a} = -\frac{12}{-4} = 3$$

The maximum value of f is given by

$$f\left(-\frac{b}{2a}\right) = f(3)$$
$$= -2(3)^2 + 12(3) - 16$$
$$= 2$$

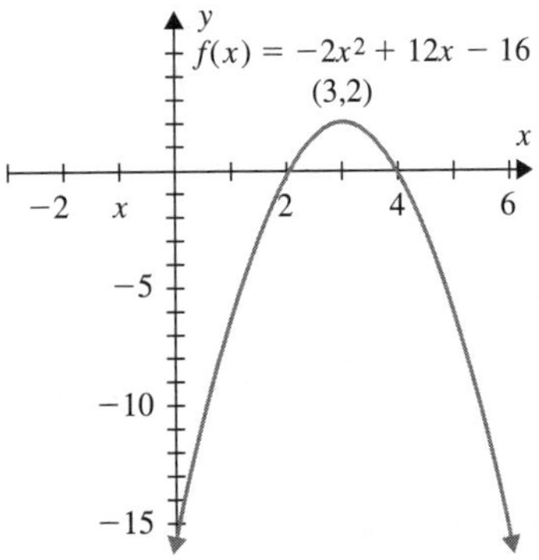

Figure 9

(b) Figure 9 shows the graph of f. From the result of part (a), we see that the function attains a maximum value of 2 at $x = 3$. ❖

Graphing Horizontal Parabolas

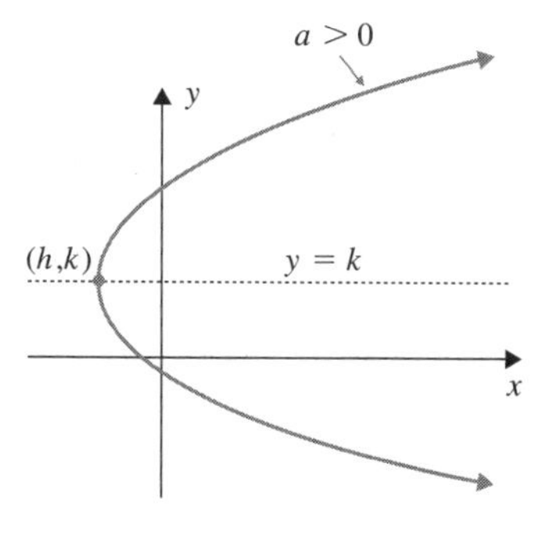

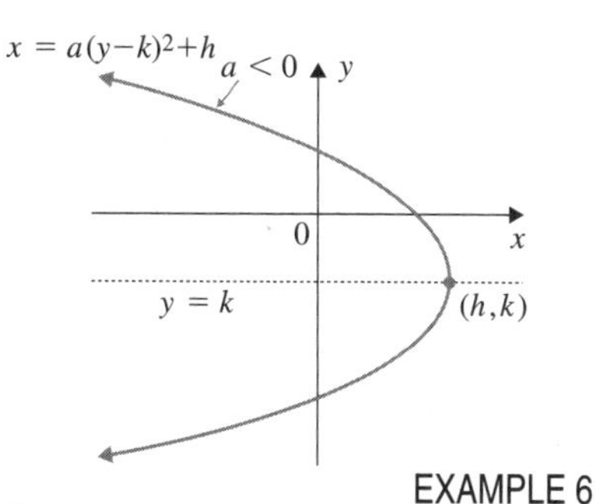

Figure 10

We have seen that the graphs of quadratic functions of the form

$$y = ax^2 + bx + c$$

are parabolas that open upward if $a > 0$, or downward if $a < 0$. By interchanging the role of x and y in the equation $y = ax^2 + bx + c$, we obtain an equation of the form

$$x = ay^2 + by + c$$

This latter equation can be written in the form

$$x = a(y - k)^2 + h$$

by completing the square. Because of symmetry, the graphs of such equations are also parabolas, which open to the right if $a > 0$ (Figure 10a), or to the left if $a < 0$ (Figure 10b). The vertex is (h, k) and the axis of symmetry is the line $y = k$ (Figure 10).

Because of the vertical line test, these equations do not represent functions.

EXAMPLE 6 Graphing Horizontal Parabolas

Sketch the graph of each parabola. Indicate the vertex and the axis of symmetry.

(a) $x = 8y^2$
(b) $x = -y^2 - 2y + 15$

Solution (a) The graph of the equation $x = 8y^2$ is a parabola that opens to the right, because $a = 8 > 0$. The vertex is at $(0, 0)$ and the axis of symmetry is the x axis or $y = 0$ (Figure 11a).

(b) By completing the square, we rewrite the equation in the form

$$x = a(y - k)^2 + h$$

as follows

$$\begin{aligned} x &= -(y^2 + 2y + 1) + 15 + 1 \\ &= -(y^2 + 2y + 1) + 16 \\ &= -(y + 1)^2 + 16 \end{aligned}$$

Thus, the vertex is at $(16, -1)$, and the parabola opens to the left. The axis of symmetry is the line $y = -1$ (Figure 11b).

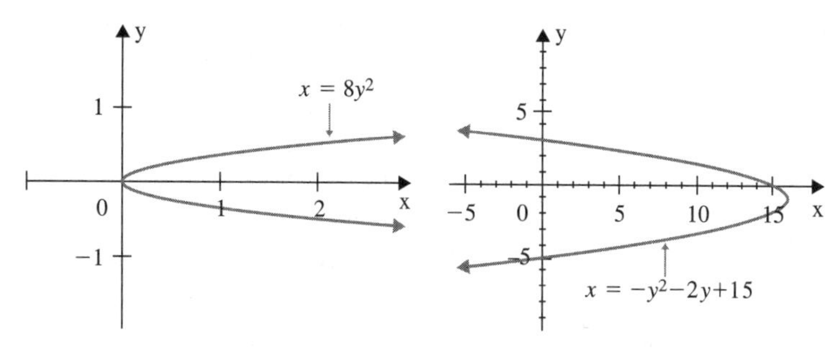

Figure 11

❖

Solving Applied Problems

The solutions of many applied problems depend on finding the extreme value of a function used to model the situation. In Example 7 we construct a quadratic function to model a problem in which we want to find the maximum value.

EXAMPLE 7 Enclosing a Rectangular Corral of Maximum Area

A rancher wishes to enclose a rectangular corral. To save fencing costs, a river is used along one side of the corral and the other three sides are formed by 3000 feet of fence. Assume that x represents the width (in feet) of two of the fenced sides, y the length (in feet) of the third fenced side, and A the area of the corral (Figure 12).

(a) Express the area A as a function of x.
(b) What are the restrictions on x?

Figure 12

(c) Find the dimensions of the corral that enclose the maximum area. What is the maximum area?
(d) Graph the function defined in part (a) with the restrictions from part (b). Discuss what happens as the width increases from 0 to 1500 feet.

Solution (a) Since the rancher plans to use 3000 feet of fence for three sides, the sketch in Figure 12 indicates that

$$2x + y = 3000$$

so

$$y = 3000 - 2x.$$

Thus the area A of the corral is given by

$A = xy$	Area = (width) · (Length)
$= x(3000 - 2x)$	Substitute for y
$= 3000x - 2x^2$	Multiply
$= -2x^2 + 3000x$	Write in standard form

So the area A is a quadratic function of the width x.

(b) Since the sum of the widths, $2x$, cannot be longer than the total amount of available fencing, it follows that

$0 < 2x < 3000$

$0 < x < 1500$ Divide each part by 2

(c) To find the maximum value of A, we first observe that

$$A = -2x^2 + 3000x = ax^2 + bx + c \text{ with } a = -2 \text{ and } b = 3000$$

So by using the results in Table 2, we find that the value of x for which the maximum value of A occurs is given by

$$x = -\frac{b}{2a}$$
$$= -\frac{3000}{2(-2)} = 750.$$

The corresponding value of y is given by

$$y = 3000 - 2x = 3000 - 2(750) = 1500.$$

Therefore the maximum value of A is given by

$$A = (750)(1500) = 1{,}125{,}000.$$

So a maximum area of 1,125,000 square feet is obtained when the corral is 750 feet wide and 1500 feet long.

(d) Next we sketch a graph of the quadratic function derived in part (a), taking into account the restrictions on x determined in part (b). The resulting graph (Figure 13) reveals that as the width x increases from 0 to 1500, the corresponding area A first increases from 0 to attain a maximum value of 1,125,000 square feet and then decreases to a value of 0 when $x = 1500$.

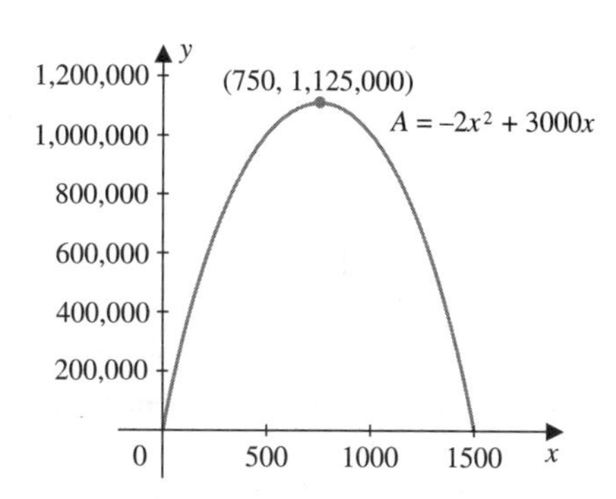

Figure 13

❖

EXAMPLE 8 Modeling Population Growth

The U.S. Census Bureau has projected the population of the United States for the years from 2000 through 2100, according to the data in Table 3.

(a) Plot a scattergram of the data. Locate the t values on the horizontal axis and the P values on the vertical axis.

(b) Use a grapher and quadratic regression to find the equation of the form

$$P = at^2 + bt + c$$

that best fits the data. Use three decimal places for a, b, and c.

(c) Sketch the graph of the best fit curve from part (b).

(d) Use the regression model to predict the population in the years 2015, 2075, and 2110. Round off the answers to the nearest thousand.

TABLE 3

Year	2000	2010	2020	2030	2040	2050	2060	2070	2080	2090	2100
t	0	10	20	30	40	50	60	70	80	90	100
U.S. Population (*P*) (nearest thousand)	275,306	299,862	324,927	351,070	377,350	403,687	432,011	463,639	497,830	533,605	570,954

(Source: U.S. Census Bureau, January 13, 2000.)

Solution (a) The scattergram for the given data is shown in Figure 14a, where P is used for the vertical axis and t is used for the horizontal axis.

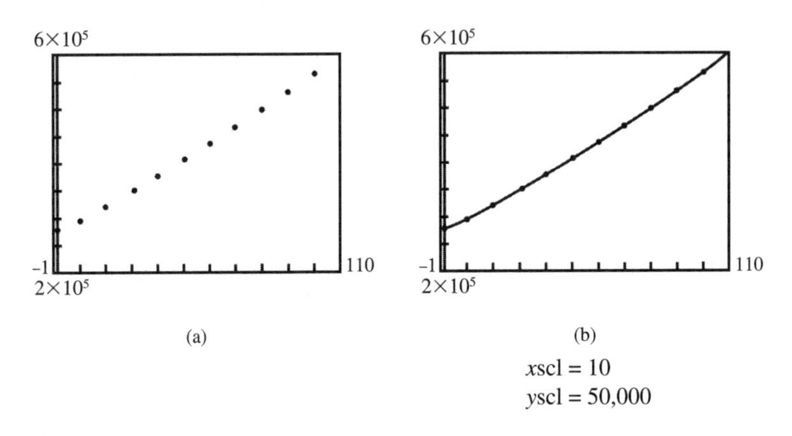

Figure 14

(b) By using a grapher, we find the quadratic regression model to be

$$P = 7.722t^2 + 2147.517t + 277436.245, \; t \geq 0$$

where the coefficients have been rounded to three decimal places.

(c) The graph of the regression equation in part (b) is shown in Figure 14b along with the scattergram of the given data.

(d) To predict the population in the year 2015, we substitute $t = 15$ into the regression equation to get 311,386. So the approximate population is predicted to be 311,386,000 in the year 2015.

Similarly, in the year 2075 we substitute $t = 75$ into the equation to get the approximate population of 481,936,000.

Finally, in the year 2110, $t = 110$ so that the approximate population is 607,099,000. ❖

At times, the same data can be modeled with different regression types of equations. For instance, in Chapter 4 we will use an exponential function as the regression model for the population data given in Table 3.

PROBLEM SET 3.1

Mastering the Concepts

1. Match each quadratic function with its graph in Figure 15.

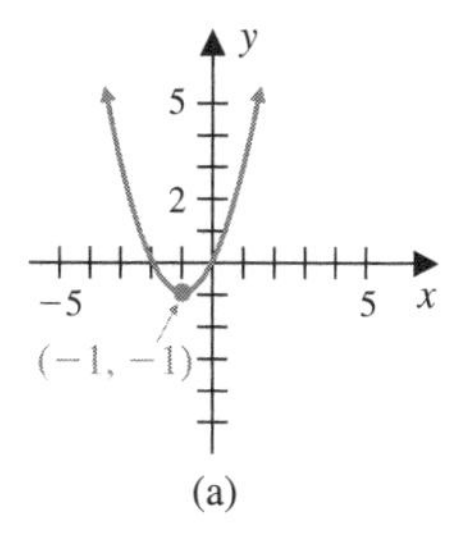

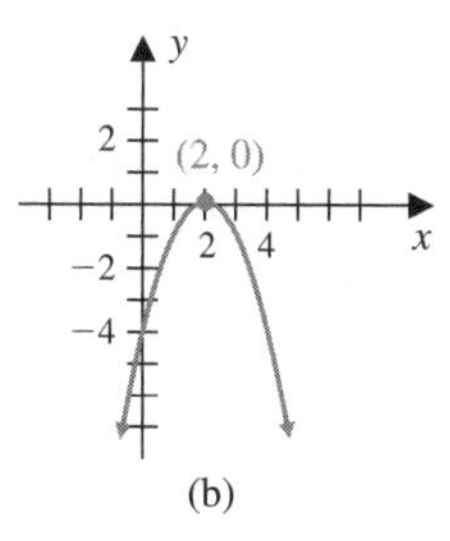

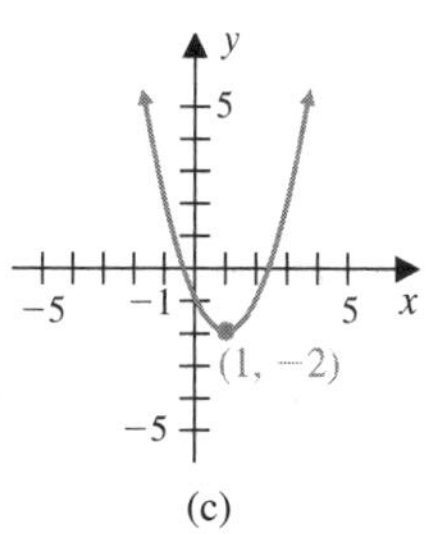

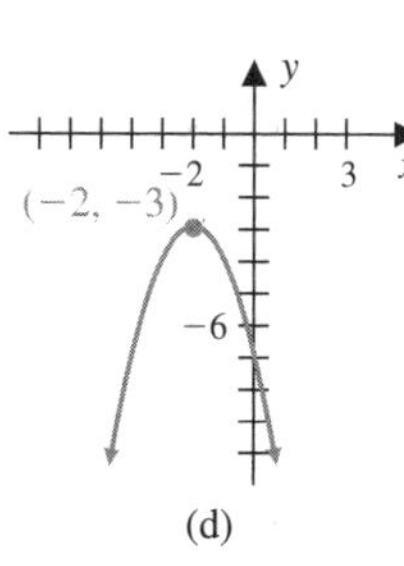

Figure 15

(a) $f(x) = -(x - 2)^2$
(b) $g(x) = (x - 1)^2 - 2$
(c) $h(x) = -(x + 2)^2 - 3$
(d) $F(x) = (x + 1)^2 - 1$

2. Describe the steps that can be used to transform the graph of $f(x) = x^2$ into the graph of g. Sketch the graph of g.

(a) $g(x) = 3x^2 + 4$
(b) $g(x) = (x + 1)^2 - 1$
(c) $g(x) = -(x + 1)^2 + 2$

In problems 3–8, use the graph of $y = x^2$ along with transformations to graph each function. Find the vertex and axis of symmetry of each parabola. Also determine the domain and range.

3. $f(x) = (x - 3)^2 + 2$
4. $g(x) = -(x + 1)^2 + 1$
5. $h(x) = -2(x + 2)^2 - 3$
6. $F(x) = 3(x - 1)^2 + 2$
7. $F(x) = \frac{1}{2}(x + 3)^2 + 1$
8. $G(x) = -\frac{1}{3}(x + 2)^2 - 1$

In problems 9–17, rewrite each function f in the form

$$f(x) = a(x - h)^2 + k$$

Find the vertex and axis of symmetry, and sketch the graph of each function. Also determine the range.

9. $f(x) = x^2 - 4x - 5$
10. $f(x) = x^2 - 10x + 20$
11. $f(x) = -x^2 - 6x - 8$
12. $f(x) = -x^2 + 5x + 11$
13. $f(x) = 3x^2 - 5x - 21$
14. $f(x) = 2x^2 - 11x + 5$
15. $f(x) = -5x^2 + 3x + 4$
16. $f(x) = -3x^2 - 6x + 5$
17. $f(x) = -\frac{1}{2}x^2 - 2x + 1$
18. By completing the square, show that $f(x) = ax^2 + bx + c$ can be written as

$$f(x) = a\left(x + \frac{b}{2a}\right)^2 + \left(c - \frac{b^2}{4a}\right)$$

In problems 19 and 20, find equations for the given parabolas.

19. (a)

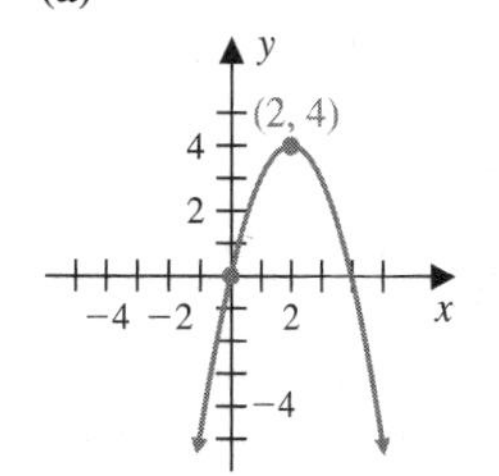

(b)

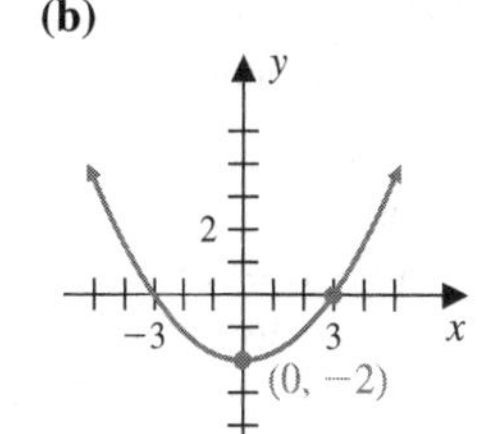

20. (a)

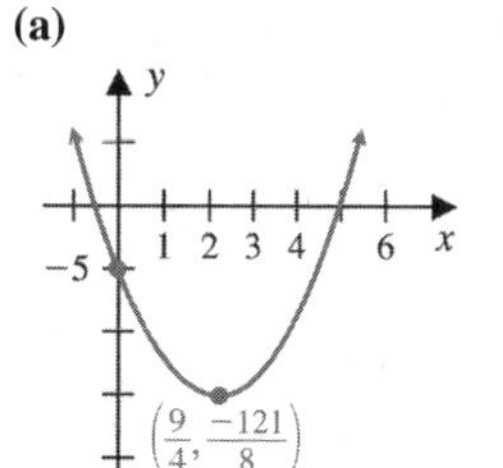

(b)

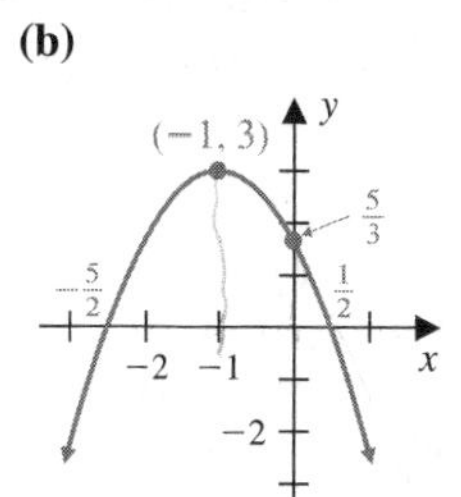

In problems 21–24, find the equation in the form

$$f(x) = a(x - h)^2 + k$$

for the quadratic function f whose graph satisfies the given conditions. Also sketch the graph of f.

21. Vertex at $(2, -2)$ and contains the point $(0, 0)$
22. Vertex at $(-1, -9)$ and contains the point $(-4, 0)$

23. Vertex at (0, 1) and contains the point (2, −3)
24. Vertex at (0, 3) and contains the point (−3, 0)

In problems 25–30, use the locations of the x intercepts and the graph of the given function f to solve the associated inequality. Express the solution in interval notation.

25. $f(x) = x^2 + 5x - 14$; $\quad x^2 + 5x - 14 > 0$
26. $f(x) = -x^2 + x + 20$; $\quad -x^2 + x + 20 \geq 0$
27. $f(x) = -2x^2 + 5x - 3$ $\quad -2x^2 + 5x \leq 3$
28. $f(x) = 2x^2 - x - 1$ $\quad 2x^2 - x \geq 1$
29. $f(x) = 6x^2 + 7x - 20$ $\quad 6x^2 + 7x - 20 \geq 0$
30. $f(x) = 2x^2 - 9x - 5$ $\quad 2x^2 - 9x - 5 < 0$

In problems 31–36, find the extreme value algebraically. Determine whether the value is a maximum or a minimum.

31. $f(x) = x^2 + 8x + 12$
32. $g(x) = 10x^2 + x - 2$
33. $h(x) = -x^2 + 4x - 5$
34. $h(x) = -9x^2 - 6x + 8$
35. $g(x) = 2x^2 + x - 1$
36. $h(x) = -2x^2 - 9x + 5$

In problems 37–44, sketch the graph of each parabola. Indicate which way the parabola opens, the vertex, and the axis of symmetry.

37. $x = (y - 4)^2 - 3$
38. $x = -(y - 2)^2 + 1$
39. $x = -5(y + 3)^2 - 6$
40. $x = (y + 1)^2 + 5$
41. $x = -y^2 + 6y - 9$
42. $x = -y^2 - 5y - 4$
43. $x = 3y^2 - 12y - 36$
44. $x = 2y^2 - 4y - 4$

Applying the Concepts

45. **Minimum Product:** The difference of two real numbers is 6.
    (a) Express the product P as a function of one of the numbers.
    (b) Find the numbers that produce a minimum value of P. What is the minimum value?

46. **Geometry:** The sum of the height y (in centimeters) and the base x (in centimeters) of a triangle is 24 centimeters.
    (a) Express the area A of the triangle as a function of x.
    (b) Find the base that produces the maximum area of the triangle. What is the maximum area?

In problems 47–58, round off the answers to two decimal places when necessary.

47. **Maximum Profit:** Suppose that the profit P (in thousands of dollars) of a manufacturing company is related to the number x (in hundreds) of full-time employees by the quadratic function

$$P = -1.25x^2 + 156.25x \qquad 0 \leq x \leq 100$$

    (a) What is the profit if the number of employees is 500? 1000? 1500?
    (b) Find the maximum profit and the corresponding number of employees.
    (c) Graph the function. Discuss the profit trend as the number of employees increases.

48. **Minimum Cost:** The cost C (in dollars) to produce x tons of an alloy used for engine blocks is modeled by the function

$$C = 2.97x^2 - 603x + 62{,}836 \qquad 0 \leq x \leq 110.$$

    (a) How much does it cost if the company produces 50 tons? 100 tons? 110 tons?
    (b) Find the minimum cost and the corresponding number of tons.
    (c) Graph the function. Discuss the cost as the number of tons increases.

49. **Maximum Height of a Baseball:** The trajectory of a baseball hit by a player at home plate is approximated by the parabola whose function is given by

$$h(x) = -0.004x^2 + x + 3.5.$$

In this model, home plate is located at the origin, $h(x)$ (in feet) is the height of the ball above ground, and x is the horizontal distance (in feet) the ball has traveled from home plate.

    (a) How high is the ball after it has traveled horizontally 50 feet? 100 feet? 200 feet?
    (b) Graph the function. Approximate the maximum height and how far the ball travels horizontally to reach this height.
    (c) If the ball is not touched in flight, how far will it have traveled horizontally when it hits the ground?

50. **Maximum Height of a Rocket:** A model rocket is launched and then it accelerates until the propellant burns out, after which it coasts upward to its highest point. The height h of the rocket (in meters) above ground level t seconds after the burnout is modeled by the function

$$h = -4.9t^2 + 343t + 2010.$$

    (a) Find the height of the rocket above ground level 10 seconds, 20 seconds, and 30 seconds after the burnout.
    (b) Graph the function and approximate the maximum height attained by the rocket. How many seconds after the burnout does it take to reach the maximum height?
    (c) How long after the burnout will it take for the rocket to reach ground level?

51. **Construction:** A lookout tower is to be constructed on the peak of a hill. A portion of a cross section of the crown of the hill is approximately parabolic in shape. If a coordinate system is superimposed on this cross section as shown in Figure 16,

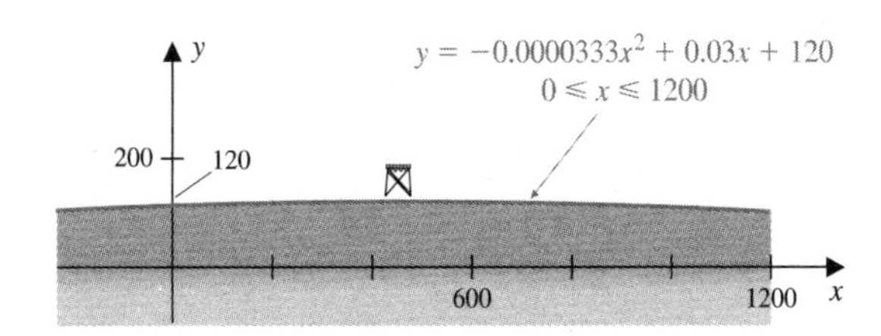

Figure 16

then the parabolic portion is modeled by the function

$$y = -0.0000333x^2 + 0.03x + 120$$

where $0 \le x \le 1200$,

where x and y are measured in feet, and y is the number of feet above sea level. (The parabolic shape is barely visible in Figure 16 due to the scale of the drawing.)

(a) Find the height of the hill above sea level when x is 200 feet, 500 feet, and 1000 feet.

(b) Graph the function and approximate the coordinates of the peak of the hill.

(c) If the observation booth of the tower is to be 150 feet above sea level, how tall should the tower be?

52. Maximum Current of a Circuit: In an electrical circuit, the power P (in watts) delivered to the load is modeled by the function

$$P = EI - RI^2$$

where E is the voltage (in volts), R is the resistance (in ohms), and I is the current (in amperes). For a circuit of 120 volts and a resistance of 12 ohms, find the current that produces the maximum power, and find the maximum power.

53. Constructing a Playground: A recreation department plans to build a rectangular playground enclosed by 1000 feet of fencing. Assume that A represents the area of the enclosure and that x and y, respectively, represent the width and length (in feet).

(a) Express the area A as a function of the width x.

(b) What are the restrictions on x?

(c) Find the dimensions of the playground that encloses the maximum area. What is this area?

(d) Graph the function defined in part **(a)**. Discuss what happens to the area as the width increases.

54. Shape of a Suspension Bridge: A suspension bridge has a main span of 1200 meters between the towers that support the parabolic main cables, and these towers extend 160 meters above the road surface (Figure 17).

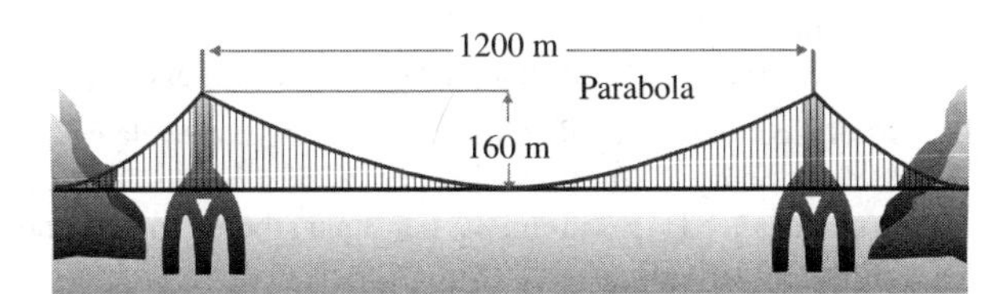

Figure 17

(a) Assuming that the main cable is tangent to the road at the center of the bridge and that the road is horizontal, find an equation of the main parabolic cable with respect to an xy coordinate system (with a vertical y axis, the x axis running along the road surface, and the origin at the center of the bridge).

(b) Graph the equation in part (a).

(c) Find the distance between the roadway and the cable at a point 300 meters from the center of the bridge.

55. Enclosure: A rancher has 450 feet of fencing to enclose three sides of a rectangular pen. A barn will be used to form the fourth side of the pen. What dimensions of the pen would enclose the maximum area? What is the maximum area?

56. Enclosure: A breeder of horses wants to fence in two grazing areas along a river, including one interior fence separation perpendicular to the river as shown in Figure 18. The river will serve as one side, and 600 yards of fencing are available. What is the largest area that can be enclosed? What are the exterior dimensions that yield the maximum area?

Figure 18

57. Architecture: A Norman-style window has the shape of a rectangle surmounted by a semicircular arch (Figure 19). Find the height y of the rectangle and the radius x of the semicircle of the Norman window with maximum area whose perimeter is $8 + 2\pi$ feet. (Use $\pi = 3.14$.)

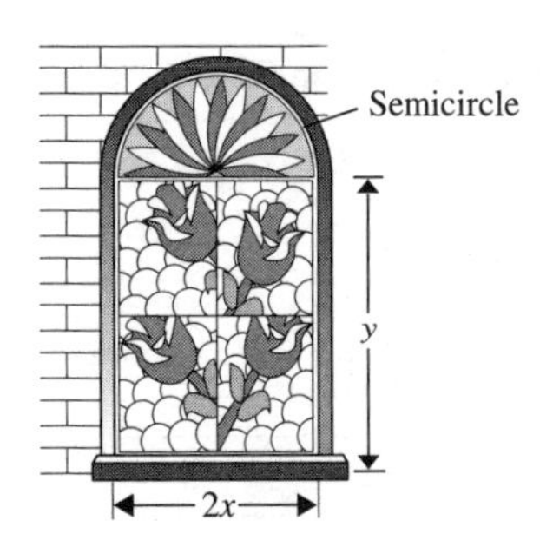

Figure 19

58. **Agriculture:** An orange grower has determined that if 120 orange trees are planted per acre, each will yield approximately 60 small boxes of oranges per tree over the growing season. The grower plans to add or delete trees in the orchard. The state agricultural service advises that because of crowding, each additional tree will reduce the average yield per tree by about 2 boxes over the growing season. How many trees per acre should be planted to maximize the total yield of oranges, and what is the maximum yield of the crop?

59. The following table gives discrete data points for pairs of values of the form (x, y).

x	y
0	0
0.5	8.5
1	26
2	30
3	16
4	0

(a) Plot a scattergram of this data. Locate the x values on the horizontal axis and the y values on the vertical axis.
(b) Use a grapher and quadratic regression to find the equation of the form

$$y = ax^2 + bx + c$$

that best fits the data. Round off a, b, and c to three decimal places.
(c) Sketch the graph of the best fit parabola from part (b).
(d) Use the regression equation to estimate the value of y if $x = 2.5$, 5, or 10. Round off the answers to one decimal place.

60. **Property Crime:** A study shows the number N (in millions) of property crimes in the United States for every other year from 1996 through 2006 is given in the following table.

Year	t	**N Crimes (in millions)**
1996	0	10.6
1998	2	11.7
2000	4	12.4
2002	6	12.7
2004	8	12.5
2006	10	12.2

(a) Plot a scattergram of this data. Locate the t values on the horizontal axis and the N values on the vertical axis.
(b) Use a grapher and quadratic regression to find a model of the form

$$N = at^2 + bt + c$$

that best fits the data. Round off a, b, and c to three decimal places.
(c) Sketch the graph of the best fit curve from part (b).
(d) Use the regression model to predict the number of property crimes in millions for the years 2011 and 2016. Round off the answers to one decimal place.

61. **Modeling Income:** The net income N (in millions of dollars) of a company during the period from 2001 through 2005 is given by the following table.

Year	t	**Income N (in millions of dollars)**
2001	0	24.8
2002	1	15.6
2003	2	16.2
2004	3	17.4
2005	4	32.2

(a) Plot a scattergram of this data. Locate the t values on the horizontal axis and the N values on the vertical axis.
(b) Use a grapher and quadratic regression to find a model of the form

$$N = at^2 + bt + c$$

that best fits the data. Round off a, b, and c to three decimal places.
(c) Sketch the graph of the best fit curve from part (b).
(d) Use the regression equation to predict the net income of the company in millions of dollars for the years 2010 and 2015. Round off to one decimal place.

62. **Fuel Economy:** The data in the following table relates the fuel economy F (in miles per gallon) of a car to different average speeds V (in miles per hour) between 40 and 80 miles per hour.

Vehicle Speed V (mph)	Fuel Economy F (mpg)
40	26.1
45	26.5
50	27.5
55	29.5
60	29.0
65	25.8
70	25.5
75	24.5
80	23.8

(a) Plot a scattergram of this data. Locate the V values on the horizontal axis and the F values on the vertical axis.

(b) Use a grapher and quadratic regression to find a model of the form

$$F = aV^2 + bV + c$$

that best fits the data. Round off a, b, and c to three decimal places.

(c) Sketch the graph of the best fit curve from part (b).

(d) Use the regression equation to estimate the fuel economy if the car speed is 52 mph, 63 mph, or 85 mph. Round off the answers to one decimal place.

Developing and Extending the Concepts

63. Decide how many x intercepts the graph of the function

$$f(x) = ax^2 + bx + c,\ a \neq 0$$

has if the *discriminant,* defined as

$$D = b^2 - 4ac$$

of the corresponding quadratic equation

$$ax^2 + bx + c = 0,$$

satisfies the given condition. Give examples of functions and their graphs to support your assertions.

(a) $D > 0$
(b) $D = 0$
(c) $D < 0$

64. Suppose that

$$f(x) = x^2 - kx + 16.$$

Determine a value for k such that the graph of f has

(a) No x intercept
(b) One x intercept
(c) Two x intercepts

In problems 65–67, use the results from problem 63 to determine how many x intercepts the graph of the function f has. Then graph the function to demonstrate the result.

65. $f(x) = -x^2 - 6x + 7$
66. $f(x) = 4x^2 + 12x + 9$
67. $f(x) = -4x^2 + 28x - 49$

68. **Calculus:** In calculus, *Simpson's rule* is used in numerical integration. Simpson's rule is based on the fact that the points $(-h, y_0)$, $(0, y_1)$ and (h, y_2) lie on the graph of a quadratic function $y = ax^2 + bx + c$, whose graph is given in Figure 20. Suppose that the shaded area A in Figure 20 is given by

$$A = \left(\frac{h}{3}\right)(2ah^2 + 6c).$$

Show that

$$A = \left(\frac{h}{3}\right)(y_0 + 4y_1 + y_2).$$

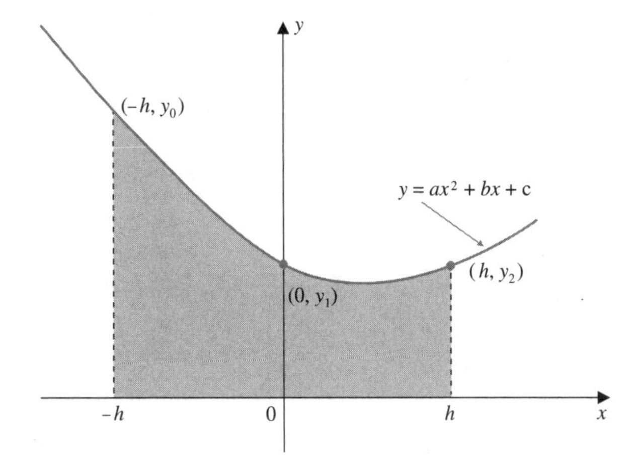

Figure 20

Interlude: Progress Check 3.1, Part 1

1. Let q denote the number of units of some product produced and sold and let $R(q)$ denote the corresponding revenue (in dollars). Suppose that $R(q)$ is a quadratic function and that selling 48 units results in the maximum revenue of \$1843.20. Find a formula for $R(q)$.

2. A rancher has 10,000 linear feet of fencing available to enclose a rectangular field and divide the field into two equal-sized pastures with an internal fence parallel to one of the external sides. Let x denote the length, in feet, of the internal fence and let $A(x)$ denote the corresponding area of the entire enclosed field. Find a formula for $A(x)$ and then determine the dimensions of each pasture that results in the maximum possible area for each pasture.

3. The owner of a retail lumber store wants to construct a fence to enclose a rectangular outdoor storage area adjacent to the store, using all of one wall of the store as part of one side of the storage area. The wall of the store is 100 feet long.

(a) Find the dimensions of the enclosure of maximum area if 340 feet of fencing is to be used. Completely justify your conclusion.

(b) Find the dimensions of the enclosure of maximum area if 240 feet of fencing is to be used (as well as all of the wall of the store). Completely justify your conclusion.

3.2 Polynomial Functions of Higher Degree

OBJECTIVES

1. Graph Functions of the Form $f(x) = a(x - h)^n + k$
2. Establish Basic Properties of Polynomial Functions
3. Relate Multiple Zeros to Graphs
4. Solve Applied Problems

Table 1 lists the types of polynomial functions that we have investigated so for.

TABLE 1

Type of Function	**Form**	**Graph**
Degree 0 (constant)	$f(x) = k\ (k \neq 0)$	Horizontal line
Degree 1 (linear)	$f(x) = mx + b\ (m \neq 0)$	Straight line
Degree 2 (quadratic)	$f(x) = ax^2 + bx + c\ (a \neq 0)$	Parabola

In the following definition, we turn our attention to polynomial functions of degree higher than two.

Definition Polynomial Function

A function f of the form

$$f(x) = a_nx^n + a_{n-1}x^{n-1} + \ldots + a_1x + a_0,\ a_n \neq 0$$

where n is a nonnegative integer, is called a **polynomial function of degree n.** The numbers $a_n, a_{n-1}, \ldots, a_1$, and a_0 are called the **coefficients** of the polynomial.

Thus $f(x) = 4x^3 - x^2 + 2x - 7$ is a polynomial function of degree 3, with coefficients 4, -1, 2, and -7.

It should be noted that $f(x) = 0$ is the **zero** polynomial function with no degree assigned to it.

Graphing Functions of the Form $f(x) = a(x - h)^n + k$

We begin by considering special kinds of polynomial functions of the form

$$f(x) = x^n, \text{ where } n \text{ is a positive integer.}$$

They are called **power functions.**
Examples of power functions are

$$f(x) = x^2, \quad g(x) = x^3, \quad h(x) = x^4 \quad \text{and} \quad F(x) = x^5$$

Recall the graphs of the special functions $f(x) = x^2$ and $g(x) = x^3$, shown in Figure 1.

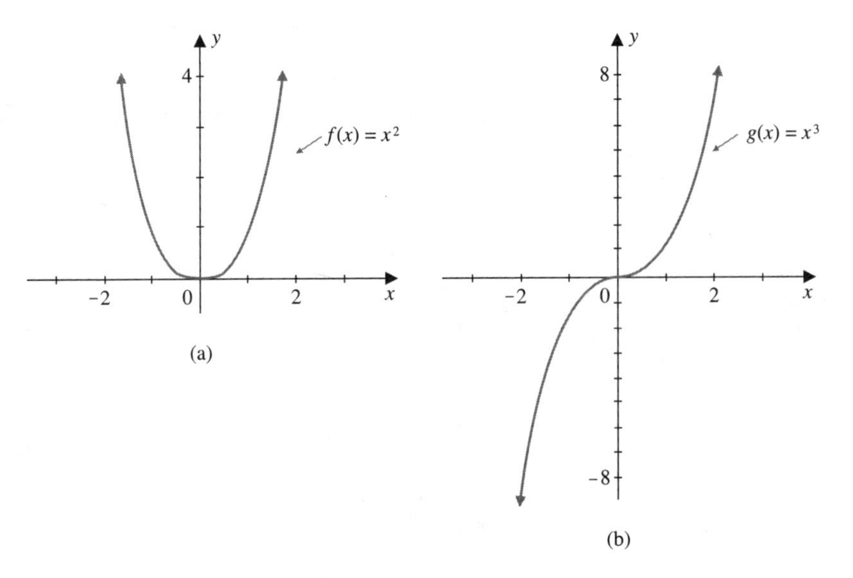

Figure 1

The first function

$$f(x) = x^2$$

is symmetric with respect to the y axis, the second function

$$g(x) = x^3$$

is symmetric with respect to the origin.

They will be used as models for the power functions with degrees greater than three.

In general, graphs of power functions

$$f(x) = x^n$$

fall into two categories according to whether n is an even or odd positive integer.

If $n \geq 4$ is an even integer, then the graphs of functions such as

$$f(x) = x^4 \quad \text{and} \quad g(x) = x^6$$

resemble the graph of $y = x^2$ (Figure 2a).

If $n \geq 5$ is an odd integer, then the graphs of functions such as

$$f(x) = x^5 \quad \text{and} \quad g(x) = x^7$$

resemble the graph of $y = x^3$ (Figure 2b).

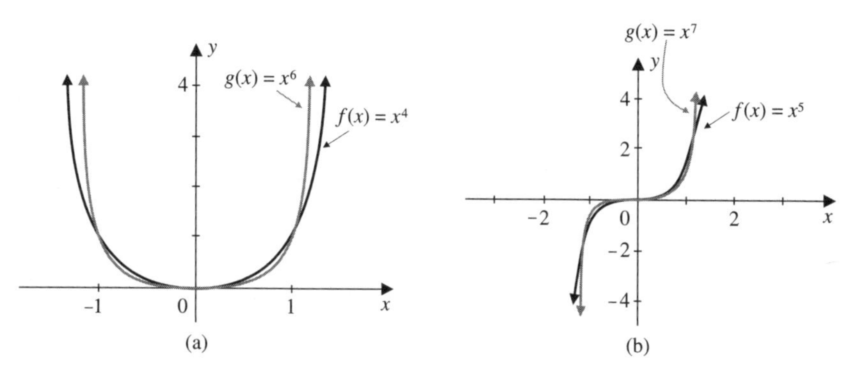

Figure 2

The graphs in Figures 1 and 2 suggest the general properties given in Table 2.

TABLE 2
Characteristics of power function graphs

$f(x) = x^n$, n a Positive Integer		
	n is Even	**n is Odd**
Symmetry with respect to	y axis, since $f(-x) = f(x)$	Origin, since $f(-x) = -f(x)$
Shape of graph	U-shaped curve; contains points $(0, 0)$, $(-1, 1)$ and $(1, 1)$; decreasing in the interval $(-\infty, 0]$ and increasing in the interval $[0, \infty)$	Contains points $(0, 0)$, $(-1, -1)$ and $(1, 1)$; increasing in interval $(-\infty, \infty)$

Transformations of the graphs of

$$y = x^n$$

can be used to sketch graphs of functions of the form

$$f(x) = a(x - h)^n + k$$

where $n \geq 2$ is an integer.

EXAMPLE 1 Using Transformations of a Power Function

Use transformations of the graph of $y = x^4$ (Figure 2a) to sketch the graph of each function.

(a) $f(x) = 3x^4$

(b) $g(x) = -3x^4$

(c) $h(x) = -3(x - 2)^4$

(d) $k(x) = -3(x - 2)^4 + 1$

Solution Each graph is obtained as follows:

(a) Vertically stretch the graph of $y = x^4$ by a factor of 3 to get the graph of

$$f(x) = 3x^4 \text{ (Figure 3a).}$$

(b) Reflect the graph of $f(x) = 3x^4$ across the x axis to obtain the graph of

$$g(x) = -3x^4 \text{ (Figure 3b).}$$

(c) Shift the graph of $g(x) = -3x^4$ horizontally, 2 units to the right to get the graph of

$$h(x) = -3(x - 2)^4 \text{ (Figure 3c).}$$

(d) Shift the graph of h vertically up 1 unit to obtain the graph of

$$k(x) = -3(x - 2)^4 + 1 \text{ (Figure 3d).}$$

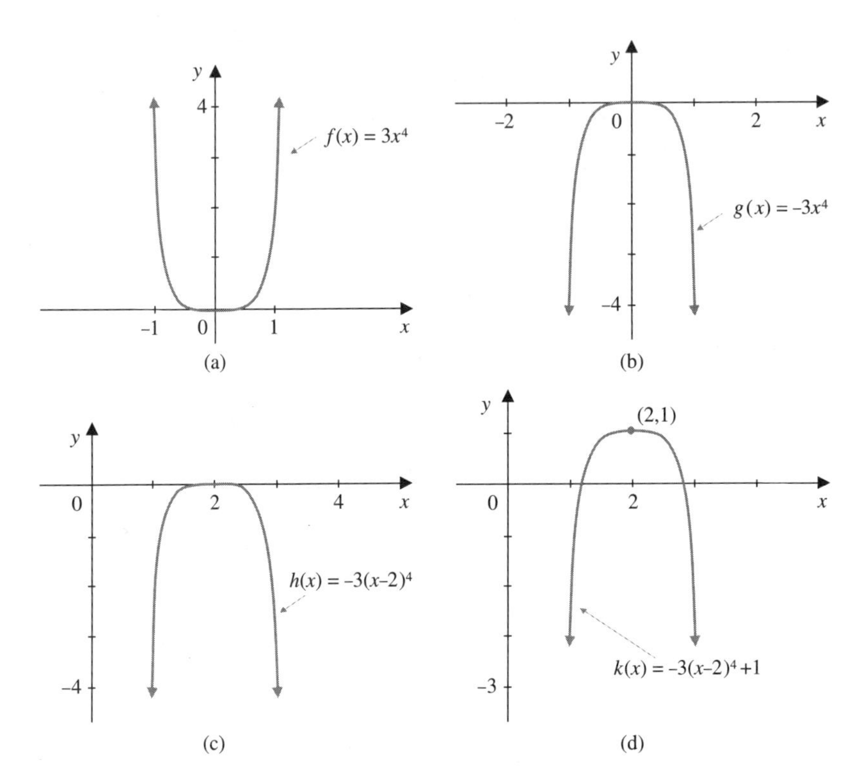

Figure 3

❖

Establishing Basic Properties of Polynomial Functions

Accurately graphing a polynomial function such as

$$f(x) = 2x^3 - 3x^2 - 3x + 2$$

that is not expressed in the form

$$f(x) = a(x - h)^n + k$$

where $n \geq 3$ is an integer, often requires advanced mathematical techniques. Whether we are graphing such a function by hand or by a grapher, it helps to understand some basic properties of polynomial functions by exploring the following questions.

1. Does the graph have gaps or breaks?
2. How many turning points does the function have? Are they high or low points?

3. How many x intercepts does the function have?
4. What happens to the graph of a polynomial function $y = f(x)$ as x gets very large or very small?

The answer to the first question is given in the following property.

Property 1 Continuity

A polynomial function is **continuous** for all real numbers.

This means that a polynomial function is defined for all real numbers, and its graph is an unbroken curve, with no *jumps, gaps,* or *holes.* It can be drawn with a pencil without having to lift it from the paper.

The second question refers to **turning points,** which are points on the graph of a polynomial function where the graph changes direction from increasing to decreasing, or vice versa.

Figure 4 displays a graph that has four turning points, labeled T_1, T_2, T_3, and T_4. This figure also shows that the graph has *peaks* at T_1 and T_3, and *valleys* at T_2 and T_4.

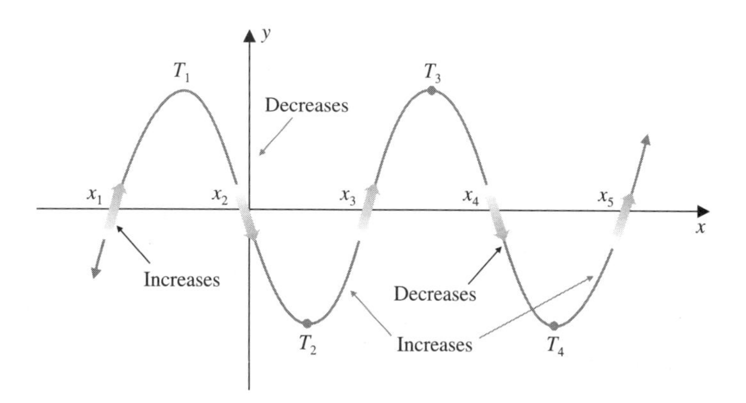

Figure 4

The following property gives us information about the number of turning points the graph of a polynomial function can have.

Property 2 Maximum Number of Turning Points

The graph of a polynomial function of degree n has at most $(n - 1)$ turning points.

The graph of a second-degree (quadratic) function has one turning point—its vertex.

This means that the total number of peaks or "high points" (called **relative maxima**) and valleys or "low points" (called **relative minima**) on the graph of a polynomial function of degree n is at most $n - 1$.

Next we turn our attention to the third question about the number of x intercepts. As Figure 4 shows, the graph of f has five x intercepts at x_1, x_2, x_3, x_4, and x_5. Since the real zeros of f are the same as the x intercepts, there are five real zeros.

In general, we have the following property.

Property 3 Maximum Number of *x* intercepts or Real Zeros

If f is a polynomial function of degree n, then the graph of f cannot have more than n x intercepts. Accordingly, a polynomial function of degree n has at most n real zeros.

For instance, consider the number of x intercepts for the graphs of the third-degree polynomial functions

$$f(x) = x^3 + x,\ g(x) = x^3 - 4x^2 + 4x \quad \text{and} \quad h(x) = x^3 - 7x - 1$$

in Figure 5.

The graph of f has one x intercept (Figure 5a); the graph of g has two x intercepts (Figure 5b); and the graph of h has three x intercepts (Figure 5c).

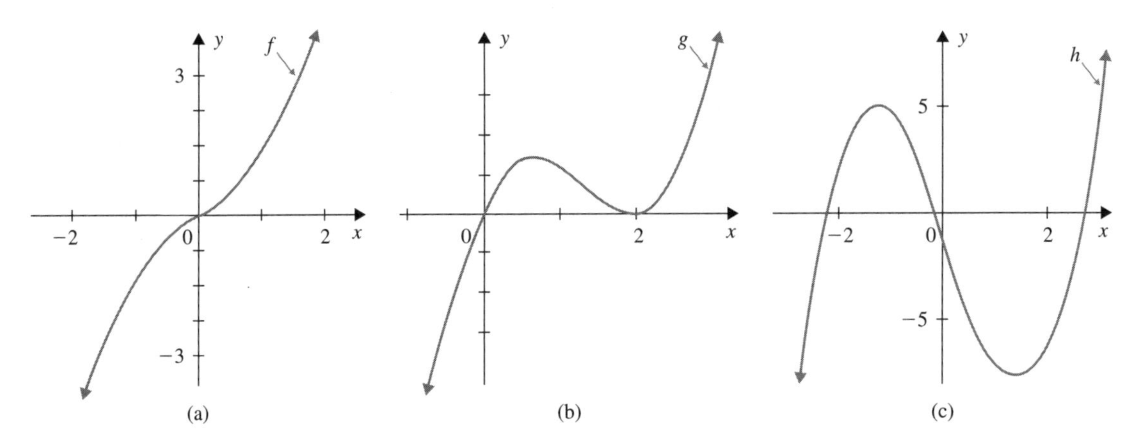

Figure 4

The fourth question regarding the *end behavior* of the graph of a polynomial function can be rephrased as follows:

What happens to the values (and graphs) of a polynomial function $y = f(x)$ as the values of $|x|$ get very large for positive or negative values of x?

Such behavior is referred to as the **limit behavior** of f.

From a numerical point of view, the data in Table 3 reveals the limit behavior of $f(x) = x^4$ in the sense that as x takes on larger and larger positive values (increases), $f(x)$ **increases without bound.** We say that $f(x)$ **approaches positive infinity as** x **approaches positive infinity,** and we describe this limit behavior symbolically by writing

$$f(x) \to +\infty \text{ as } x \to +\infty.$$

TABLE 3
Limit behavior pattern of $f(x) = x^4$

x	-100	-10	-5	-1	0	1	5	10	100
$f(x) = x^4$	100,000,000	10,000	625	1	0	1	625	10,000	100,000,000

Similarly, we observe that $f(x)$ increases without bound as x takes on smaller and smaller negative values (decreases). So we say that $f(x)$ **approaches positive infinity as x approaches negative infinity.** Accordingly, we write

$$f(x) \to +\infty \quad \text{as} \quad x \to -\infty$$

This numerical pattern is confirmed by examining the graph of $f(x) = x^4$ in Figure 2a on page 167. As the x values get larger (or smaller), the y values on the graph get bigger and bigger without restriction.

Similarly, from a graphical point of view, the graph in Figure 1b on page 166 shows that $g(x) = x^3$ increases without bound as positive values of x increase. Also, $g(x)$ decreases without bound as negative values of x decrease. So we describe the limit behavior of g by writing, respectively,

$$g(x) \to +\infty \quad \text{as} \quad x \to +\infty \quad \text{and} \quad g(x) \to -\infty \quad \text{as} \quad x \to -\infty$$

To understand the limit behavior of a polynomial function that is not a power function, let us compare the graphs of

$$f(x) = x^3 + x^2 - 2x - 1 \quad \text{and} \quad g(x) = x^3$$

given in Figures 6a and 6b, respectively.

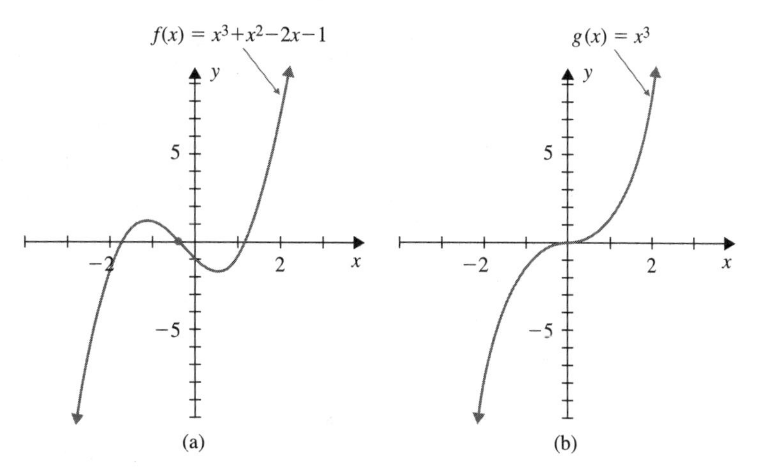

Figure 6

The graph of g increases without bound; that is, $g(x) \to +\infty$ as $x \to \infty$, and the same is true of the graph of f. Also, the graph of g decreases without bound; that is, $g(x) \to -\infty$ as $x \to -\infty$, and the same is true of the graph of f.

Thus, the *far left* and *far right* behavior of the polynomial function f can be discovered by using the far left and far right behavior of its leading term x^3.

In a sense, the leading term x^3 in the polynomial function f "dominates" all other terms the farther away we get from the origin. Whereas the graphs of the functions f and g are not similar near the origin, the *limit behaviors* are the same.

In general, the direction and shape of the graph of a polynomial function at the *far right* and *far left* can be determined by examining its *leading term* as explained in the following property.

Property 4 Limit Behavior

The **limit behavior** of a polynomial function

$$f(x) = a_n x^n + a_{n-1} x^{n-1} + \ldots + a_1 x + a_0,\ a_n \neq 0$$

is the same limit behavior as the function defined by its *leading term,* $g(x) = a_n x^n$.

We refer to the function g as the *limit behavior model of f.* Table 4 gives two additional examples of the limit behavior property.

TABLE 4

Examples of limit behavior models

Given Function	Limit Behavior Model	Limit Behavior Comparison	Graphical Display
$f(x) = 2x^3 - 3x^2 - 3x + 2$	$g(x) = 2x^3$	$g(x) \to +\infty$ as $x \to +\infty$ $g(x) \to -\infty$ as $x \to -\infty$ and so $f(x) \to +\infty$ as $x \to +\infty$ $f(x) \to -\infty$ as $x \to -\infty$	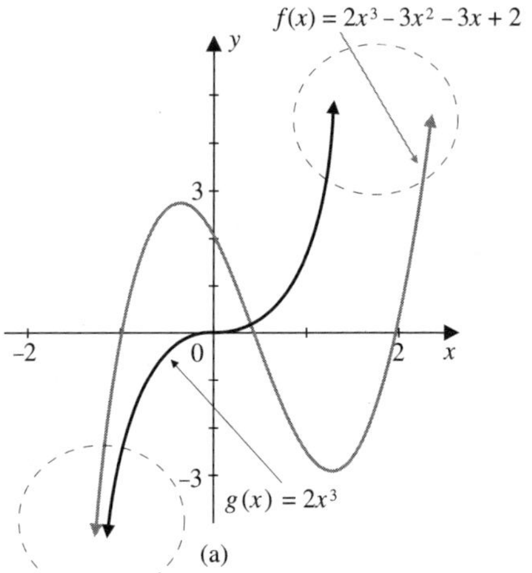 $f(x) = 2x^3 - 3x^2 - 3x + 2$ $g(x) = 2x^3$ (a)
$h(x) = -2x^4 + 2x^2$	$g(x) = -2x^4$	$g(x) \to -\infty$ as $x \to +\infty$ $g(x) \to -\infty$ as $x \to -\infty$ and so $h(x) \to -\infty$ as $x \to +\infty$ $h(x) \to -\infty$ as $x \to -\infty$	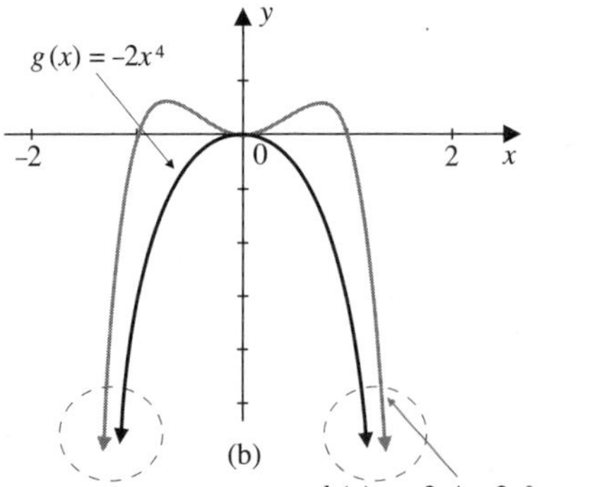 $g(x) = -2x^4$ $h(x) = -2x^4 + 2x^2$ (b)

To understand algebraically why f and g in the first example in Table 4 have the same limit behavior, consider

$$f(x) = 2x^3 - 3x^2 - 3x + 2.$$

Factor $f(x)$ as follows and compare with $g(x)$:

$$f(x) = 2x^3\left(1 - \frac{3/2}{x} - \frac{3/2}{x^2} + \frac{1}{x^3}\right)$$

As $x \to +\infty$, the term $\frac{3/2}{x}, \frac{3/2}{x^2}$, and $\frac{1}{x^3}$ approach zero. So that in turn

$$f(x) \text{ approaches } 2x^3(1 - 0 - 0 + 0) = 2x^3 = g(x).$$

Graphing can be done more accurately if we know the above four properties as the next example illustrates.

EXAMPLE 2 Analyzing the Graph of a Polynomial Function

Figure 7 shows the graph of the function

$$y = f(x) = x^3 - 4x^2 + x + 6 = (x + 1)(x - 2)(x - 3)$$

obtained by point-plotting and using continuity and the fact that the limit behavior of f is the same as that of $y = x^3$

(a) Find the x intercepts and the zeros of f.
(b) How many turning points are there?
(c) Use the graph to solve the inequality.

$$x^3 - 4x^2 + x + 6 \le 0$$

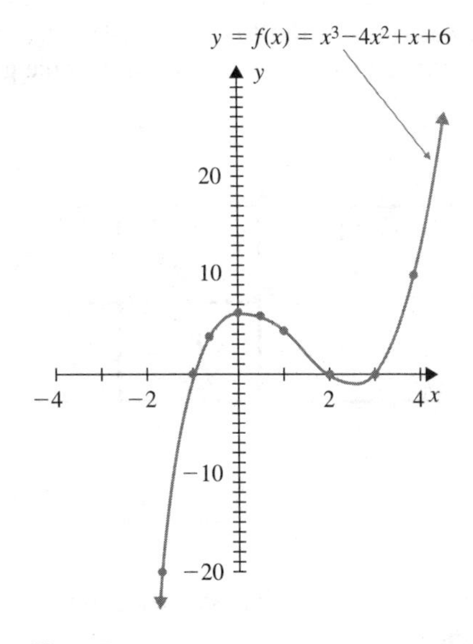

x	y
−2	−20
−1	0
−0.5	4.375
0	6
0.5	5.625
1	4
2	0
3	0
4	10

Figure 7

The factorization made it easy to find the x intercepts. Later, we will learn how to locate x intercepts for polynomial functions that are difficult to factor.

Solution (a) Since the function is written in a factored form, it is not difficult to find the x intercepts. By setting $y = 0$ and then solving for x, we have

$$x^3 - 4x^2 + x + 6 = (x + 1)(x - 2)(x - 3) = 0$$

so that,

$x = -1, x = 2, x = 3$ are the three x intercepts. The real zeros of f are the same as the x intercepts, namely, -1, 2, and 3.

(b) The graph has two turning points, the maximum indicated in Property 2.

(c) Next we solve the inequality

$$x^3 - 4x^2 + x + 6 \leq 0$$

From part (a), the locations of the x intercepts are -1, 2, and 3. The y values of points on the graph represent the values of the expression

$$x^3 - 4x^2 + x + 6.$$

So the portions of the graph on or below the x axis indicate the values of x that satisfy the inequality $y \leq 0$. Thus by reading the graph, we discover that the solution set of the inequality consists of all values of x such that $x \leq -1$ or $2 \leq x \leq 3$, that is, all values of x in the intervals $[-\infty, -1]$ or $[2, 3]$. ❖

The four properties above help us anticipate and verify the general behavior of the graph of a polynomial function even when we use a grapher.

For instance, suppose we use a grapher to graph the function

$$f(x) = x^3 - 4x^2 + x + 6$$

given in Example 2, without knowing the factored form.

By using the WINDOW specified in Figure 8a and without anticipating the behavior of the function, it would appear that the graph of f is as shown in Figure 8b.

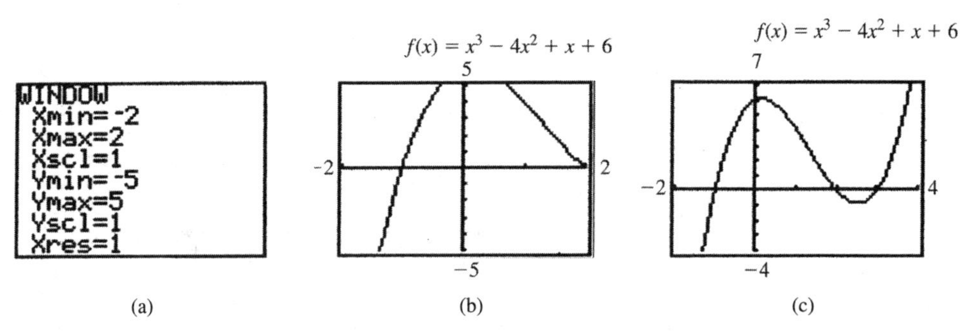

Figure 8

Since the graph of f is supposed to be continuous, the Ymax choice needs to be modified.

Also, the limit behavior model for f is

$$g(x) = x^3$$

(Figure 6b on page 171). Consequently, the limit behavior for f should be

$$f(x) \to +\infty \quad \text{as} \quad x \to +\infty$$

and

$$f(x) \to -\infty \quad \text{as} \quad x \to -\infty$$

Since Figure 8b does not include the appropriate limit behavior of f for large positive values of x, we need to change the Xmax choice. Figure 8c displays one possible WINDOW selection that yields a more accurate graph of f, showing both the continuity and limit behavior. More confidence in the accuracy of the graph is gained by observing that this third-degree polynomial function has the maximum number of turning points, $3 - 1 = 2$, and the maximum number of x intercepts.

Relating Multiple Zeros to Graphs

Consider the factored form of the polynomial function f in Example 2, given by

$$f(x) = x^3 - 4x^2 + x - 6 = (x + 1)(x - 2)(x - 3)$$

We note that $f(x) = 0$ whenever $x + 1 = 0$ or $x - 2 = 0$ or $x - 3 = 0$; that is, any zero of $f(x)$ must be a zero of one of the factors of $f(x)$. Since the number of *different* linear factors of f is 3, we say the function f has zeros -1, 2, and 3, each of "multiplicity 1."

When a polynomial function f is written in factored form, the same factor, say, $x - c$, may occur more than once, in which case, c is referred to as a *repeated* or *multiple zero* of f.

For instance, $f(x) = (x - 5)^2$ has zero 5 with a multiplicity of 2.

In general, we have the following definition:

Definition Multiplicity of a Zero

If $f(x) = (x - c)^s \cdot Q(x)$, $Q(c) \neq 0$, and s is a positive integer, then we say that c is a **zero of multiplicity** s of the function f.

Thus the function

$$f(x) = (x + 2)^3 (x - 1)^4 (x + 1)^2$$

has three different zeros, -2, 1, and -1, with the multiplicities given in the following table.

Zero	Multiplicity
−2	3
−1	4
−1	2

If c is a real zero of $f(x)$ of multiplicity s, then $f(x)$ has the factor $(x - c)^s$, and the graph of f has an x intercept at c. The general behavior of the graph of f at c depends on whether s is an odd or even integer.

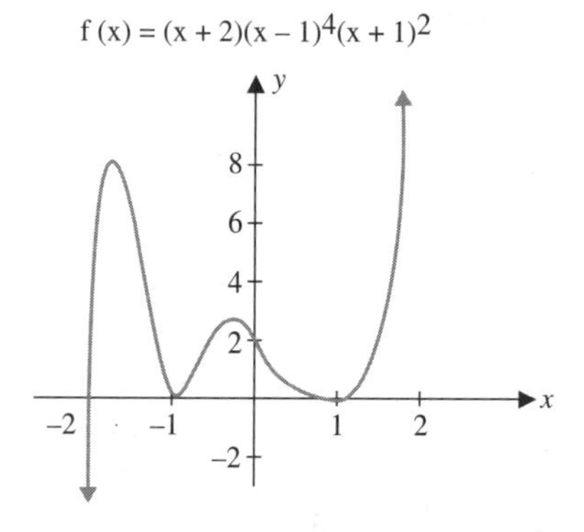

Figure 9

For example, Figure 9 shows the graph of the function

$$f(x) = (x + 2)(x - 1)^4(x + 1)^2$$

It has zeros -2, -1 and 1, with multiplicities 1, 2, and 4 respectively. By examining the graph of f, we observe the following features.

1. The graph of f crosses the x axis at $x = -2$, a zero of odd multiplicity, 1.
2. The graph of f only touches, but does not cross the x axis at $x = -1$, a zero of even multiplicity, 2.
3. The graph of f only touches, but does not cross the x axis at $x = 1$, a zero of even multiplicity, 4.

This illustration is generalized as follows:

Even-Odd Multiplicity Rule Graph Features

If c is a real zero of even multiplicity of a polynomial function f, then the graph of f touches the x axis at $x = c$ and bounces back.

If c is a real zero of odd multiplicity of a polynomial function f, then the graph of f crosses the x axis at $x = c$.

EXAMPLE 3 Applying the Even-Odd Multiplicity Rule

Use the even-odd multiplicity rule to determine where the graph of

$$f(x) = x(x - 4)^2(x - 2)^3$$

crosses the x axis and where the graph touches the x axis and bounces back.

Solution Table 5 identifies the real zeros of f along with their multiplicities and corresponding graphical features as indicated by the even-odd multiplicity rule.

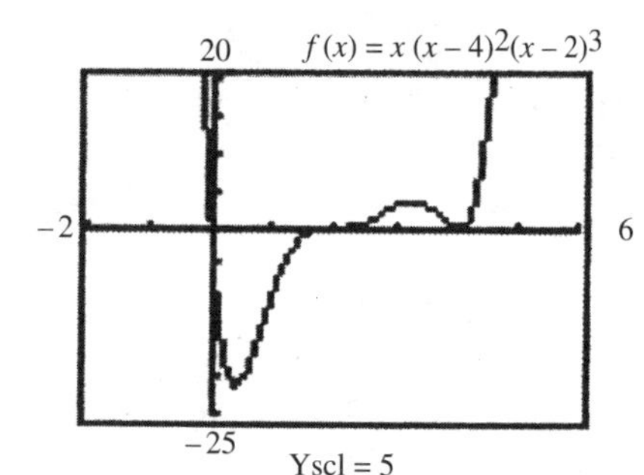

Figure 10

TABLE 5
$f(x) = x(x - 4)^2(x - 2)^3$

Real Zero	Multiplicity	Graphical Features
0	1 (odd)	Crosses x axis at $x = 0$
2	3 (odd)	Crosses x axis at $x = 2$
4	2 (even)	Touches x axis at $x = 4$ and bounces back

❖

Note that the graph of f (Figure 10), which was obtained by using a grapher, contains the features listed in Table 5.

Solving Applied Problems

Higher-degree polynomial functions are used in mathematical modeling, as illustrated in the next example.

EXAMPLE 4 **Examining a Growth Rate Model**

A biologist studying bacteria growth in a culture determines that the growth rate $f(t)$ (in grams per hour) of bacteria during the first 30 hours of an experiment is given by the model

$$f(t) = -0.003(t - 19)(t + 5)(t + 0.75),\ 0 \le t \le 30$$

where t is the elapsed time in hours.

(a) What are the bacteria growth rates at 8 hours, 11.8 hours, 15 hours, and 29 hours? Round off the answers to two decimal places.
(b) Indicate the time when the growth rate is zero and what the growth rate is initially, that is, when $t = 0$. Round off the answers to two decimal places.
(c) Sketch the graph of this model.
(d) During what time interval is the growth rate positive? Negative?

Solution (a) The bacteria growth rates at 8 hours, 11.8 hours, 15 hours, and 29 hours, rounded to two decimal places, are obtained by performing the following computations.

$$\begin{aligned} f(8) &= -0.003(8 - 19)(8 + 5)(8 + 0.75) \\ &= 3.75 \\ f(11.8) &= -0.003(11.8 - 19)(11.8 + 5)(11.8 + 0.75) \\ &= 4.55 \\ f(15) &= -0.003(15 - 19)(15 + 5)(15 + 0.75) \\ &= 3.78 \\ f(29) &= -0.003(29 - 19)(29 + 5)(29 + 0.75) \\ &= -30.35 \end{aligned}$$

This means that at 8 hours, at 11.8 hours, and at 15 hours, the bacteria are growing at rates of 3.75 grams per hour, 4.55 grams per hour, and 3.78 grams per hour, respectively. However, the growth rate of -30.35 grams per hour at 29 hours indicates that the number of bacteria is declining.

(b) The growth rate is zero when $f(t) = 0$, that is, when

$$t = 19,\ t = -5,\ \text{and}\ t = -0.75.$$

We only use the positive value, $t = 19$, because the given restriction indicates that $0 \le t \le 30$. Thus the growth rate is zero at 19 hours.
Notice that when $t = 0$ we get

$$\begin{aligned} f(0) &= -0.003(-19)(5)(0.75) \\ &= 0.21 \end{aligned}$$

rounded to two decimal places.
This means the initial growth rate is 0.21 gram per hour.

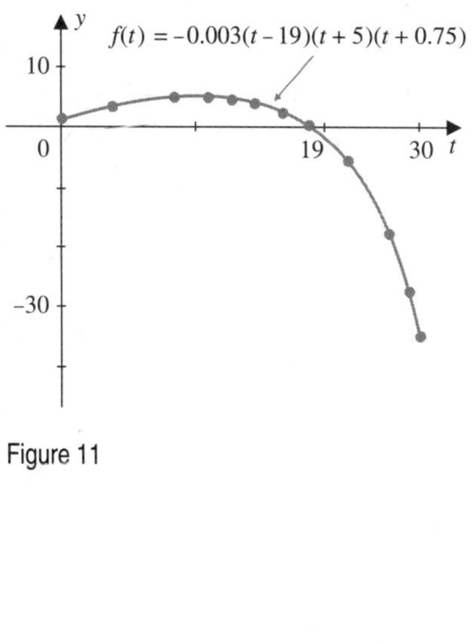

(c) By plotting points for $0 \le t \le 30$, including the data obtained in part (a), and connecting them with a smooth continuous curve, we obtain the graph of the model (Figure 11).

Figure 11

(d) The growth rate is positive whenever $f(t) > 0$. As shown in Figure 11, this occurs where the graph is above the t axis, that is, when

$$0 \le t < 19.$$

Similarly, the growth rate is negative when $f(t) < 0$, that is, when

$$19 < t \leq 30.$$

❖

Note that there is no systematic way to locate a maximum (high) or a minimum (low) point on the graph of a higher-degree polynomial function with the tools we have learned. However, by using programmed MIN and MAX finding features of a grapher, we are able to approximate the locations of any extreme points.

EXAMPLE 5 Fabricating a Box

The liner for an outside window flower box is to be made by cutting equal squares from the corners of a rectangular sheet of perforated aluminum and turning up the sides. Assume the aluminum sheet is 40 inches by 16 inches and the length of each square is x inches.

(a) Express the volume V as a function of x.
(b) What are the restrictions on x?
(c) Use a grapher to sketch the graph of the function in part (a) incorporating the restrictions from part (b).
(d) Use a grapher to estimate the value of x that gives the maximum volume. Also, find the maximum volume. Round off the answers to one decimal place.

Solution (a) Figure 12 displays a physical model of the situation. From the sketches we see that the height h of the liner is x inches, the width w is $16 - 2x$ inches, and the length l is $40 - 2x$ inches. So the volume V of the liner is given by

$V = lwh$	Formula from geometry
$= (40 - 2x)(16 - 2x)x$	Substitute
$= 4x(20 - x)(8 - x)$	Factor

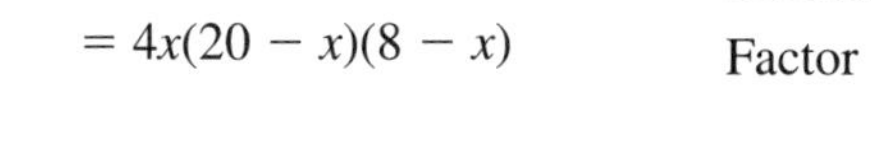

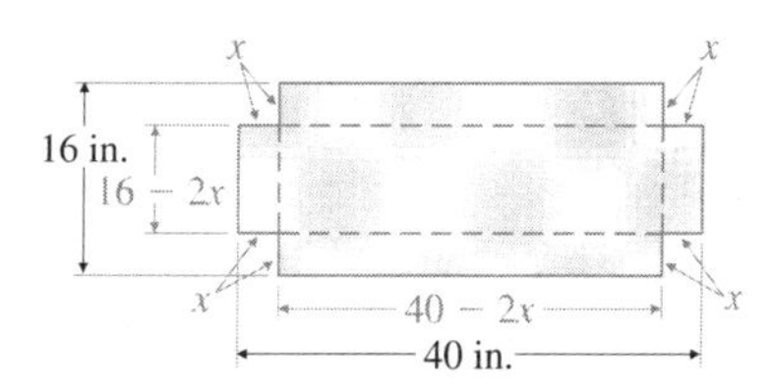

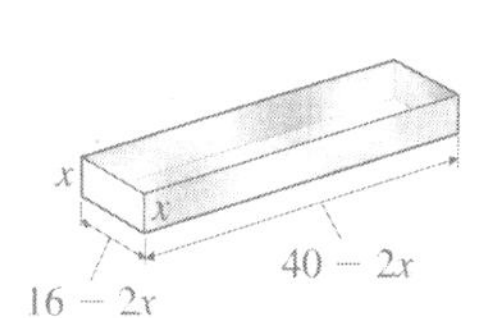

Figure 12

Thus, V is expressed as a function of x.

(b) Each physical dimension must be positive, so $x > 0$.

Also $40 - 2x > 0$ and $16 - 2x > 0$

$40 > 2x$	$16 > 2x$
$20 > x$	$16 > 2x$

Thus $0 < x < 8$.

(c) Figure 13 shows a viewing window of the graph of

$$V = 4x(20 - x)(8 - x) \text{ for } 0 < x < 8.$$

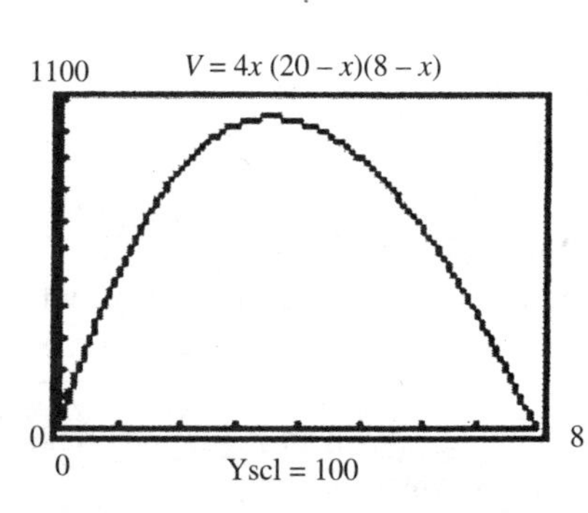

Figure 13

(d) By using the features of a grapher we find that the maximum value of V is about 1039.5 cubic inches, and it is obtained when x is approximately 3.5 inches.

PROBLEM SET 3.2

Mastering the Concepts

In problems 1 and 2, first describe the symmetry and general shape of the graph of each function. Then sketch the graph.

1. **(a)** $f(x) = x^8$
(b) $h(x) = x^9$

2. **(a)** $f(x) = x^{10}$
(b) $h(x) = x^{11}$

In problems 3–14, use transformations of the graph of an appropriate power function to sketch the graph of the given function.

3. $g(x) = \frac{1}{3}x^4$

4. $h(x) = -5x^4$

5. $f(x) = -2x^5$

6. $g(x) = \frac{1}{3}x^4$

7. $h(x) = x^5 + 3$

8. $f(x) = x^6 - 7$

9. $f(x) = -\frac{1}{3}(x+2)^5$

10. $h(x) = 4(x-4)^5$

11. $g(x) = 2(x-1)^4 + 3$

12. $f(x) = \frac{1}{3}(x+2)^5 + 2$

13. $h(x) = -4(x-2)^3 + 1$

14. $g(x) = -\frac{1}{2}(x+3)^4 - 2$

In problems 15–20, use the graph of f to determine the limit behavior of $f(x)$ as:

(a) $x \to +\infty$ **(b)** $x \to -\infty$

Indicate the number of turning points on each graph, and determine whether the graph has low and high points. Also, discuss the continuity and the number of x intercepts of each graph.

15.

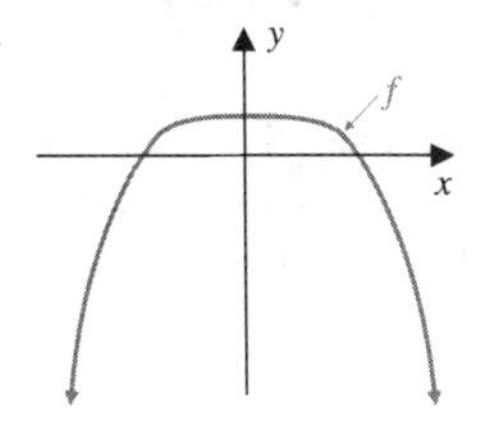

16.

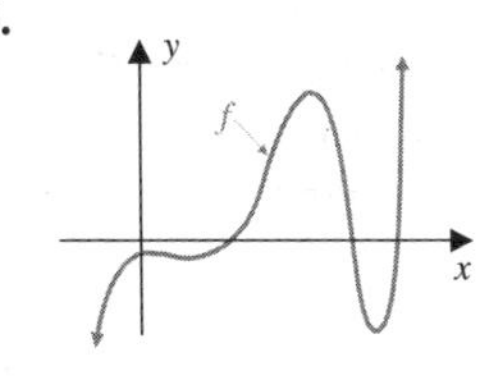

17.

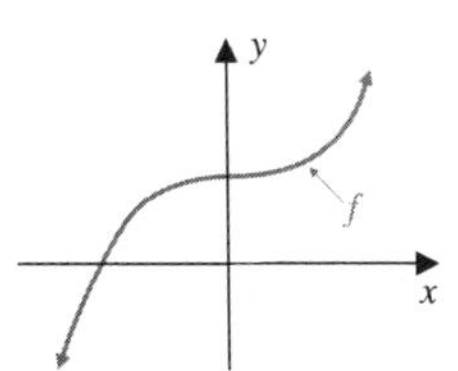

18.

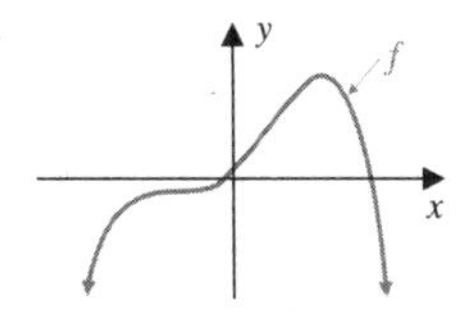

19.

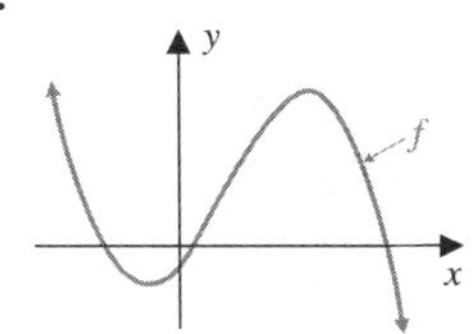

20.

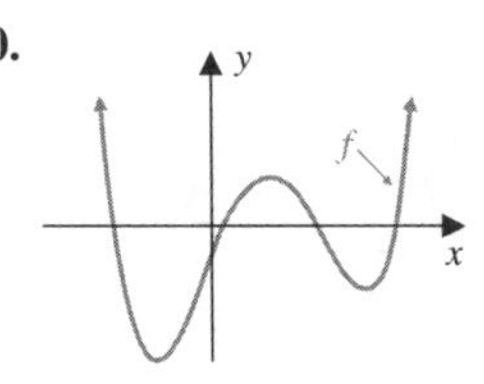

In problems 21 and 22, graph the limit behavior model function and then use the result to match the given polynomial function with its graph.

21. **(a)** $f(x) = -x^3 + 2x - 5$
(b) $g(x) = x^4 - 8x^3 + 20x^2 - 16x + 8$

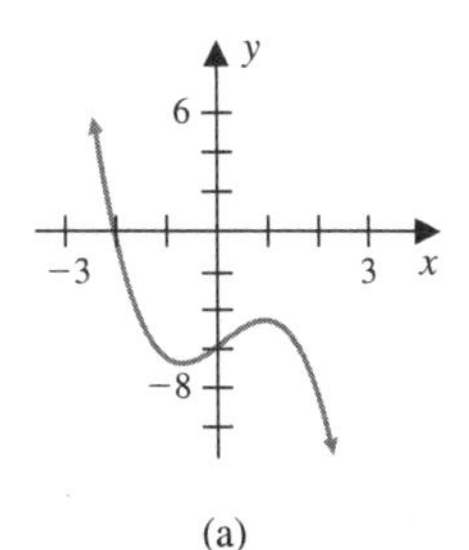

(a)

(b)

22. **(a)** $h(x) = x^3 + 3x^2 - 2x - 5$
(b) $F(x) = -2x^4 - 8x^3 - 12x^2 - 8x + 1$

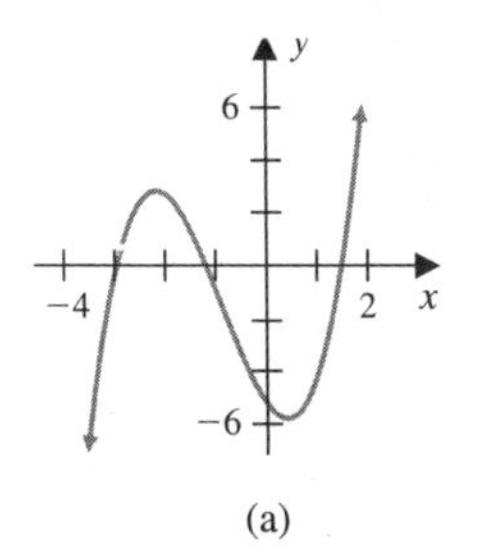

(a)

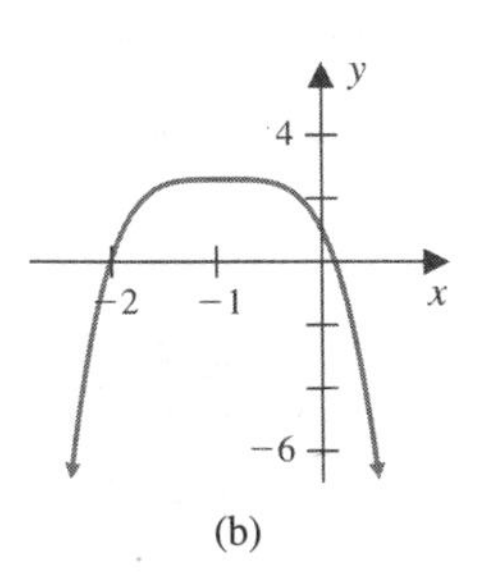

(b)

In problems 23 and 24, graph the given function for the specified WINDOW selections. Explain why the selection is inappropriate.

23. $f(x) = x^3 - 6x^2 + 11x - 6$

(a)

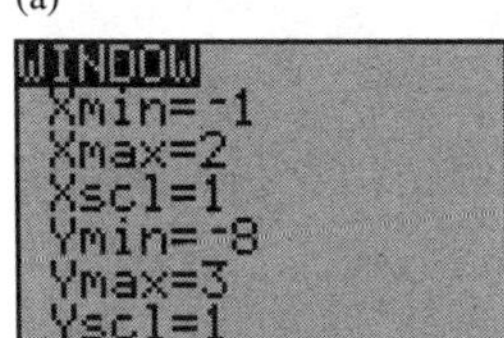

(b)

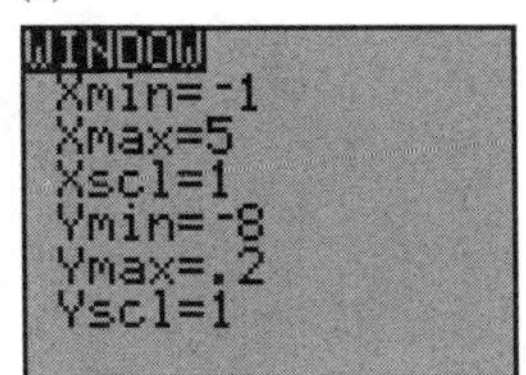

24. $h(x) = -x^4 + 2x + 3$

(a)

WINDOW
Xmin=-1
Xmax=2
Xscl=1
Ymin=-8
Ymax=3
Yscl=1

(b)

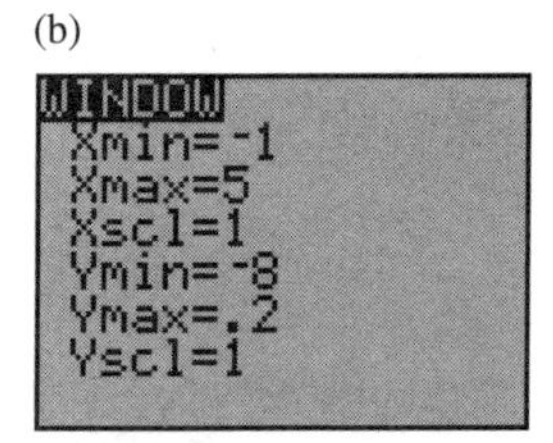

In problems 25–30, use the limit behavior model of the given polynomial function f to determine the limit behavior of $f(x)$:

(a) As $x \to +\infty$
(b) As $x \to -\infty$

Then use a grapher to sketch a graph of f by selecting an appropriate viewing window. Indicate the number and types of turning points.

25. $f(x) = x^4 - x^2 - 20$

26. $f(x) = x^4 - 4x^2 - 3x + 12$

27. $f(x) = x^3 - 8x^2 + 16x$

28. $f(x) = -x^3 - 6x^2$

29. $f(x) = -2x^4 + 2x^3 - x + 3$

30. $f(x) = x(x + 1)(x - 2)$

In problems 31–36, graph each function. Find the exact locations of the x intercepts and the zeros of f of the graph, and then use the graph to solve the associated inequality.

31. $f(x) = x(x - 1)(x + 2)$; $x(x - 1)(x + 2) > 0$

32. $h(x) = x^3 - x^2 - 2x$
$= x(x - 2)(x + 1)$;
$x^3 - x^2 - 2x \le 0$

33. $f(x) = x^3 - x^2 - 12x$
$= x(x - 4)(x + 3)$;
$x^3 - x^2 - 12x < 0$

34. $g(x) = -2x^4 + 2x^2$
$= -2x^2(x + 1)(x - 1)$;
$-2x^4 + 2x^2 > 0$

35. $f(x) = (2x - 5)(x - 1)^2$; $(2x - 5)(x - 1)^2 \ge 0$

36. $f(x) = 2x^4 - 3x^3 - 12x^2 + 7x + 6$
$= (2x + 1)(x - 1)(x - 3)(x + 2)$
$2x^4 - 3x^3 - 12x^2 + 7x + 6 > 0$

In problems 37–40, find all zeros of f and indicate the multiplicity of each zero. Also, use the even-odd multiplicity rule to determine where the graph of f crosses the x axis, and where it touches the x axis and bounces back.

37. $f(x) = (x - 5)^2 (x + 2)^3 (x + 5)$

38. $f(x) = x(x - 4)^3 (x + 2)^2$

39. $f(x) = x(x + 3)^2 (x - 4)^3$

40. $f(x) = x(x + 2)^2 (x - 1)^3$

Applying the Concepts

41. Average Temperature: The average hourly temperature in the month of January, in degrees Fahrenheit, in a midwestern city is modeled by the function

$$f(t) = -0.1(t + 6)(t - 2)(t - 11) \quad 0 \le t \le 12$$

where $t = 0$ represents 8:00 AM and $t = 12$ represents 8:00 PM.

(a) What is the average temperature at 11:00 AM, 3:00 PM, and 6:00 PM? Round off the answers to two decimal places.
(b) Indicate when the temperature is zero.
(c) Sketch the graph of this model.
(d) During what hours of the day is the temperature above zero? Below zero?

42. Population Rate: The population growth rate $f(t)$ (in thousands per year) of a certain region during a 9-year period is given by the model

$$f(t) = -0.04(t - 7)(t + 9)(t + 0.8) \quad 0 \le t \le 9$$

where t is the elapsed time in years.

(a) What is the population growth rate at 5 years and 9 years? Round off the answers to two decimal places.
(b) Indicate when the growth rate is zero.
(c) Sketch the graph of this model.
(d) During what time intervals is the growth rate positive? Negative?

In problems 43–46, round off the answers to two decimal places.

43. Unemployment Rate: A study conducted by a state unemployment commission determined the percentage of the workforce expected to be unemployed in the state over a 4-year period beginning in January 2000. The unemployment R (in percent) over that period is approximated by the model

$$R = 0.0002t^3 - 0.0147t^2 + 0.216t + 5.9$$

where t is the number of months, with $t = 0$ representing the beginning of the 4-year period.

(a) What are the restrictions on t?
(b) Use a grapher to graph this function over the restricted interval.

(c) Use a grapher to determine when the unemployment is expected to be the lowest over this period. What is the lowest expected rate of unemployment?

(d) What is the highest unemployment rate expected over this period of time? When is it expected to occur?

44. **Drug Dosage:** Data collected from a study conducted by medical researchers show the amount A of glucose (measured in tenths of a percent) in a patient's bloodstream at time t (in minutes) during a stress test. The amount A is approximated by the model

$$A = 0.1t^3 - 2.1t^2 + 12t + 2 \quad 1.5 < t \leq 11$$

(a) Use a grapher to graph this function.

(b) Use a grapher to determine the maximum amount of glucose and when it occurs.

(c) What is the minimum amount of glucose, and when does it occur?

45. **Package Design:** An open-top box is to be constructed by removing equal squares, each of length x inches, from the corners of a piece of cardboard that measures 20 inches by 20 inches and then folding up the sides.

(a) Draw a sketch of the situation and then express the volume V of the box as a function of x.

(b) What are the restrictions on x?

(c) Use a grapher to sketch the graph of the function from part (a) incorporating the restrictions from part (b).

(d) Use a grapher to determine the value of x for which the volume is a maximum. Also find the maximum volume.

46. **U.S. Postal Package Design:** A closed box with a square end x inches on a side is to be mailed. The U.S. Postal Service will accept the box for domestic shipment only if the length L of the box plus its girth ($4x$) is at most 108 inches (Figure 14). Suppose that the length plus girth of a box is exactly 108 inches.

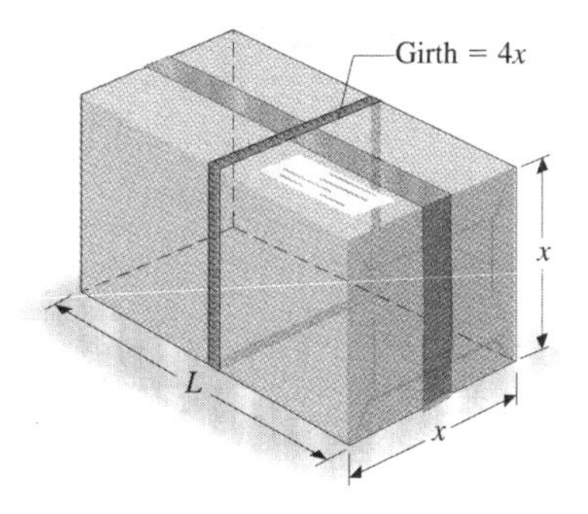

Figure 14

(a) Express the volume V of the box as a function of x.

(b) What are the restrictions on x?

(c) Use a grapher to sketch the graph of the function from part (a) incorporating the restrictions from part (b).

(d) Use a grapher to determine the value of x for which the volume is a maximum. Also find the maximum volume.

Developing and Extending the Concepts

47. Compare the limit behaviors of the functions

$$f(x) = ax^4 \quad \text{and} \quad g(x) = bx^4$$

if:

(a) $ab > 0$

(b) $ab < 0$

48. Compare the limit behaviors of the functions

$$f(x) = ax^3 \text{ and } g(x) = bx^3$$

if:

(a) $ab > 0$

(b) $ab < 0$

49. Examine the graph of $y = x^3$. For $x < 0$ the shape of the curve is **concave down** (cupped downward), and for $x > 0$ the shape is **concave up** (cupped upward). The point on the graph where a curve changes concavity is said to be a **point of inflection.** For $y = x^3$, the point of inflection is (0, 0). With this in mind, graph each given function. For what values of x is the curve concave down? For what values of x is the curve concave up? Where is the point of inflection?

(a) $f(x) = 2x^3 + 1$

(b) $g(x) = -(x - 4)^3 + 1$

50. Graph

$$f(x) = x^3$$

and

$$g(x) = |x^3|$$

on the same coordinate system. How can the graph of g be obtained from the graph of f? In general, explain how the graph of

$$y = |f(x)|$$

can be geometrically obtained from the graph of the polynomial function $y = f(x)$. When does

$$|f(x)| = f(x)?$$

In problems 51–54, display a general sketch of the limit behavior of the graph of the polynomial function

$$f(x) = a(x - h)^n + k, n > 0$$

under the given conditions.

51. $a > 0$ and $n \geq 1$ is an odd integer.

52. $a < 0$ and $n \geq 1$ is an odd integer.

53. $a > 0$ and $n \geq 2$ is an even integer.

54. $a < 0$ and $n \geq 2$ is an even integer.

In problems 55 and 56, complete the following table:

$x =$	-10^3	-10^2	-10	-1	1	10	10^2	10^3
$f(x)$								
$g(x)$								
$R(x)$								

What happens to the values of $R(x)$ as

$$x \to -\infty$$

and as

$$x \to +\infty?$$

How do the data in the table show that g is the limit behavior model for f?

55. $f(x) = x^3 + 2x^2 - 5x + 1$

$$= x^3\left(1 + \frac{2}{x} - \frac{5}{x^2} + \frac{1}{x^3}\right)$$

$$= g(x) \cdot R(x)$$

56. $f(x) = 2x^4 - x^2 + x - 7$

$$= 2x^4\left(1 - \frac{1}{2x^2} + \frac{1}{2x^3} - \frac{7}{2x^4}\right)$$

$$= g(x) \cdot R(x)$$

❖ In problems 57 and 58, each table gives discrete points for ordered pairs (x, y).

(a) Plot a scattergram of each set of data. Locate the x values on the horizontal axis and the y values on the vertical axis.

(b) Use a grapher and cubic regression to find a model of the form

$$y = ax^3 + bx^2 + cx + d$$

that best fits the data. Round off a, b, c, and d to four decimal places.

(c) Sketch the graph of the best fit curve from part (b).

(d) Use the regression equation to estimate the value of y if x is 27, 35, or 45. Round off the answers to the nearest integer.

57.

x	y
21	33
28	41
33	45
38	49
40	51

58.

x	y
12	53
18	48
23	44
28	42
30	38

OBJECTIVES

1. Use the Remainder Theorem
2. Factor Polynomials
3. Relate Factors, Zeros, Roots, and x Intercepts
4. Determine the Number and Types of Zeros
5. Find Rational Zeros

3.3 Real Zeros of Polynomials

So far, we have introduced some properties of polynomial functions, and have seen their usefulness in analyzing graphs. To advance our ability to find the real zeros or x intercepts of these functions, we must first review some basic concepts in algebra related to long division and synthetic division (see Appendix II).

For instance, we can either use long division or synthetic division to divide

$$f(x) = 2x^4 - 5x^2 + 3x - 11 \text{ by } x - 2$$

Long Division

$$
\begin{array}{rl}
\text{Quotient} \rightarrow & 2x^3 + 4x^2 + 3x + 9 \\
\text{Divisor} \rightarrow x - 2 \,\big)\!\! & 2x^4 + 0x^3 - 5x^2 + 3x - 11 \leftarrow \text{Polynomial} \\
 & \underline{2x^4 - 4x^3} \\
 & 4x^3 - 5x^2 \quad \leftarrow \text{Subtract} \\
 & \underline{4x^3 - 8x^2} \\
 & 3x^2 + 3x \quad \leftarrow \text{Subtract} \\
 & \underline{3x^2 - 6x} \\
 & 9x - 11 \leftarrow \text{Subtract} \\
 & \underline{9x - 18} \\
\text{Remainder} \rightarrow & 7 \leftarrow \text{Subtract}
\end{array}
$$

Synthetic Division

Polynomial Coefficients

$$
\begin{array}{r|rrrrr}
2 & 2 & 0 & -5 & 3 & -11 \\
 & & 4 & 8 & 6 & 18 \\
\hline
 & 2 & 4 & 3 & 9 & \boxed{7}
\end{array}
\leftarrow \text{Remainder}
$$

Quotient Coefficients

The result of this division may be written in the *multiplicative form*

$$2x^4 - 5x^2 + 3x + 11 = (x - 2)(2x^3 + 4x^2 + 3x + 9) + 7$$

$$\underset{\text{Polynomial}}{\uparrow} \qquad \underset{\text{Divisor}}{\uparrow} \qquad \underset{\text{Quotient}}{\uparrow} \qquad \underset{\text{Remainder}}{\uparrow}$$

This observation leads to the following generalization:

The Division Algorithm Result

If $f(x)$ and $D(x)$ are nonconstant polynomials such that the degree of $f(x)$ is greater than or equal to the degree of $D(x)$, then there exist unique polynomials $Q(x)$ and $R(x)$ such that

$$f(x) = D(x) \cdot Q(x) + R(x)$$

where the degree of $R(x)$ is less than the degree of $D(x)$ [$R(x)$ may be 0]. The expression $D(x)$ is called the **divisor,** $f(x)$ is the **dividend,** $Q(x)$ is the **quotient,** and $R(x)$ is the **remainder.**

As we shall see, being able to divide a polynomial function by a linear polynomial will facilitate the search for zeros of higher-degree polynomial functions.

Using the Remainder Theorem

A useful special case of the division algorithm occurs if the divisor is of the form $x - c$, where c is a real number. When this is the situation, we could use synthetic division to determine the quotient and remainder. The result is the form

$$f(x) = (x - c) \cdot Q(x) + R$$

where $Q(x)$ is the quotient and R is the remainder.

Now, if we let $x = c$, it follows that

$$\begin{aligned} f(c) &= (c - c) \cdot Q(c) + R \\ &= 0 \cdot Q(c) + R \\ &= R \end{aligned}$$

For example, if we divide $f(x) = 3x^3 - 2x + 1$ by $x - 2$, we obtain

$$f(x) = (x - 2)\,(3x^2 + 6x + 10) + 21.$$

Now if we substitute 2 for x into the latter expression we get

$$f(2) = (2 - 2)\,[3(2^2) + 6(2) + 10] + 21 = 0 + 21 = 21$$

Thus, we observe that $f(2) = 21$, which is the same value as the remainder when $f(x)$ is divided by $x - 2$.

In general, we have the following theorem.

The Remainder Theorem

If a polynomial $f(x)$ of degree $n \geq 1$ is divided by $x - c$, then the remainder $R = f(c)$.

EXAMPLE 1 Evaluating a Polynomial by Using the Remainder Theorem

Use the remainder theorem to find $f(5)$ if $f(x) = 3x^3 - x^2 + 2x - 18$.

Solution First, we use synthetic division to divide $3x^3 - x^2 + 2x - 18$ by $x - 5$ to get

$$\begin{array}{r|rrrr} 5 & 3 & -1 & 2 & -18 \\ & & 15 & 70 & 360 \\ \hline & 3 & 14 & 72 & 342 \end{array}$$

It follows that $f(5) = 342$, which is the remainder from dividing $f(x)$ by $x - 5$. 

Factoring Polynomials

Suppose we divide a polynomial function f by $x - c$ to obtain the result.

$$f(x) = (x - c) \cdot Q(x) + R,$$

where $Q(x)$ is the quotient and $R = f(c)$ is the remainder

If $f(c) = 0$, then the remainder is 0, so that

$$f(x) = (x - c) \cdot Q(x);$$

that is, $x - c$ is a factor of $f(x)$.

Conversely, if $x - c$ is a factor of $f(x)$, then

$$f(x) = (x - c) \cdot Q(x)$$

and by the remainder theorem, $f(c) = 0$.

Thus we have the following theorem

The Factor Theorem

If $f(x)$ is a polynomial of degree $n \geq 1$ and $f(c) = 0$, then $x - c$ is a factor of the polynomial $f(x)$; conversely, if $x - c$ is a factor of $f(x)$, then $f(c) = 0$.

The factor theorem, in conjunction with synthetic division, enables us to determine if an expression of the form $x - c$ is a factor of a polynomial; and, if it is, we find the other factor. This idea is illustrated in the next example.

EXAMPLE 2 Using the Factor Theorem to Determine Factors of a Polynomial

Let $f(x) = x^3 + 5x^2 + 5x - 2$. Use the factor theorem and synthetic division to determine if $x + 2$ is a factor of $f(x)$. If it is, find the factorization.

Solution We have $x + 2 = x - (-2)$, which has the form $x - c$ with $c = -2$. Using synthetic division to divide $f(x)$ by $x + 2$, we get

$$\begin{array}{r|rrrr} -2 & 1 & 5 & 5 & -2 \\ & & -2 & -6 & 2 \\ \hline & 1 & 3 & -1 & 0 \end{array}$$

It follows that $f(-2) = 0$, so by the factor theorem, we know $x + 2$ is a factor of $f(x)$. The synthetic division indicates that the factorization is

$$x^3 + 5x^2 + 5x - 2 = (x + 2)(x^2 + 3x - 1).$$ ❖

Relating Factors, Zeros, Roots, and x Intercepts

The relationship between zeros, roots, factors, and x intercepts is fundamental to the study of polynomials. For instance, by examining the factorization in Example 2, given by

Figure 1

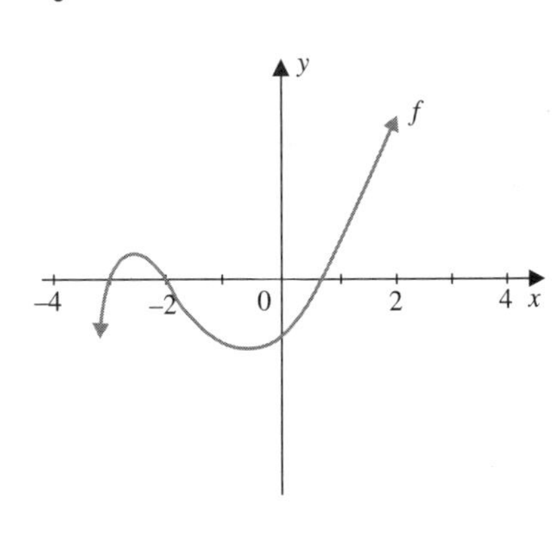

$$f(x) = x^3 + 5x^2 + 5x - 2 = (x + 2)(x^2 + 3x - 1),$$

we see that -2 is a root of the equation

$$x^3 + 5x^2 + 5x - 2 = 0.$$

We also see that -2 is a zero of the function

$$f(x) = x^3 + 5x^2 + 5x - 2.$$

It follows that -2 is one of the x intercepts of the graph (Figure 1).

In general, the following statements are equivalent for any polynomial function f.

Equivalent Statements

If f is a polynomial function and if c is a real number, then the following statements are equivalent.

1. The number c is zero of f.
2. The number c is a solution or root of the equation $f(x) = 0$.
3. $x - c$ is a factor of $f(x)$.
4. The number c is an x intercept for the graph of f.

EXAMPLE 3 Relating Factors, Zeros, Roots, and x Intercepts

Let $f(x) = x^3 + 2x^2 - 5x - 6$

(a) Use the factor theorem to show that $x + 3$ is a factor of $f(x)$, and then write $f(x)$ in complete factored form.
(b) Find the zeros of f.
(c) Find the roots of $x^3 + 2x^2 - 5x - 6 = 0$.
(d) Find the x intercepts of the graph of $f(x) = x^3 + 2x^2 - 5x - 6$.

Solution (a) Since $x + 3 = x - (-3)$, we use $c = -3$, to get

$$f(c) = f(-3) = (-3)^3 + 2(-3)^2 - 5(-3) - 6 = 0$$

Hence, by the factor theorem $x + 3$ is a factor of $f(x)$. Using synthetic division to divide $f(x)$ by $x + 3$, we obtain

$$\begin{array}{r|rrrr} -3 & 1 & 2 & -5 & -6 \\ & & -3 & 3 & 6 \\ \hline & 1 & -1 & -2 & 0 \end{array}$$

$$\text{so that } f(x) = x^3 + 2x^2 - 5x - 6$$
$$= (x + 3)(x^2 - x - 2)$$

Since $x^2 - x - 2 = (x - 2)(x + 1)$, it follows that

$$f(x) = (x + 3)(x^2 - x - 2)$$
$$= (x + 3)(x - 2)(x + 1)$$

Consequently, the complete factored form of $f(x)$ is given by

$$f(x) = (x + 3)(x - 2)(x + 1).$$

(b) Because of the equivalent statements, -3, -1, and 2 are the zeros of $f(x)$.

(c) Again we use the equivalent statements to conclude that the roots of the equation $f(x) = x^3 + 2x^2 - 5x - 6 = 0$ are -3, -1, and 2.

(d) Finally the equivalent statements indicate that the x intercepts of the graph of f are -3, -1, and 2 (Figure 2). ❖

Figure 2

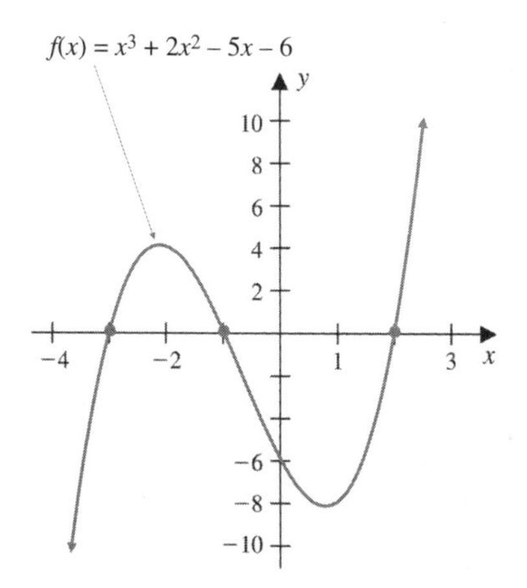

Determining the Number and Types of Zeros

In view of the equivalency between the zeros, the roots, the x intercepts, and factors of a polynomial function, we can use a graph to help predict the number and types of real zeros. For example, the graph of the function

$$f(x) = x^3 + 2x^2 - 5x - 6$$

in Figure 2 reveals that there are three real zeros, two negative and one positive. In addition, because of the equivalency, the graph in itself indirectly shows the function f has three linear factors.

To gain information about the number of real zeros for a polynomial in the real number system from an algebraic point of view, we observe that according to the factor theorem, if c_1 is a real zero of a polynomial function f of degree n, $n \geq 1$, then $f(x)$ can be factored as

$$f(x) = a(x - c_1) \cdot q_1(x)$$

where a is the leading coefficient of $f(x)$ and $q_1(x)$ is of degree $n - 1$.

If c_2 is another real zero of f, then $f(x)$ can be factored as

$$f(x) = a(x - c_1)(x - c_2) \cdot q_2(x)$$

where q_2 is of degree $n - 2$.

Continuing this process, we observe that the factored form of $f(x)$ cannot have more than n linear factors, since q_n would be of degree zero, and hence constant.

Since each linear factor produces a real zero, we are led to the following result:

Property Maximum Number of Zeros

If f is a polynomial function of degree n, then f cannot have more than n zeros, not necessarily all different.

Since each real zero corresponds to an x intercept, we have established Property 3 introduced in Section 3.2, which states that the graph of a polynomial function of degree n has at most n x intercepts.

EXAMPLE 4 **Finding a Polynomial with Prescribed Zeros**

Find a polynomial function f that has degree 3, zeros -1, 3, and 5, and satisfies $f(1) = 4$.

Solution By the factor theorem, $f(x)$ has factors $x + 1$, $x - 3$, and $x - 5$. Hence, $f(x)$ has the form:

$$f(x) = a(x + 1)(x - 3)(x - 5)$$

for some number a.

Since the above property assures us that there can be no more than three zeros, it follows that there can only be three linear factors for $f(x)$. Because $f(1) = 4$, it follows that

$$4 = a(1 + 1)(1 - 3)(1 - 5)$$

$$4 = 16a$$

$$a = \frac{4}{16} = \frac{1}{4}.$$

Hence,

$$f(x) = \frac{1}{4}(x + 1)(x - 3)(x - 5)$$

If we multiply this factored form, we obtain the polynomial

$$f(x) = \frac{1}{4}x^3 - \frac{7}{4}x^2 + \frac{7}{4}x + \frac{15}{4}.$$

Finding Rational Zeros

In Section 3.2, we pointed out that it is generally difficult to find the zeros of polynomial functions of higher degree. Here, we will adopt an algebraic strategy to find *rational zeros,* that is, rational numbers that are zeros, if they exist for polynomial functions with rational coefficients. We shall see that a grapher is helpful in carrying out this algebraic strategy. To understand the technique, we begin by examining the factored form of the following polynomial function.

$$\begin{aligned} f(x) &= 24x^3 - 22x^2 - 5x + 6 \\ &= (4x - 3)(2x + 1)(3x - 2) \end{aligned}$$

The zeros of f are found by solving the equation

$$f(x) = (4x - 3)(2x + 1)(3x - 2) = 0$$

to obtain the rational zeros, 3/4, $-1/2$, and 2/3.

We observe the following relationship between the rational zeros and the coefficients of $f(x)$:

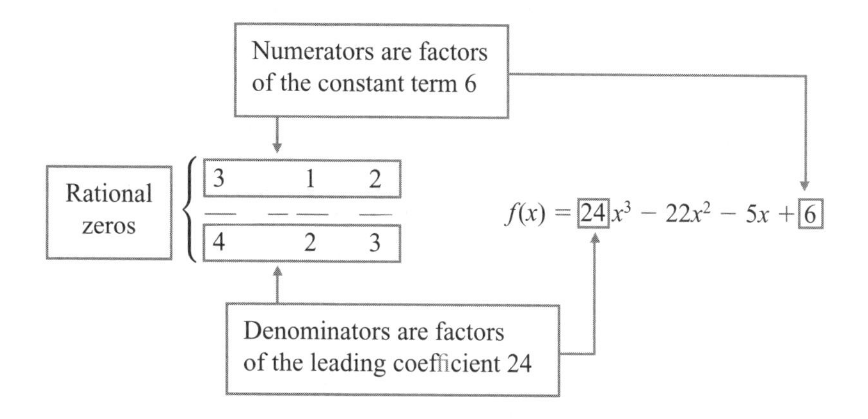

This observation leads us to the next theorem.

Rational Zero Theorem

Assume all the coefficients in the polynomial function

$$f(x) = a_n x^n + a_{n-1} x^{n-1} + \ldots + a_1 x + a_0$$

are integers, where $a_n \neq 0$ and $a_0 \neq 0$. If f has a rational zero p/q (where p and q are integers) in lowest terms, then

p must be a factor of a_0 and q must be a factor of a_n

It is important to understand the following points about the rational zero theorem:

1. The polymonial must have integer coefficients.
2. Acceptable factors for a_0 and a_n include -1 and 1.
3. The theorem does not guarantee the existence of rational zeros; indeed, there may not be any rational zero.

The rational zero theorem leads to the following strategy for finding all possible rational zeros of a polynomial function with integer coefficients.

Strategy for Finding Rational Zeros of a Polynomial Function

If p/q is a rational zero of f, then by the factor theorem, $x - \frac{p}{q}$ *is a factor of f(x).*

Step 1 List all factors of the constant term a_0 in the polynomial. (This list provides us with all possible values of p.) List all factors of the coefficient a_n of the term of the highest degree in the polynomial. (This list provides us with all possible values of q.)

Step 2 Use the lists from step 1 to determine all possible rational numbers p/q in reduced form. These rational numbers are the only possible rational zeros of f.

Step 3 Test the values produced in step 2. If none of them is a root of $f(x) = 0$, we conclude that f has no rational zeros.

As we shall see, shortcuts and variations of these steps are sometimes possible.

EXAMPLE 5 Finding Rational Zeros

Use the rational zero theorem to find the rational zeros of

$$f(x) = x^3 - 2x^2 - x + 2$$

and then use the results to factor $f(x)$.

Solution Assume that p/q is a rational zero of f and p/q is reduced to lowest terms. We apply the above strategy as follows:

Step 1. The factors of $a_0 = 2$ provide the possibilities for p, which are

$$p: \pm 1, \pm 2.$$

The possible values for q are the factors of $a_3 = 1$, which are

$$q: \pm 1$$

Step 2. The possible values for p/q are

$$\frac{p}{q}: \pm 1, \pm 2.$$

We test the $\frac{p}{q}$ values by using synthetic division. If -1 is selected, then

$$\begin{array}{r|rrrr} -1 & 1 & -2 & -1 & 2 \\ & & -1 & 3 & -2 \\ \hline & 1 & -3 & 2 & 0 \end{array}$$

so that -1 is a rational zero, and we obtain the factorization

$$x^3 - 2x^2 - x + 2 = (x + 1)(x^2 - 3x + 2)$$

The remaining zeros of f can be found by determining the roots of the so called *reduced equation*, $x^2 - 3x + 2 = 0$, to get

$$\begin{aligned} x^2 - 3x + 2 &= 0 \\ (x - 2)(x - 1) &= 0 \\ x = 2 \quad \text{or} \quad x &= 1 \end{aligned}$$

Hence, the rational zeros of f are -1, 1, and 2. By the factor theorem,

$$f(x) = (x + 1)(x - 1)(x - 2)$$

If we had first tested 1 or 2 in Step 3, the final result would turn out to be the same.

If the rational zero theorem produces several possibilities for rational zeros (Step 2 in the above strategy), then testing each possibility can become rather tedious. At times the testing process can be shortened by first examining the graph of $y = f(x)$

to determine likely candidates for zeros. The use of the rational zero theorem in conjunction with a graph is illustrated in the next example.

EXAMPLE 6 **Finding Rational Roots of an Equation**

Find all rational roots of the equation

$$4x^3 - 13x = 6$$

Use a graph to determine likely candidates for the rational roots.

Solution The given equation is equivalent to the equation $4x^3 - 13x - 6 = 0$. However, the solutions of this latter equation are the same as the zeros of the polynomial function

$$f(x) = 4x^3 - 13x - 6.$$

We apply the three-step strategy on page 191 to find the zeros of f.
Assume p/q in reduced form, is a rational zero of f.

Step 1. The possible values of p and q are:

$$p: \pm 1, \pm 2, \pm 3, \pm 6$$
$$q: \pm 1, \pm 2, \pm 4$$

Step 2. The possible values of p/q are:

$$\frac{p}{q}: \pm 1, \pm 2, \pm 3, \pm 6, \pm\frac{1}{2}, \pm\frac{3}{2}, \pm\frac{1}{4}, \pm\frac{3}{4}$$

The graph of f in Figure 3 shows that there are zeros near 2, between -2 and -1, as well as between -1 and 0 so the only feasible, candidates for zeros are

$$2, \frac{-3}{2}, \text{ and } -\frac{1}{2}.$$

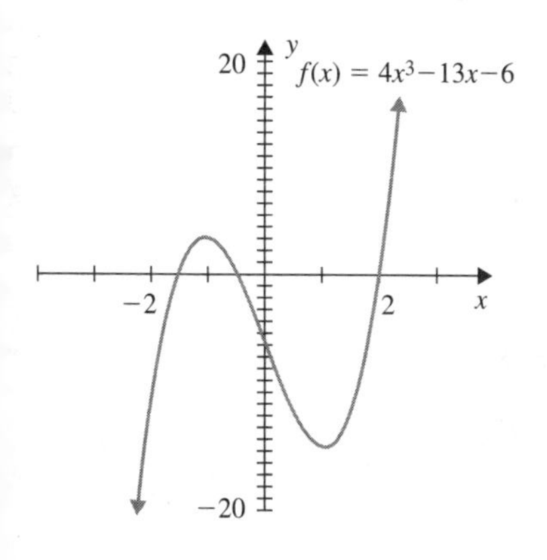

Figure 3

We begin by testing 2. Using synthetic division, we get

$$\begin{array}{r|rrrr} 2 & 4 & 0 & -13 & -6 \\ & & 8 & 16 & 6 \\ \hline & 4 & 8 & 3 & \vert\ 0 \end{array}$$

Since the reminder $f(2) = 0$, 2 is a zero of f, $x - 2$ is a factor of $f(x)$, and

$$f(x) = 4x^3 - 13x - 6 = (x - 2)(4x^2 + 8x + 3)$$

Any solution of the reduced equation $4x^2 + 8x + 3 = 0$ will also be a zero of f. Solving the reduced equation, we get

$$4x^2 + 8x + 3 = (2x + 3)(2x + 1) = 0$$

$$\begin{array}{r|r} 2x + 3 = 0 & 2x + 1 = 0 \\ x = -\dfrac{3}{2} & x = -\dfrac{1}{2} \end{array}$$

Therefore, the zeros of f, which are the same as the roots of the given equation, are $-\frac{3}{2}, -\frac{1}{2}$, and 2. ❖

Suppose a polynomial equation $f(x) = 0$ has noninteger rational coefficients. Then we need to make an adjustment in order to apply the rational zero theorem, since it requires integer coefficients. The next example illustrates one technique for doing this.

EXAMPLE 7 Using an Equivalent Form to Find Rational Zeros

(a) Find the rational zeros of the polynomial function

$$f(x) = \frac{2x^4}{3} - \frac{7x^3}{3} - \frac{10x^2}{3} + 11x + 6$$

(b) Use the results from part (a) to factor $f(x)$.

Solution (a) The zeros of f are the same as the zeros of the function

$$F(x) = \frac{1}{3}(2x^4 - 7x^3 - 10x^2 + 33x + 18)$$

obtained by factoring out $\frac{1}{3}$. Thus the zeros of f are the same as the zeros of

$$g(x) = 2x^4 - 7x^3 - 10x^2 + 33x + 18$$

Now we can apply the rational zero theorem to the function g by assuming that $\frac{p}{q}$, in reduced form, is a rational zero of g.

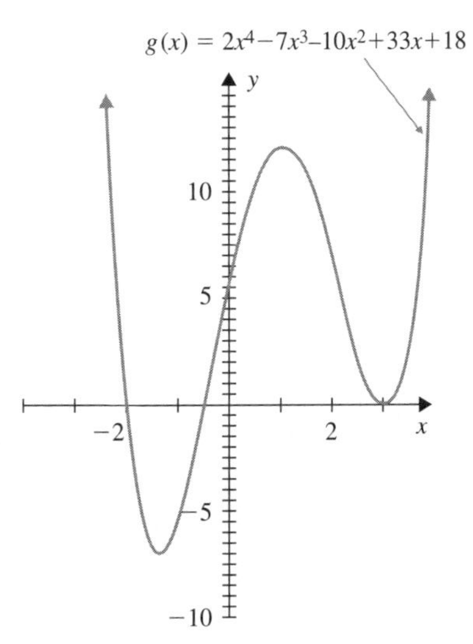

Figure 4

The possible values for p and q are:

$$p: \pm 1, \pm 2, \pm 3, \pm 6, \pm 9, \pm 18$$
$$q: \pm 1, \pm 2$$

As a result, any rational zero of $g(x)$ must occur in the list:

$$\frac{p}{q}: \pm 1, \pm 2, \pm 3, \pm 6, \pm 9, \pm 18, \pm\frac{1}{2}, \pm\frac{3}{2}, \pm\frac{9}{2}$$

Examining the graph of $y = g(x)$ in Figure 4, we see there are zeros "near" -2 and 3, and a zero between -1 and 0. Thus, the only likely candidates are

$$-2, -\frac{1}{2}, \text{ and } 3.$$

Using synthetic division we test -2, to get

$$\begin{array}{r|rrrrr} -2 & 2 & -7 & -10 & 33 & 18 \\ & & -4 & 22 & -24 & -18 \\ \hline & 2 & -11 & 12 & 9 & 0 \end{array}$$

Thus we have

$$g(x) = (x + 2)(2x^3 - 11x^2 + 12x + 9)$$

Synthetic division is again used to test a potential zero 3, in the reduced equation $2x^3 - 11x^2 + 12x + 9 = 0$ to get

$$\begin{array}{r|rrrr} 3 & 2 & -11 & 12 & 9 \\ & & 6 & -15 & -9 \\ \hline & 2 & -5 & -3 & 0 \end{array}$$

So 3 is also a zero. Thus we have the factorization

$$g(x) = (x + 2)(x - 3)(2x^2 - 5x - 3).$$

All the remaining zeros must be roots of the reduced polynomial equation $2x^2 - 5x - 3 = 0$. Upon factoring this latter equation we get

$$2x^2 - 5x - 3 = (2x + 1)(x - 3) = 0$$

so that

$$x = -1/2 \text{ and } x = 3 \text{ are also zeros.}$$

Therefore, the zeros of g, and consequently of f, are

$$-2, -\frac{1}{2}, \text{ and } 3,$$

where 3 has a multiplicity of 2.

(b) Since

$$f(x) = \frac{2}{3}x^4 - \frac{7}{3}x^3 - \frac{10}{3}x^2 + 11x + 6$$

is of degree 4 and has zeros

$$-2, -1/2, \text{ and } 3 \text{ (with multiplicity 2)}$$

it follows that

$$\begin{aligned} f(x) &= \frac{1}{3}(2x^4 - 7x^3 - 10x^2 + 33x + 18) \\ &= \frac{1}{3}g(x) \\ &= \frac{1}{3}(x + 2)(x - 3)^2(2x + 1). \end{aligned}$$

❖

Approximating Irrational Zeros

The technique we have learned so far for finding the zeros of polynomial functions has limitations. For instance, consider the function

$$f(x) = x^3 + 3x^2 - 2x - 5$$

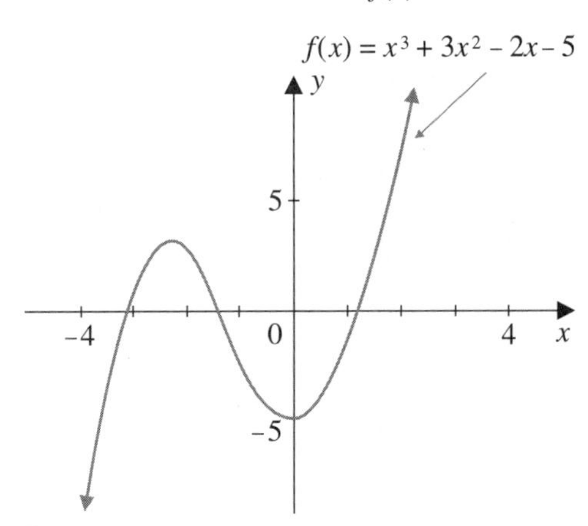

Figure 5

The graph of f given in Figure 5 shows that there are three x intercepts. This means, of course, that there are three real number zeros for f. To determine if any of these zeros is a rational number, we apply the rational zero theorem as follows:

If $\frac{p}{q}$ is a rational zero, then the possibilities for

$$p \text{ are } \pm 1, \pm 5$$

and those for

$$q \text{ are } \pm 1$$

so that the possibilities for

$$\frac{p}{q} \text{ are } \pm 1, \pm 5.$$

Upon testing these values, we find that none of them is a zero for f, so we conclude that the real zeros (x intercepts of the graph) must be irrational numbers. We will explore ways of approximating such values.

Numerical techniques for approximating the values of real zeros are based on the following property.

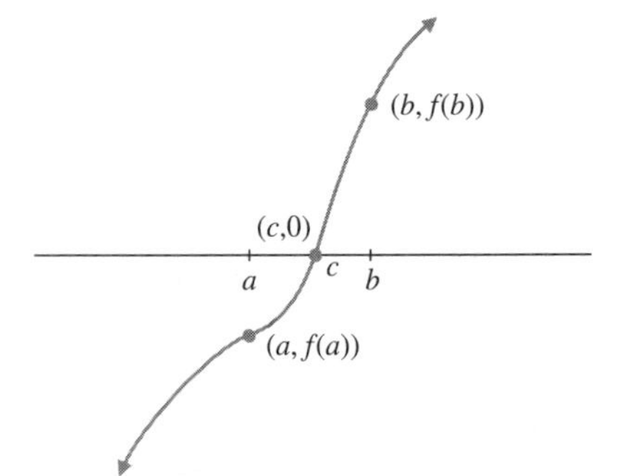

Figure 6

Property Intermediate-Value Property

Suppose that f is a polynomial function with real coefficients. If the values of $f(a)$ and $f(b)$ have opposite algebraic signs, where $a > b$, then there is at least one number c in the interval (a, b) such that c is a zero of f, that is, $f(c) = 0$.

This property is illustrated in Figure 6, which shows part of the graph of a polynomial function f that changes sign on an interval $[a, b]$. Because the graph is a continuous or unbroken curve, it must cross the x axis at least at one point, say $(c, 0)$, between the points $(a, f(a))$ and $(b, f(b))$.

In other words, there must be a real zero in the interval (a, b).

EXAMPLE 8 Using the Intermediate-Value Property

Use the intermediate-value property to show that

$$f(x) = x^3 + 3x^2 - 2x - 5$$

has a real zero in the interval (1, 2). Notice that the graph of f shown in Figure 5 on page 196 indicates there is a zero between 1 and 2.

Solution We find that

$$f(1) = -3 \text{ and } f(2) = 11.$$

Since

$$f(1) < 0 < f(2),$$

we conclude from the intermediate-value property that f has at least one zero in the interval (1, 2). ❖

If the intermediate-value property identifies an interval that contains a real zero of a polynomial function with rational coefficients, we can apply the rational zero theorem to find its value if it is a rational number. If the zero is an irrational number, we use an approximating method to estimate its value.

One method of approximating the irrational zeros is based on the recursive process illustrated in Figure 7, where portions of the graph of the function $y = f(x)$ are displayed.

Each bisection point (midpoint of the interval) is getting closer to c, where f(c)=0.

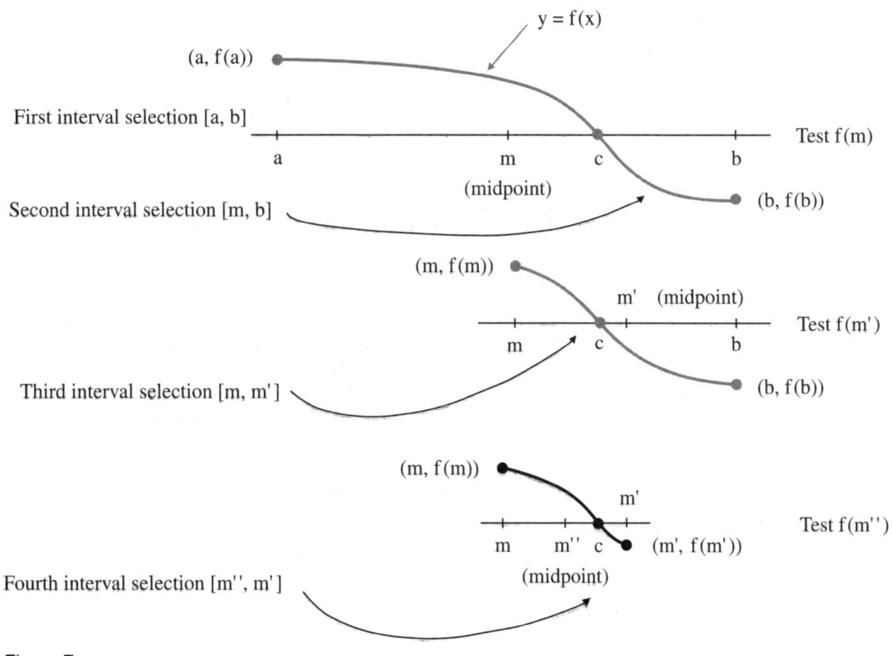

Figure 7

The actual algorithm for this recursive process, called the **bisection method,** proceeds as follows:

The Bisection Method

Assume that f is a polynomial function with real coefficients.

Step 1 Find values for a and b where $a < b$ and the values $f(a)$ and $f(b)$ change signs once, either from positive to negative or from negative to positive. (A graph of the function is helpful here.)

Step 2 Determine the midpoint m of interval $[a, b]$, namely,

$$m = (a + b)/2$$

Step 3 From the intervals $[a, m]$ and $[m, b]$, select the interval where the values of the given function f at the endpoints have opposite algebraic signs (one is positive and the other is negative).

Step 4 Repeat steps 2 and 3 on the interval selected above. Continue this process and evaluate $f(m)$ in each repetition until the value of $f(m)$ differs from 0 by less than an amount decided on at the start.

It is common to express the desired accuracy of the approximation of a zero in terms of an absolute value inequality. For instance, if we are asked to find an approximation c for a real zero such that $|f(c)| \leq 0.0005$, then we continue the approximating process until we find a value c such that

$$-0.0005 \leq f(c) \leq 0.0005.$$

EXAMPLE 9 Using the Bisection Method

Use the bisection method to approximate c, the zero of the function

$$f(x) = x^3 + 3x^2 - 2x - 5$$

in the interval (1, 2), to four decimal places, so that $|f(c)| < 0.005$.

Solution We apply the bisection method as follows:

Step 1. From Example 8, we know that there is a zero between 1 and 2, since $f(1)$ and $f(2)$ have opposite signs.

Step 2. The midpoint of interval $\left[1, 2\right]$ is $\dfrac{(1+2)}{2} = 1.5$ and $f(1.5) = 2.125$

Step 3. Since $f(1) = -3$ and $f(1.5) = 2.125$, we select interval $[1, 1.5]$ to do the next bisection.

Step 4. The midpoint of interval $\left[1, 1.5\right]$ is $\dfrac{(1+1.5)}{2} = 1.25$ and

$f(1.25) = -0.8594$ (approx.)

Next, we select interval $[1.25, 1.5]$ because $f(1.25)$ is negative and $f(1.5)$ is positive. The midpoint is 1.375 and

$$f(1.375) = 0.5215 \text{ (approx.)}$$

Then we select interval [1.25, 1.375] because $f(1.25)$ is negative and $f(1.375)$ is positive. The midpoint is 1.3125 and

$$f(1.3125) = -0.1960 \text{ (approx.)}.$$

Continuing this process will eventually yield more accurate approximations of the zero of f between 1 and 2. The progression of improved approximations is shown in Table 1. Since $f(1.3302) = 0.0016 < 0.005$, we conclude that 1.3302 is the desired approximation for the zero

TABLE 1

Interval	**Midpoint m (four decimal places)**	**$f(m)$ (four decimal places)**
[1, 2]	1.5000	2.1250
[1, 1.5]	1.2500	−0.8594
[1.25, 1.5]	1.3750	0.5215
[1.25, 1.375]	1.3125	−0.1960
[1.3125, 1.375]	1.3438	0.1564
[1.3125, 1.3438]	1.3282	−0.0210
[1.3282, 1.3438]	1.3360	0.0673
[1.3282, 1.3360]	1.3321	0.0231
[1.3282, 1.3321]	1.3302	0.0016

The bisection method is easy to understand but tedious to carry out.

Another way to approximate real zeros is to use a grapher. From the graph in Figure 8, we see that there is a zero of

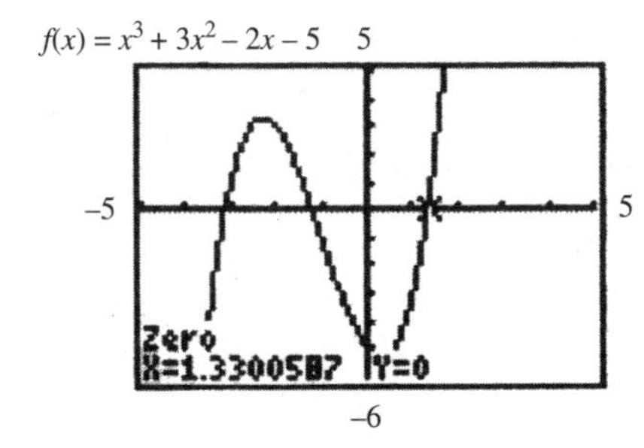

Figure 8

$$f(x) = x^3 + 3x^2 - 2x - 5$$

between 1 and 2. The value of this zero is the same as the value of the x intercept between 1 and 2. To locate the x intercept we use the ROOT -finding or ZERO -finding feature on a grapher to obtain the approximation 1.3300587. Other irrational zeros of this polynomial that can also be found by this method are −3.128419 and −1.20164.

Solving Applied Problems

Many applications of mathematics involve solving polynomial equations. As we know, finding the solution of a polynomial equation is equivalent to finding the zeros of a function. The next example illustrates an application of the use of this equivalency. It also stresses the importance of carefully interpreting the results of a mathematical model.

EXAMPLE 10 Modeling the Dimensions of a Container

A large container for filling bags of low-nitrogen fertilizer is to be constructed by attaching an inverted cone to the bottom of a right circular cylinder of radius r. The total depth of the container is 12 feet and the radius is the same as the height of the cone, as shown in Figure 9.

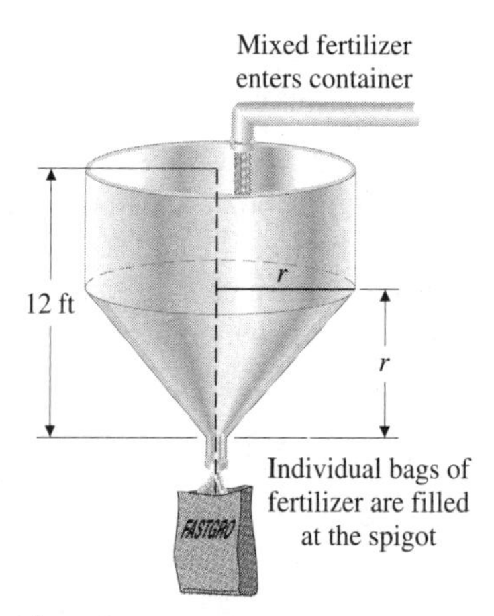

Figure 9

(a) Express the volume V of the container as a function of r. Also, find any restrictions on r.

(b) Suppose that the container is to have a capacity of 90π cubic feet. What is the radius?

Solution (a) The volume V of the container is the sum of the volumes of the cylinder and the cone. Since the height of the cone is r feet, the height of the cylinder is $12 - r$ feet. So the volume V is given by

$$V = (\text{Volume of cylinder}) + (\text{Volume of cone})$$

$$= \pi r^2(12 - r) + \frac{1}{3}\pi r^3 \qquad \begin{cases} \text{Volume of cylinder} = \pi r^2 h; \\ \text{Volume of cone} = \frac{1}{3}\pi r^2 h \end{cases}$$

$$= 12\pi r^2 - \pi r^3 + \frac{1}{3}\pi r^3$$

$$= -\frac{2}{3}\pi r^3 + 12\pi r^2$$

Thus, the equation

$$V = -\left(\frac{2}{3}\right)\pi r^3 + 12\pi r^2$$

defines V as a function of r.

Since r and V represent physical dimensions, $r > 0$ and $V > 0$.

However, since the total depth of the container is 12 feet and r is the same as the height of the cone, then r is at most 12 feet, that is, $r < 12$. So we conclude that $0 < r < 12$.

(b) Using the function found in part (a), we let $V = 90\pi$, so we need to solve the equation

$$90\pi = -\frac{2}{3}\pi r^3 + 12\pi r^2.$$

This equation is equivalent to

$$r^3 - 18r^2 + 135 = 0.$$

Solving the latter equation is equivalent to finding the zeros of the function

$$f(r) = r^3 - 18r^2 + 135.$$

The rational zero theorem leads to the possible rational zeros:

$$\frac{p}{q}: \pm 1, \pm 3, \pm 5, \pm 9, \pm 15, \pm 27, \pm 45, \pm 135$$

However, we established in part (a) that $0 < r < 12$, so the only feasible potential zeros from this list are 1, 3, 5, and 9. After testing these possibilities, we get $f(3) = 0$, so 3 is a zero of f and $r - 3$ is a factor of $f(r)$. In fact

$$f(r) = (r - 3)(r^2 - 15r - 45).$$

Solving the reduced equation $r - 15r - 45 = 0$ by using the quadratic formula yields two additional zeros of f:

$$\frac{15 - 9\sqrt{5}}{2} \text{ and } \frac{15 + 9\sqrt{5}}{2}$$

Hence, the real zeros of f are

$$3, \frac{15 - 9\sqrt{5}}{2} \left(\text{approx. } -2.56\right), \quad \text{and} \quad \frac{15 + 9\sqrt{5}}{2} \left(\text{approx. } 17.56\right).$$

However, r cannot equal $(15 - 9\sqrt{5})/2$ because this number is negative. Furthermore, r cannot equal $(15 + 9\sqrt{5})/2$ since we have already established that $r < 12$. Consequently, the redius r must be 3 feet. ❖

EXAMPLE 11 Modeling a Suspension Bridge

Figure 10 shows the main span of a suspension bridge. Assume the relationship between the length c of the supporting cable between two vertical towers and the sag x, the vertical distance from the top of either tower to the lowest point of the supporting cable at or above the roadway, is given by the formula

$$c = l + \frac{8x^2}{3l} - \frac{32x^4}{5l^3}$$

where l is the horizontal distance of the *span* between the towers.

The Mackinac Bridge, one of the world's longest suspension bridges, connects the upper and lower peninsulas of Michigan. The distance between its main towers is 3800 feet, the top of each main tower is 552 feet above water level, and the roadway is 199 feet above water, as shown in Figure 11.

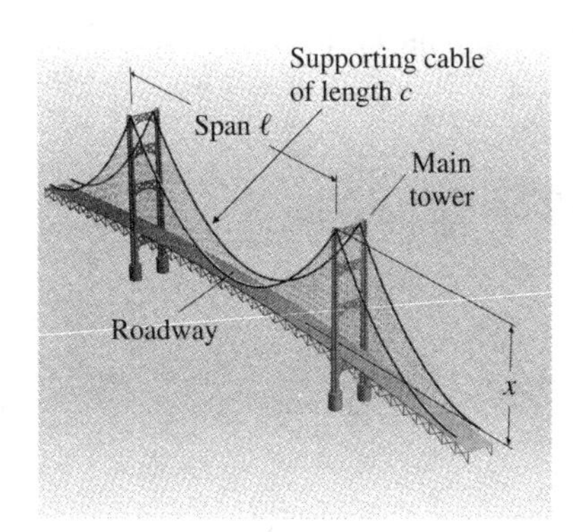

Figure 10

(a) Express the length of the supporting cable c as a function of x, the sag.
(b) Determine the restrictions on x.

(c) Use a grapher to find how high above the roadway the lowest point of the cable is, if the supporting cable is 3884 feet long. Round off the answer to one decimal place.

Solution (a) Since l is given to be 3800, we substitute this value into the above formula to get

$$c = 3800 + \frac{8x^2}{3(3800)} - \frac{32x^4}{5(3800)^3}$$
$$= 3800 + \frac{x^2}{1425} - \frac{x^4}{857{,}375 \cdot 10^4}$$

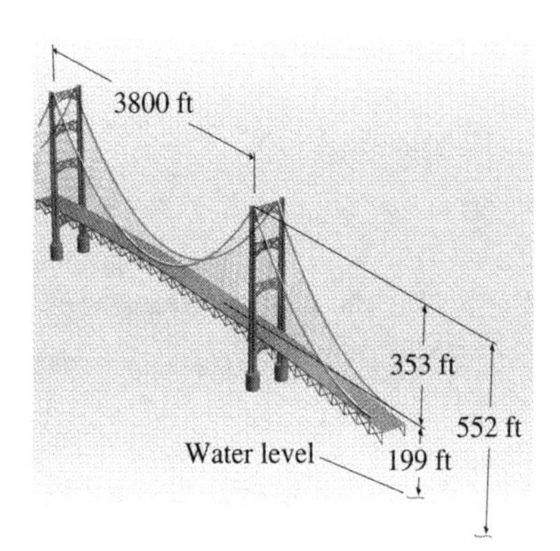

Figure 11

This latter equation expresses c as a function of x.

(b) Since the lowest point of the supporting cable is at or above the roadway, then the sag x must be less than or equal to the tower's height above the roadway. Upon examining the given data, which are shown in Figure 11, it follows that $0 < x < 353$.

(c) We begin by finding the sag x, given that the cable is 3884 feet long. To determine x, we substitute $c = 3884$ into the function in part (a) to get

$$3884 = 3800 + \frac{x^2}{1425} - \frac{x^4}{857{,}375 \cdot 10^4}$$
$$0 = -84 + \frac{x^2}{1425} - \frac{x^4}{857{,}375 \cdot 10^4}$$

The solutions of this latter equation are the same as the zeros of the function f defined by

$$f(x) = -84 + \frac{x^2}{1425} - \frac{x^4}{857{,}375 \cdot 10^4}, \text{ where } 0 < x \le 353$$

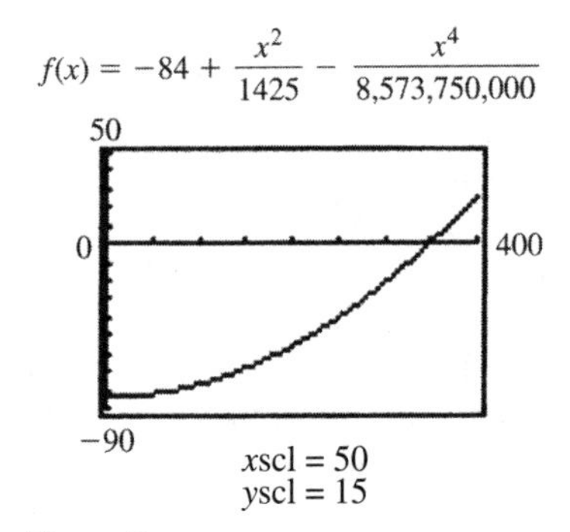

Figure 12

Now we use a grapher to sketch a graph of f (Figure 12). It clearly shows that there is an x intercept in the interval (0, 400). By using the zero finding feature we get an approximate value of 349.5 for the x intercept.

Thus f has a zero with value of about 349.5, making the sag approximately 349.5 feet. It follows that the lowest point of the cable is about $353 - 349.5 = 3.5$ feet above the roadway. ❖

PROBLEM SET 3.3

Mastering the Concepts

In problems 1 and 2, use the remainder theorem to find the value of each function.

1. **(a)** $f(-1)$ for $f(x) = 4x^3 - 2x^2 + 7x - 1$
(b) $f(2)$ for $f(x) = 8x^3 - x^2 - x + 5$
(c) $f(3)$ for $f(x) = 3x^4 - 2x^3 + 4x^2 + 3x + 7$

2. **(a)** $f(-2)$ for $f(x) = 2x^5 + 3x^4 - x^2 + x - 4$
(b) $f(-3)$ for $f(x) = 6x^4 + 10x^2 + 7$.
(c) $f(3/5)$ for $f(x) = 15x^4 + 5x^3 + 6x^2 - 2$

3. Use the factor theorem to determine whether each binomial is a factor of the polynomial $f(x) = x^3 - 28x - 48$. If it is, find the factorization.
(a) $x + 2$
(b) $x + 4$
(c) $x + 6$
(d) $x - 6$

4. Use the factor theorem to determine whether each binomial is a factor of the polynomial $2x^4 - 9x^3 - 34x^2 - 9x + 14$. If it is, find the factorization.
(a) $x - 7$
(b) $x + 2$
(c) $x - (1/2)$
(d) $x - 1$

In problems 5–8, use the factor theorem to show the binomial is a factor of $f(x)$, and then factor $f(x)$.

5. $f(x) = x^3 - 6x^2 + 11x - 6;\ x - 1$
6. $f(x) = x^3 + 6x^2 - 11x - 16;\ x + 1$
7. $f(x) = 2x^4 + 3x - 26;\ x + 2$
8. $f(x) = 6x^3 - 25x^2 - 29x + 20;\ x - 5$

In problems 9 and 10 use the factor theorem to show that c is a zero of f.

9. $f(x) = 4x^3 - 9x^2 - 8x - 3;\ c = 3$
10. $f(x) = x^3 + x^2 - 12x + 12;\ c = 2$

In problems 11 and 12, find a polynomial function that has degree 3 and has the given zero with the specified value of f.

11. Zeros, $1, 3, -3$, and $f(2) = 5$
12. Zeros, $-1, 2, 3$, and $f(-2) = 4$

In problems 13–16, list all real zeros of each polynomial function. Also, specify the x intercepts, and find the real roots of the equation $f(x) = 0$.

13. $f(x) = x(x^2 - 4)(x + 3)$
14. $f(x) = x(x^2 - 1)(x^2 - 9)$
15. $f(x) = (x^2 - x - 2)(x^2 + 1)$
16. $f(x) = (x^2 + x - 6)(x^2 + 4)$

In problems 17–22, use the rational zero theorem to find the rational zeros of the function. Use the results to factor the function expression.

17. $f(x) = x^3 - x^2 - 4x + 4$
18. $g(x) = x^3 + 2x - 12$
19. $h(x) = 5x^3 - 12x^2 + 17x - 10$
20. $f(x) = x^3 - x^2 - 14x + 24$
21. $f(x) = x^5 - 5x^4 + 7x^3 + x^2 - 8x + 4$
22. $h(x) = 4x^5 - 23x^3 - 33x^2 - 17x - 3$

In problems 23–26, use the rational zero theorem to find all rational roots of each equation.

23. $x^3 + 2x^2 - 3x - 6 = 0$
24. $x^3 + 3x^2 - 6x - 18 = 0$
25. $x^4 - 24x^2 - 25 = 0$
26. $x^4 - 8x^3 - 10x^2 + 72x + 9 = 0$

In problems 27–30, use an equivalent form and the rational zero theorem to find the rational zeros of the function. Use the rational zeros to factor the function expression.

27. $f(x) = x^4 - x^3 - \frac{25}{4}x^2 + \frac{x}{4} + \frac{3}{2}$

28. $g(x) = \frac{2}{5}x^4 - x^3 - \frac{8}{5}x^2 + 5x - 2$

29. $g(x) = \frac{2}{3}x^3 + \frac{x^2}{15} - \frac{7}{15}x + \frac{2}{15}$

30. $f(x) = \frac{2}{3}x^4 - \frac{5}{2}x^2 + \frac{5}{6}x + 1$

In problems 31 and 32, use an equivalent form and the rational zero theorem to solve each equation.

31. $x^3 + \frac{x^2}{2} - 12x - 6 = 0$

32. $2x^3 - 3x^2 - \frac{16}{5}x + \frac{12}{5} = 0$

In problems 33–38, use the rational zero theorem, along with the graph of the function, to find the rational zeros. Use the results to factor the function expression.

33. $h(x) = 4x^4 - 4x^3 - 7x^2 + 4x + 3$
34. $g(x) = x^4 - 9x^2 + 20$
35. $f(x) = x^4 - 3x^3 - 12x - 16$
36. $h(x) = 2x^3 + x^2 + x$

37. $f(x) = x^3 - \frac{7}{2}x^2 + 3x + \frac{5}{2}$

38. $g(x) = x^3 - \frac{x^2}{2} - 4x + 2$

In problems 39–44, use the intermediate-value property to show that the function f has a zero in the given interval.

39. $f(x) = x^3 - 3x^2 + 4x - 5$; $[1, 3]$
40. $f(x) = x^3 - 4x^2 + 3x + 1$; $[2, 4]$
41. $f(x) = 3x^3 - 10x + 9$; $[-3, -2]$
42. $f(x) = 2x^3 + 6x^2 - 8x + 2$; $[-5, -4]$
43. $f(x) = x^4 - 4x^3 + 10$; $[0.9, 2.1]$
44. $f(x) = x^4 + 6x^3 - 18x^2$; $[2.1, 2.2]$

In problems 45–50, use the bisection method to approximate the value of the zero c on the specified interval to four decimal places so that $|f(c)| < 0.005$.

45. $f(x) = x^3 - 3x + 1$, $(0, 1)$
46. $g(x) = x^4 + 2x^3 + 2x^2 - 4x - 8$, $(1, 2)$
47. $h(x) = x^3 - 6x^2 + 3x + 13$, $(-2, -1)$
48. $h(x) = -x^3 + 3x + 1$, $(1, 2)$
49. $h(x) = 23x - x^3$, $(4, 5)$
50. $g(x) = 2x^5 - 4x^4 + x^2 - 10$, $(2, 3)$

In problems 51–54, use a grapher to sketch the graph of each function, and to approximate the value c of each real zero so that $|f(c)| \leq 0.0005$. Round off the answers to four decimal places, if necessary. Also, approximate the solution of the associated inequality. Round off to four decimal places.

51. $f(x) = x^3 - 3x - 2$; $x^3 - 3x > 2$
52. $f(x) = x^3 - 3x^2 + 5$; $x^3 + 5 \leq 3x^2$
53. $f(x) = x^4 - 3x^2 + 2$; $x^4 + 2 \leq 3x^2$
54. $f(x) = x^3 + 4x^2 - 3x - 10$; $x^3 + 4x^2 \leq 3x + 10$

Applying the Concepts

55. **Radar Enclosure:** A *radome* is a geometric solid that is formed by removing a section from a sphere so that it has a flat base, as shown in Figure 13. Such structures are used to enclose radar antennas for protection from rain, wind, and snow. The volume V of a radome of radius r and height x is given by the formula

$$V = \pi r x^2 - \frac{\pi}{3}x^3$$

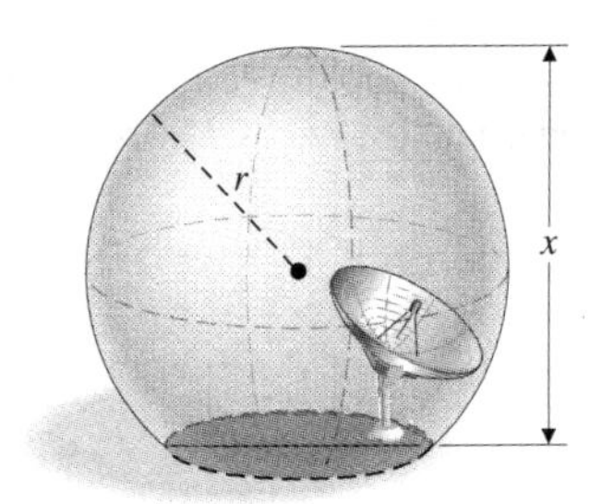

Figure 13

(a) Suppose that a radome has a radius of 21 feet and height of at most 40 feet. Express the volume V as function of the height x.
(b) What are the restrictions on x?
(c) Find the height x of the radome if it encloses a volume of $11{,}664\pi$ cubic feet.

56. **Percent of Income:** Suppose that the relationship between the percent of population x (expressed as a decimal) and the percent of the total income y (expressed as a decimal) of a certain community is modeled by the function

$$y = 2.7x^3 - 2.4x^2 + 0.7x$$

(a) What are the restrictions on x?
(b) What percent of the total income does 40% of the population in the community account for?
(c) What percent of the population accounts for 20% of the total income of the community?

57. **Storage:** A cistern to be fabricated in the shape of a right circular cylinder with a hemisphere at the bottom is to have a depth of 30 feet (Figure 14).

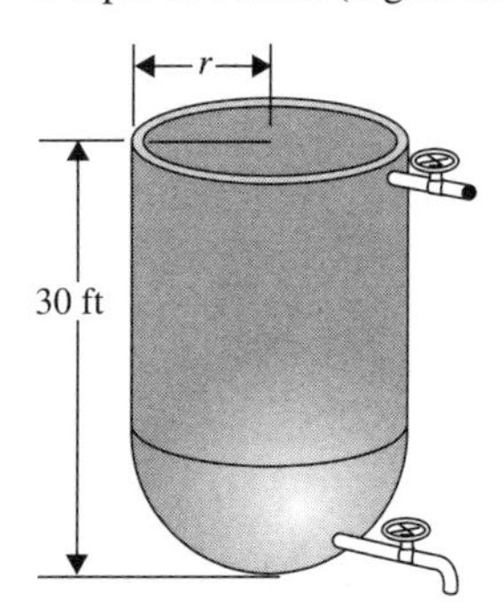

Figure 14

(a) Express the volume V of the cistern as a function of r, the radius of the cylinder.
(b) Determine the restrictions on r.
(c) What is the radius if the volume is 1008π cubic feet?

58. **Construction:** A rectangular toy chest has a square base that is 2 feet by 2 feet and a height of 1 foot. Suppose that each dimension of the chest is increased by x feet, where x is at most 2 feet.
(a) Express the volume V of the new chest as a function of x.
(b) What is the value of x if the volume of the new chest is 4.5 times the volume of the old one?

In problems 59–62, round off the answers to two decimal places.

59. **Rate of Inflation:** Suppose that a typical family in the United States will save approximately s percent of its income per year if the rate of inflation is x percent, where s is modeled by the function

$$s = -0.001x^3 + 0.06x^2 - 1.2x + 12, 0 \leq x \leq 10.$$

Use a grapher to approximate what the rate of inflation would be if the average saving percent was 10 percent, 8 percent, or 5 percent.

60. **Hot Air Balloon:** Atmospheric pressure P (in pounds per square inch) is approximated by the function

$P = 15 - 6h + 1.2h^2 - 0.16h^3, 0 \leq h \leq 3.5$ where height h is in thousands of feet above sea level. Use a grapher to approximate the height of a hot air balloon if the atmospheric pressure at that height is 10 pounds per square inch, 5 pounds per square inch, or 3.4 pounds per square inch.

61. **The Golden Gate Bridge Cable:** The Golden Gate Bridge in San Francisco has twin supporting main towers extending 525 feet above the road surface. The two main towers are 4200 feet apart.

(a) Draw a sketch of the situation.

(b) Use the formula in Example 11 to express the length of the supporting cable c between the two towers as a function of the sag x.

(c) Determine the restrictions on x.

(d) Use a grapher to find how high the lowest point of the cable is above the roadway if the length of the supporting cable is 4363 feet.

62. **Radius of a Propane Tank:** A steel propane gas storage tank is to be constructed in the shape of a right circular cylinder with a hemisphere attached at each end. The total length of the tank is 12 feet (Figure 15).

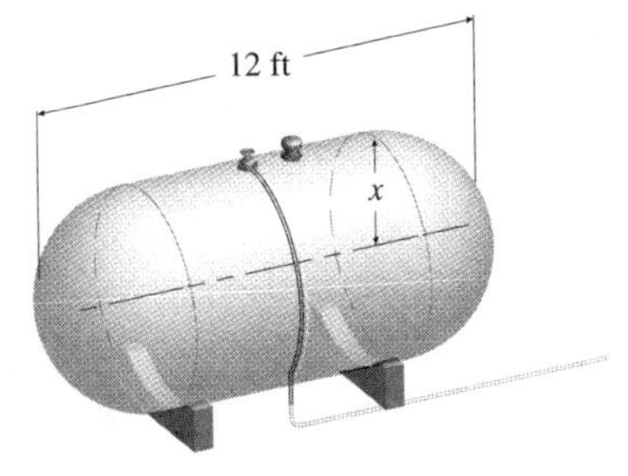

Figure 15

(a) Express the volume V of the tank as a function of the radius x (in feet) of the cylinder.

(b) What are the restrictions on x?

(c) Use a grapher to determine the radius so that the resulting volume is 260 cubic feet.

Developing and Extending the Concepts

In problems 63–68, indicate whether the statement is true or false. If the statement is false, give an example to disprove it; if it is true, justify it with an explanation.

63. If the graph of a polynomial function

f has three x intercepts, then the degree of $f(x)$ is 3.

64. The numbers -1, 0, 2, and 3 cannot all be zeros of a polynomial function of degree 3.

65. If $\frac{2}{3}$ is a rational zero of the polynomial function $g(x) =$

$a_n x^n + a_{n-1} x^{n-1} + \ldots + a_1 x + a^0$ and the coefficients are all integers, then a_0 is an even integer.

66. If a polynomial function f has three rational zeros, then the graph of f has exactly three x intercepts.

67. Suppose that f is a polynomial function with real coefficients. If the values of $f(a)$ and $f(b)$ are both positive, where $a < b$, then f has no zero between a and b.

68. Suppose that f is a polynomial function with real coefficients. If the values of $f(a)$ and $f(b)$ have opposite algebraic signs, where $a < b$, then there is exactly one zero between a and b.

OBJECTIVES

1. Graph Functions of the Form $R(x) = \dfrac{a}{(x-h)^n} + k$
2. Locate Vertical and Horizontal Asymptotes
3. Graph Nonreduced Rational Functions
4. Solve Applied Problems

3.4 Rational Functions

Just as rational numbers are defined in terms of quotients of integers, *rational functions* are defined in terms of quotients of polynomials.

> A function R of the form
>
> $$R(x) = \frac{p(x)}{q(x)}$$
>
> where $p(x)$ and $q(x)$ are polynomials and $q(x) \neq 0$, is called a **rational function.**

Examples of rational functions are

$$R(x) = \frac{3}{x},\ R(x) = \frac{2x}{x^2-4},\ R(x) = \frac{3x-1}{x^2+4},\ \text{and } R(x) = \frac{x^2-4}{x-2}.$$

The **domain** of a rational function $R(x) = p(x)/q(x)$ consists of all real numbers x for which $q(x) \neq 0$ since division by 0 is not defined.

Table 1 identifies the domains of the above functions.

TABLE 1

Function	Domain
1. $R(x) = \dfrac{3}{x}$	All real numbers except $x = 0$.
2. $R(x) = \dfrac{2x}{x^2-4}$	All real numbers except $x = -2$ and $x = 2$ because $x^2 - 4 = 0$ when $x = -2$ or 2.
3. $R(x) = \dfrac{3x-1}{x^2+4}$	All real numbers since $x^2 + 4 \neq 0$ for any real number x.
4. $R(x) = \dfrac{x^2-4}{x-2}$	All real numbers except $x = 2$ because $x - 2 = 0$ when $x = 2$.

Graphing Functions of the Form $R(x) = \dfrac{a}{(x-h)^n} + k$

Our investigation of the graphs of rational functions begins with a study of functions of the form

$$R(x) = \frac{1}{x^n}.$$

This type of function is not defined when $x = 0$.

Understanding the numerical behavior of $R(x)$ for values of x "near" 0, is useful in sketching the graph of R.

To illustrate this behavior, consider the rational function

$$f(x) = \frac{1}{x}.$$

The domain of f includes all real numbers except $x = 0$. So the graph has no y intercept. The data in Table 2 suggest that as x takes on positive values closer and closer to 0 (that is, as *x approaches 0 from the right*), the positive function values increase without bound.

TABLE 2

x	1000	100	10	1	0.1	0.01	0.001	0.0001	0
$f(x) = 1/x$	0.001	0.01	0.1	1	10	100	1000	10,000	$f(0)$ is not defined

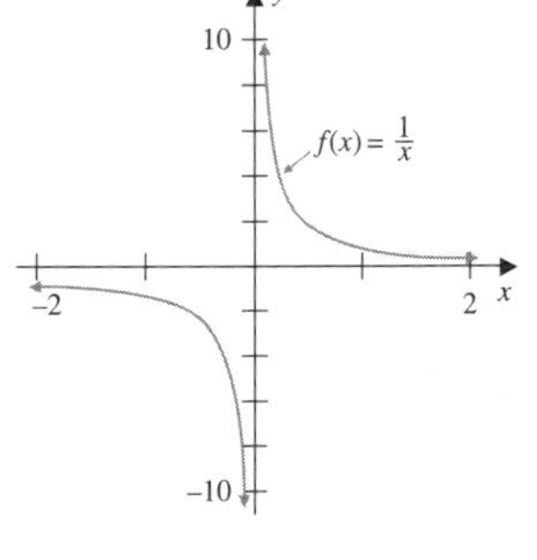

Figure 1

We describe this behavior symbolically by writing

$$f(x) \to +\infty \quad \text{as} \quad x \to 0^+.$$

Correspondingly, the graph of

$$f(x) = 1/x.$$

(Figure 1) gets "higher and higher" and approaches the y axis, but never touches it, as x approaches 0 *from the right.*

Similarly, as Table 3 suggests, the negative function values $f(x)$ decrease without bound as negative x values get closer and closer to 0 (that is, as *x approaches 0 from the left*), and we write

$$f(x) \to -\infty \text{ as } x \to 0^-.$$

TABLE 3

As $x \to 0^-$, $f(x) \to -\infty$

x	-1000	-100	-10	-1	-0.1	-0.01	-0.001	-0.0001
$f(x) = 1/x$	-0.001	-0.01	-0.1	-1	-10	-100	-1000	$-10{,}000$

This means that as x approaches 0 from the left, the graph of $f(x) = 1/x$ gets "lower and lower" and approaches the y axis, but never touches it, as shown in Figure 1.

Because of the behavior as x approaches 0 from either the right or the left, we say that the graph of $f(x) = 1/x$ approaches the y axis *asymptotically* and the y axis (which has the equation $x = 0$) is referred to as a *vertical asymptote.*

More formally, we have the following definition:

Definition Vertical Asymptote

The line $x = k$, where k is constant, is called a **vertical asymptote** of the graph of a function f if

$$f(x) \to -\infty \quad \text{or} \quad f(x) \to +\infty$$

either as $x \to k^-$(from the left) or as $x \to k^+$ (from the right).

Figure 2 contains illustrations of some graphs that exhibit vertical asymptotic behavior.

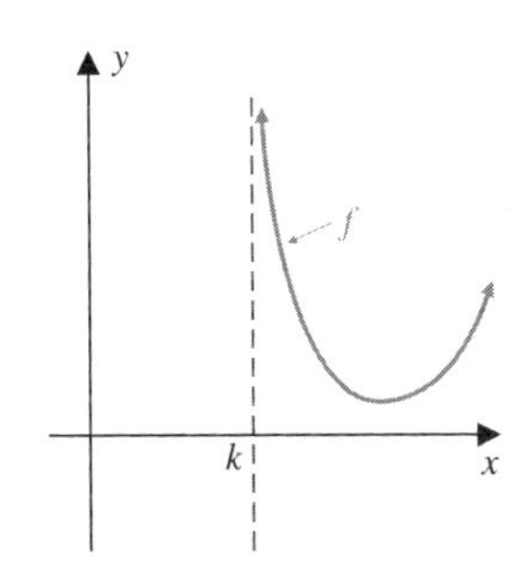
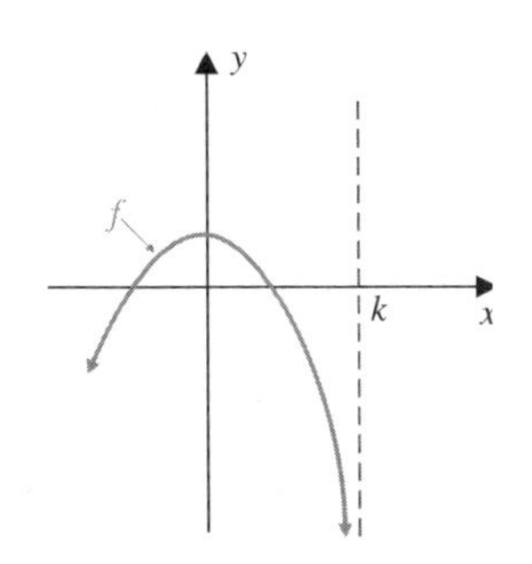
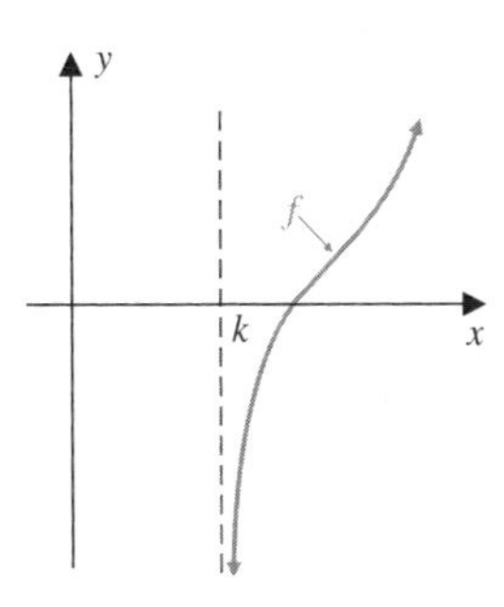

(a) $f(x) \to +\infty$ as $x \to k^-$ (b) $f(x) \to +\infty$ as $x \to k^+$ (c) $f(x) \to -\infty$ as $x \to k^-$ (d) $f(x) \to -\infty$ as $x \to k^+$

Figure 2

Let us reexamine the function $f(x) = 1/x$ on page 212 to determine its limit behavior. The graph in Figure 1 indicates that the function

$$f(x) = \frac{1}{x}.$$

has the following limit behavior:

$$f(x) \to 0 \text{ (through positive values) as } x \to +\infty$$

$$f(x) \to 0 \text{ (through negative values) as } x \to -\infty$$

In other words, the graph gets closer and closer to the x axis either as

$$x \to +\infty \text{ or as } x \to -\infty.$$

Because of this limit behavior, the x axis (which has the equation $y = 0$) is referred to as a *horizontal asymptote* of the graph of f.

In general, we have the following definition:

Definition Horizontal Asymptote

The line $y = k$, where k is a constant, is called a **horizontal asymptote** for the graph of a function f if

$$f(x) \to k \text{ as } x \to -\infty$$

or

$$f(x) \to k \text{ as } x \to +\infty.$$

Figure 3 contains illustrations of some graphs that exhibit horizontal asymptotic behavior.

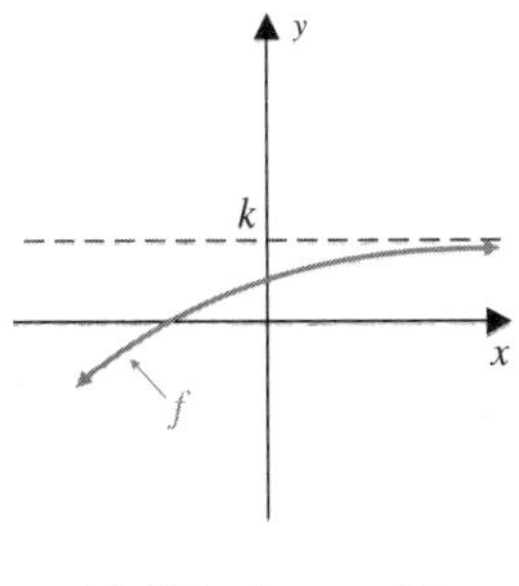
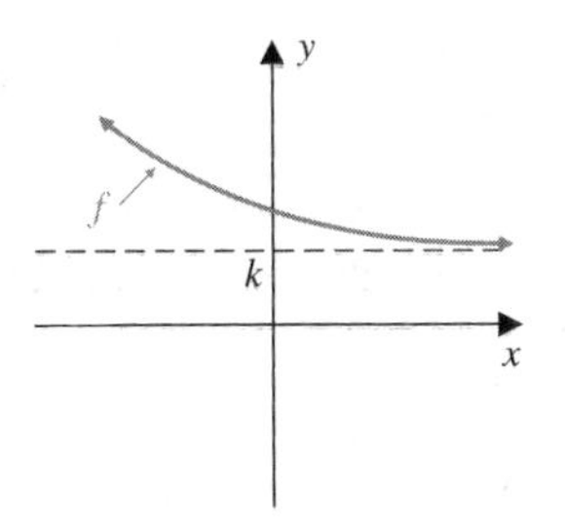
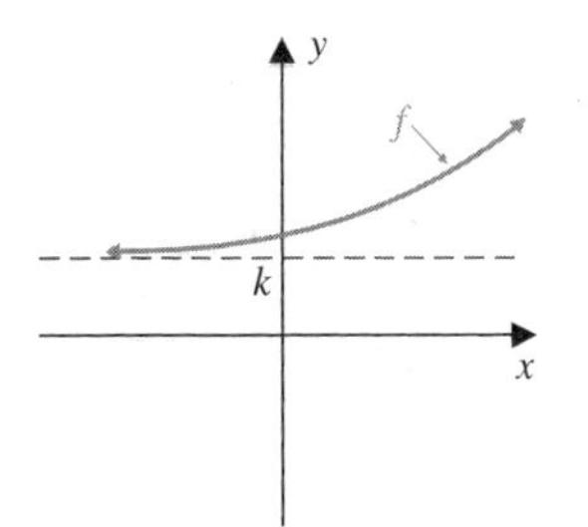
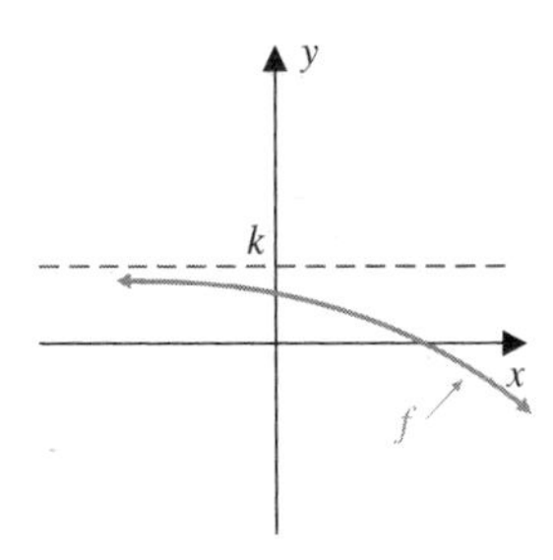

(a) $f(x) \to k$ as $x \to +\infty$ (b) $f(x) \to k$ as $x \to +\infty$ (c) $f(x) \to k$ as $x \to -\infty$ (d) $f(x) \to k$ as $x \to -\infty$

Figure 3

EXAMPLE 1 Graphing $g(x) = \dfrac{1}{x^2}$

Examine the numerical behavior of

$$g(x) = \frac{1}{x^2}$$

as $x \to 0^+$. Also, describe the limit behavior as $x \to +\infty$. Then use this information to sketch the graph of g.

Solution The domain of g includes all real numbers except $x = 0$, so the graph has no y intercept. The data in Table 4 suggests that

$$g(x) \to +\infty \text{ as } x \to 0^+$$

so the y axis is a vertical asymptote. Also, the data in Table 4 help us to conclude that g has the following limit behavior:

$$g(x) \to 0 \text{ (through positive values) as } x \to +\infty$$

TABLE 4
As $x \to 0^+$, $g(x) \to +\infty$

x	1000	100	10	1	0.1	0.01	0.001
$g(x) = 1/x^2$	0.000001	0.0001	0.01	1	100	10,000	1,000,000

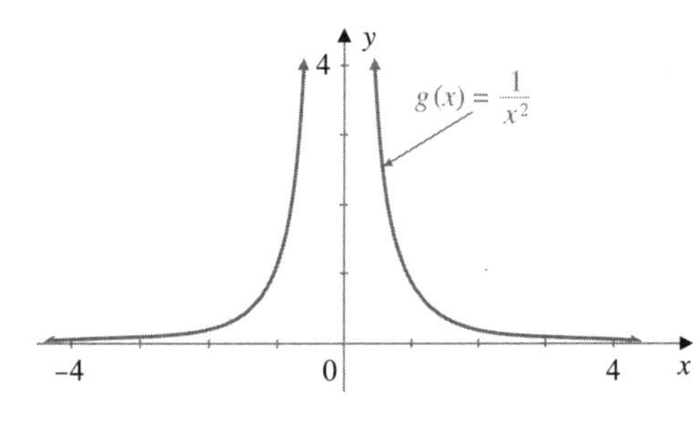

Figure 4

Consequently, the x axis is a horizontal asymptote of the graph of the function g.

Since, $g(-x) = 1/(-x)^2 = 1/x^2 = g(x)$, it follows that g is an even function and its graph is symmetric with respect to the y axis. By using the asymptotic behavior and symmetry, we get the graph of $g(x) = 1/x^2$ (Figure 4). Note that there is no x intercept since $1/x^2 = 0$ has no solution. ❖

In general, graphs of functions of the form

$$R(x) = \frac{1}{x^n} \text{ where } n \geq 1 \text{ is a positive integer}$$

fall into two categories according to whether n is an even or odd positive integer.

If $n \geq 2$ is an even integer, then the graphs of functions such as

$$f(x) = \frac{1}{x^4} \text{ and } g(x) = \frac{1}{x^6} \text{ (Figure 5a)}$$

resemble the graph of $y = 1/x^2$ shown in Figure 4

If $n \geq 1$ is an odd integer, then the graphs of functions such as

$$f(x) = \frac{1}{x^3} \text{ and } g(x) = \frac{1}{x^5} \text{ (Figure 5b)}$$

resemble the graph of $y = 1/x$ shown in Figure 1 on page 212.

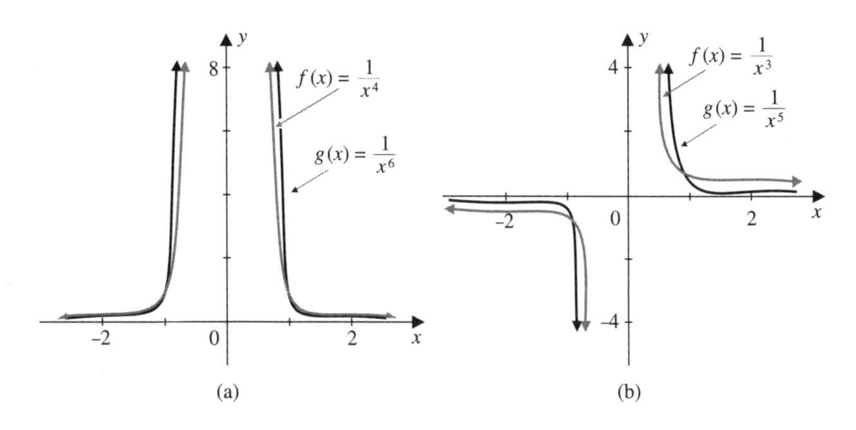

Figure 5

In general, the graphs of $R(x) = 1/x^2$ have the characteristics given in Table 5.

TABLE 5
Characteristics of rational functions $R(x) = 1/x^n$, where $n \geq 1$ is a positive integer

Property	n is Even	n is Odd
Vertical asymptote	y axis (equation is $x = 0$)	y axis (equation is $x = 0$)
Horizontal asymptote	x axis (equation is $y = 0$)	x axis (equation is $y = 0$)
Symmetry	y axis, since $R(-x) = R(x)$	Origin, since $R(-x) = -R(x)$
Intercepts	None	None

Transformations of the graph of $R(x) = 1/x^n$ can be used to graph rational functions of the form

$$R(x) = \frac{a}{(x - h)^n} + k$$

where $n \geq 1$ is a positive integer.

EXAMPLE 2 Using Transformations to Graph Rational Functions

Use transformations of the graph of either $y = 1/x$ or $y = 1/x^2$ to graph each function. Identify the asymptotes.

(a) $f(x) = \dfrac{1}{x - 2}$ (b) $g(x) = \dfrac{1}{x^2} + 2$

Solution

(a) The graph of f can be obtained by shifting the graph of $y = 1/x$ horizontally 2 units to the right (Figure 6a). Note that the vertical asymptote is shifted from the y axis to the line $x = 2$, but the horizontal asymptote remains the x axis.

(b) The graph of g can be obtained by shifting the graph of $y = 1/x^2$ vertically upward 2 units (Figure 6b). Note that the vertical asymptote is still the y axis, but the horizontal asymptote is shifted 2 units up from the x axis to become the line $y = 2$.

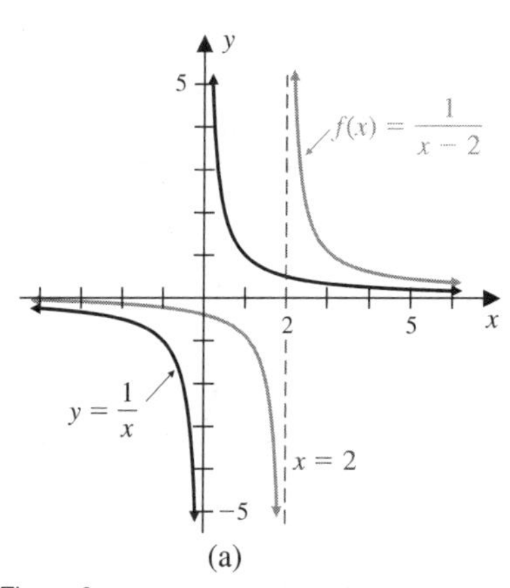

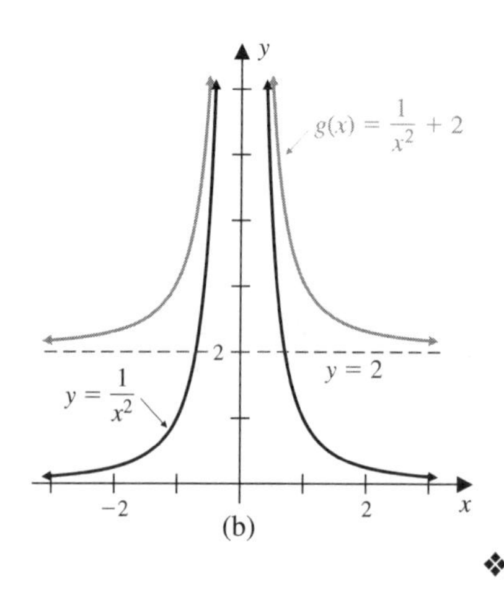

Figure 6

❖

If the degree of $p(x)$ is equal to the degree of $q(x)$ for the rational function $R(x) = \frac{p(x)}{q(x)}$, then the graph of R has a horizontal asymptote. To find this horizontal asymptote, we may use long division to rewrite $R(x)$ in the form

$$R(x) = \frac{a}{(x-h)^n} + k$$

then proceed to sketch the graph by using transformations. The next example illustrates the technique.

EXAMPLE 3 Graphing a Rational Function

(a) Use long division to rewrite the given function in the form $f(x) = \frac{a}{x-h} + k$ and then use transformations of graphs to graph the function

$$f(x) = \frac{3x+17}{x+5}$$

Identify the asymptotes of the graph.

(b) Find the domain and x intercepts of f.

(c) Solve the inequality $\frac{2}{x+5} > -3$.

Solution (a) After dividing $3x + 17$ by $x + 5$, we rewrite $f(x)$ as

$$f(x) = \frac{2}{x+5} + 3$$

Thus the graph of f can be obtained by shifting the graph of $y = \frac{1}{x}$ horizontally 5 units to the left, vertically scaling by a factor of 2, and shifting 3 units vertically upward (Figure 7). Note that the horizontal asymptote is $y = 3$ because of the vertical shift and the vertical asymptote is $x = -5$ because of the horizontal shift.

(b) The domain of f consists of all real numbers except -5, that is,

$$(-\infty, -5) \text{ or } (-5, \infty).$$

The x intercept of f is obtained by setting $f(x) = 0$ and then solving for x to get

$$3x + 17 = 0 \text{ or } x = -\frac{17}{3}.$$

(c) To solve $2/(x + 5) > -3$, we begin by rewriting the inequality in the equivalent form

$$\frac{2}{x+5} + 3 > 0$$

Figure 7

in which 0 is on one side of the inequality.

Since

$$\frac{2}{x+5} + 3 = \frac{3x+17}{x+5} = f(x)$$

we can rewrite the latter inequality as $f(x) > 0$.

Thus we need to find all values of x that yield positive values of $f(x)$; that is, we need to find all values of x that yield points on the graph of f that are above the x axis.

The graph of f in Figure 7 is above the x axis when $x < -17/3$ and when $x > -5$. Therefore, the solution of the original inequality is $x < -17/3$ or $x > -5$. In interval notation the solution includes all numbers in the interval $(-\infty, -17/3)$ or the interval $(-5, \infty)$. ❖

Locating Vertical and Horizontal Asymptotes

To get an accurate graph of a rational function, it is helpful to first identify its asymptotes by examining the equation that defines the function. So far our investigation of the form

$$R(x) = \frac{1}{x^n}$$

indicates the following behavior:

For any fraction $p(x)/q(x)$, as the values of the denominator $q(x)$ get closer to 0 and the values of the numerator $p(x)$ approach a value different from 0, the cumulative effect is that $p(x)/q(x)$ will either increase or decrease without bound, depending on whether the function values are positive or negative. This behavior provides the basis for determining the vertical asymptotes of a rational function from its equation.

Property 1 Locating Vertical Asymptotes

A rational function

$$R(x) = \frac{p(x)}{q(x)}$$

has a vertical asymptote at the line $x = c$ if c is a zero of q and not a zero of p, that is, if $q(c) = 0$ but $p(c) \neq 0$.

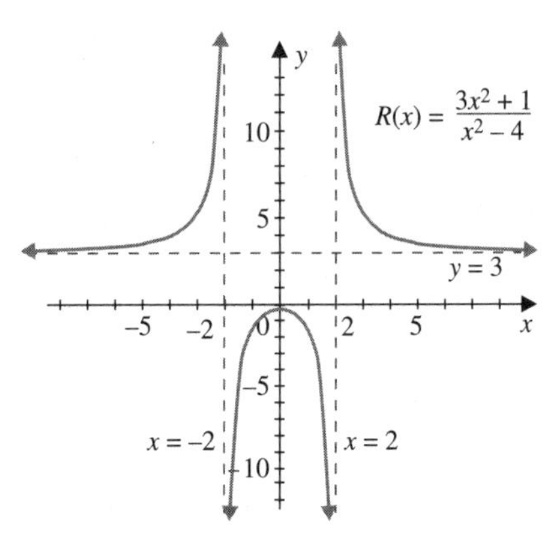

Figure 8

For instance, Figure 8 shows the graph of the function

$$R(x) = \frac{3x^2 + 1}{x^2 - 4}, \text{ which is of the form } \frac{p(x)}{q(x)}.$$

The graph of R has vertical asymptotes at the lines

$$x = -2 \text{ and } x = 2,$$

because $q(-2) = 0$ and $q(2) = 0$ but $p(-2) \neq 0$ and $p(2) \neq 0$.

Let us turn our attention to the limit behavior of the function

$$R(x) = \frac{3x^2 + 1}{x^2 - 4}$$

as $x \to +\infty$.

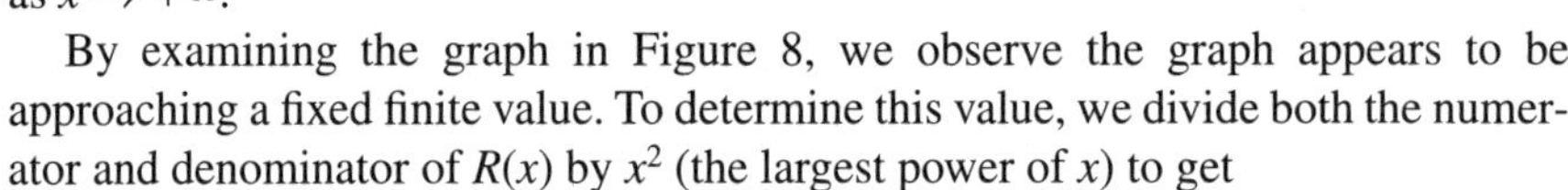

By examining the graph in Figure 8, we observe the graph appears to be approaching a fixed finite value. To determine this value, we divide both the numerator and denominator of $R(x)$ by x^2 (the largest power of x) to get

$$R(x) = \frac{3x^2 + 1}{x^2 - 4} = \frac{\frac{3x^2}{x^2} + \frac{1}{x^2}}{\frac{x^2}{x^2} - \frac{4}{x^2}} = \frac{3 + \frac{1}{x^2}}{1 - \frac{4}{x^2}}$$

When $|x|$ gets very large, the values of $\frac{1}{x^2}$ and $\frac{4}{x^2}$ approach zero; that is

$$\text{as } |x| \to +\infty, \frac{1}{x^2} \to 0 \text{ and } \frac{4}{x^2} \to 0$$

so that

$$R(x) \to \frac{3 + 0}{1 - 0} = 3, \text{ that is, 3 is the limit value.}$$

So the line $y = 3$ is a horizontal asymptote.

In general, to determine the limit behavior of the graph of a rational function when $|x| \to \infty$, we first divide both the numerator and the denominator by x raised to its highest power. Then we examine the latter form to help locate any horizontal asymptote. The result depends on the relative degrees of the polynomials in the numerator and the denominator, as indicated in the following property:

Property 2 Locating Horizontal Asymptotes

Assume R is a rational function (in reduced form) defined by

$$R(x) = \frac{p(x)}{q(x)} = \frac{a_m x^m + a_{m-1}x^{m-1} + \ldots + a_1 x + a_0}{b_n x^n + b_{n-1}x^{n-1} + \ldots + b_1 x + b_0}$$

where $m \geq 0$ and $n > 0$ are integers; and and $a_m \neq 0$ and $b_n \neq 0$.

1. If $m < n$, then the line $y = 0$ (the x axis) is the horizontal asymptote.
2. If $m = n$, then the line $y = a_m/b_n$ is the horizontal asymptote.
3. If $m > n$, then there is no horizontal asymptote.

EXAMPLE 4 Locating a Horizontal Asymptote

Locate the horizontal asymptote of

$$g(x) = \frac{3x}{2x+1}$$

and display this asymptotic behavior graphically.

First Solution Because $m = n = 1$, it follows from Property 2, that the line $y = \frac{3}{2}$ is a horizontal asymptote of the graph of g.

The graph in Figure 9 shows the following limit behaviors.

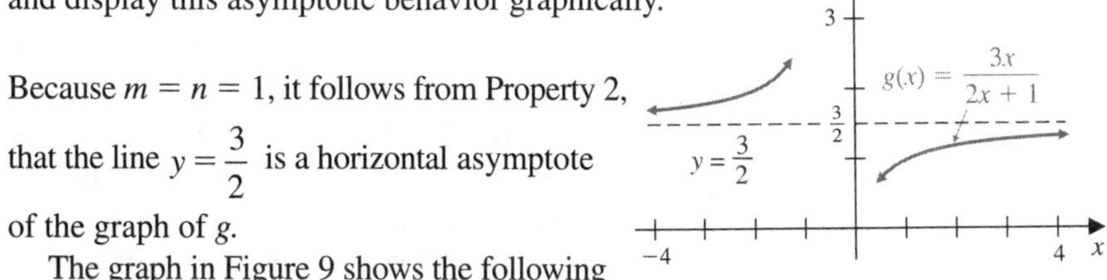

Figure 9

$$g(x) \to \frac{3}{2} \text{ as } x \to +\infty \text{ and } g(x) \to \frac{3}{2} \text{ as } x \to -\infty$$

Alternate Solution We divide both the numerator and the denominator by x (the largest power of x) to get

$$g(x) = \frac{3x}{2x+1} = \frac{\frac{3x}{x}}{\frac{2x}{x} + \frac{1}{x}} = \frac{3}{2 + \frac{1}{x}}$$

When the values of $|x|$ get very large, the values of $\frac{1}{x}$ approach 0, that is

$$\text{as } |x| \to +\infty, 1/x \to 0$$

so

$$g(x) \to \frac{3}{2+0} = \frac{3}{2}$$

That is,

$$g(x) \to \frac{3}{2} \text{ as } x \to +\infty \text{ and } g(x) \to \frac{3}{2} \text{ as } x \to -\infty$$

Thus, the line $y = 3/2$ is the horizontal asymptote of the graph of g (Figure 9). ❖

Graphing Nonreduced Rational Functions

A rational function is said to be in *reduced form* if its numerator and denominator have no common factor. The next example investigates graphs of rational functions that are not in reduced form.

EXAMPLE 5 Graphing Nonreduced Rational Functions

Graph the given functions and find any asymptotes.

(a) $R(x) = \dfrac{x^2 - 4}{x - 2}$

(b) $h(x) = \dfrac{x^2 - 2x}{x^2 - 4x + 4}$

Solution (a) By writing the numerator of R in factored form, we get the reduced form

$$R(x) = \frac{x^2 - 4}{x - 2} = \frac{\cancel{(x-2)}(x+2)}{\cancel{x-2}} = x + 2 \quad \text{where } x \neq 2$$

Note that the function $y = x + 2$ is *not* the same as the function R, because they have different domains. The graph of $y = x + 2$ is a straight line that includes the point (2, 4) (Figure 10a), whereas the graph of R is the same straight line except that it has a *hole* at the point (2, 4) (Figure 10b). In this situation, 2 is a zero of both the numerator and denominator of R. However, it turns out that there is no vertical asymptote at $x = 2$, only a hole.

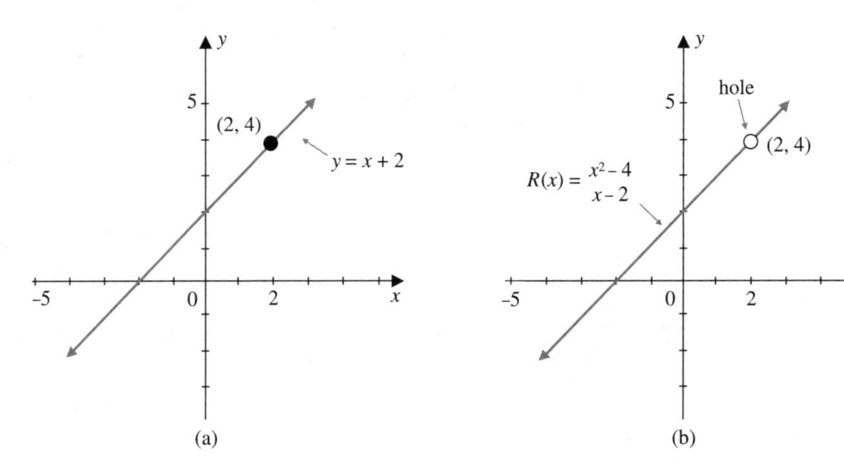

Figure 10

(b) By writing both the numerator and denominator of $h(x)$ in factored form, we get the reduced form

$$h(x) = \frac{x^2 - 2x}{x^2 - 4x + 4}$$

$$= \frac{x\cancel{(x-2)}}{\cancel{(x-2)}(x-2)} = \frac{x}{x-2}, \quad \text{where } x \neq 2..$$

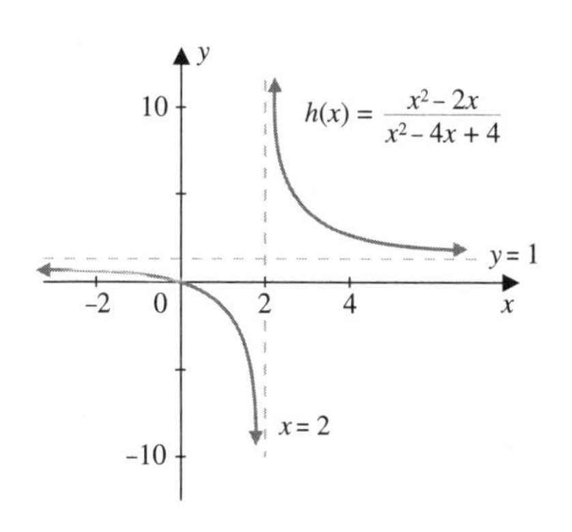

Figure 11

Note that the function

$$y = \frac{x}{(x-2)} = \frac{2}{x-2} + 1$$

is the same as function h. The graph is obtained by using transformations of the graph of $y = \frac{1}{x}$ (Figure 11). In this situation, 2 is a zero of both the numerator and denominator of h, and it turns out that there is a vertical asymptote at $x = 2$. ❖

The graphs of more complicated rational functions such as

$$f(x) = \frac{x^2}{x^2 - 3} \text{ and } g(x) = \frac{3x^2 - 12}{x^2 + 2x - 3}$$

are examined extensively in calculus. For now, we can use a grapher to sketch such functions.

When a grapher is used to sketch the graph of a rational function the result may not correctly represent the behavior of the function near a vertical asymptote. The grapher may erroneously connect separate branches of the graph across a vertical asymptote. For instance, the viewing window of the graph of

$$f(x) = \frac{x-1}{x^2 - x - 6}$$

shown in Figure 12 incorrectly displays a graph that includes curves that appear to be the vertical asymptotes. Actually, the grapher is erroneously connecting separate branches of the graph. Inaccuracies such as this one may occur near vertical asymptotes, depending on the selected viewing window.

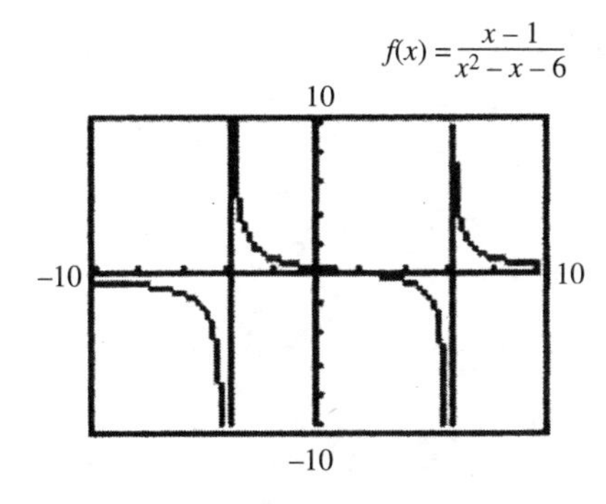

Figure 12

Solving Applied Problems

Trends that occur in real-world situations are sometimes modeled by rational functions.

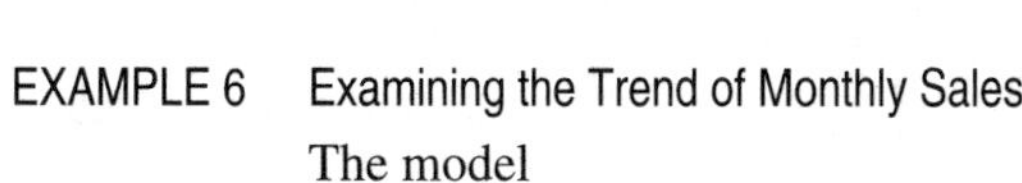

EXAMPLE 6 Examining the Trend of Monthly Sales

The model

$$s(t) = \frac{2000(1 + t)}{2 + t}$$

relates the monthly sales s (in dollars) of a certain new product to the time t (in months), beginning 1 month after the product has been introduced.

(a) What are the restrictions on t?
(b) Locate the horizontal asymptote for s.
(c) Use the graph of s, taking into account the restrictions on t, and the horizontal asymptote to determine the trend of the monthly sales as time elapses.

Solution (a) We use $t \geq 1$, because the model is applicable at the beginning of the second month.

(b) Since

$$s(t) = \frac{2000(1+t)}{2+t} = \frac{2000t + 2000}{t+2} = \frac{2000 + \frac{2000}{t}}{1 + \frac{2}{t}}$$

it follows that as

$$t \to +\infty,\ s(t) \to 2000;$$

that is,

the line $s = 2000$ is a horizontal asymptote.

(c) After plotting a few points, recognizing that $s = 2000$ is a horizontal asymptote, and taking into account that $t \geq 1$, we obtain the graph shown in Figure 13. Notice that even though $t = -2$ is a vertical asymptote for the function for s (when it has no restrictions), it will not show in this graph because $t \geq 1$.

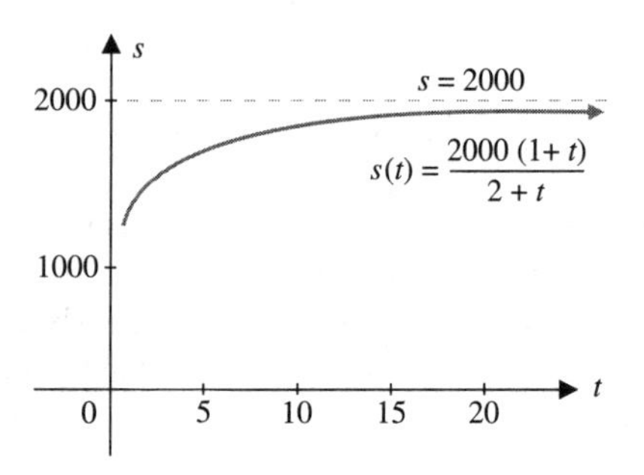

Figure 13

Upon examining the graph, we observe that as $t \to +\infty$ (the values of t are increasing), $s(t) \to 2000$ (from below). So as time elapses, the monthly sales trend will approach but not exceed \$2000. We refer to \$2000 as the *limiting value* of the monthly sales. ❖

PROBLEM SET 3.4

Mastering the Concepts

In problems 1 and 2, graph each function. Identify the asymptotes and demonstrate the results numerically with a table of values.

1. **(a)** $F(x) = \dfrac{1}{x^6}$

(b) $H(x) = \dfrac{1}{x^5}$

2. **(a)** $G(x) = \dfrac{1}{x^8}$

(b) $h(x) = \dfrac{1}{x^7}$

In problems 3–6, use the given graph of each function to identify the vertical and horizontal asymptotes. Describe the asymptotic behaviors of the function.

3.

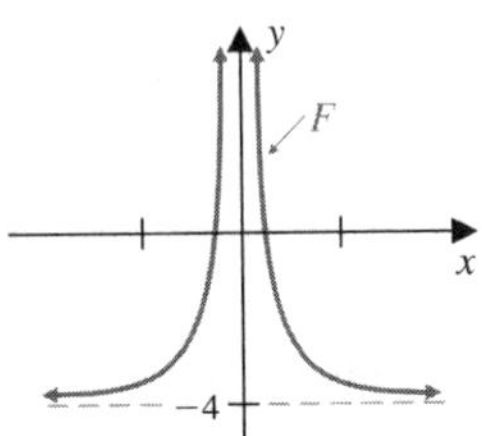

4.

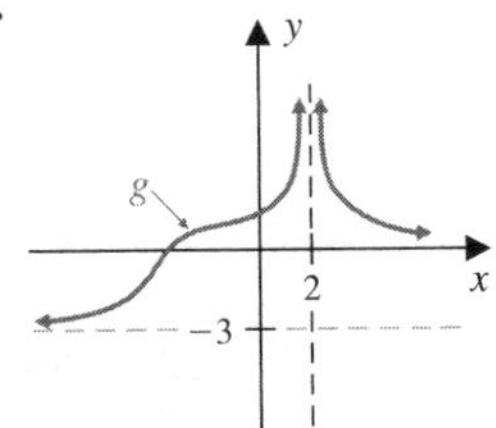

5.

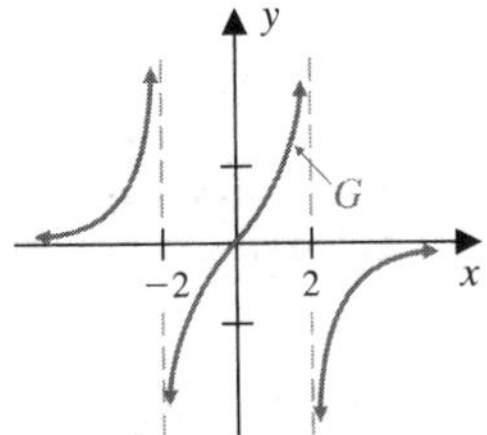

6.

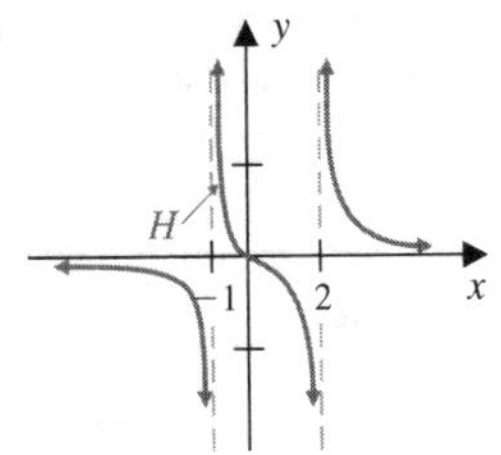

In problems 7–12, describe a sequence of transformations that will transform the graph of either $y = 1/x$ or $y = 1/x^2$ into the graph of each function. Sketch the graph of each function, and identify the asymptotes in each case. Also, find the domain and x intercepts of the graph of f.

7. **(a)** $f(x) = \dfrac{4}{x+3}$

(b) $g(x) = \dfrac{1}{x} + 3$

8. **(a)** $g(x) = \dfrac{1}{(x-3)^2}$

(b) $h(x) = \dfrac{4}{x^2} - 1$

9. **(a)** $f(x) = \dfrac{-3}{(x+1)^2}$

(b) $g(x) = \dfrac{2}{x-1} + 3$

10. **(a)** $h(x) = \dfrac{3x+2}{x}$

(b) $f(x) = \dfrac{-2}{(x+1)^2} - 1$

11. **(a)** $f(x) = \dfrac{4}{(x+3)^2} - 1$

(b) $g(x) = \dfrac{2x^2 + 16x + 33}{(x+4)^2}$

[*Hint:* Use long division first.]

12. **(a)** $g(x) = \dfrac{1}{x-2} + 3$

(b) $h(x) = \dfrac{3x+7}{x+2}$

[*Hint:* Use long division first.]

In problems 13 and 14, find any vertical asymptotes. Graph each function and locate any holes in the graph.

13. **(a)** $f(x) = \dfrac{x^2 - 1}{x-1}$

(b) $g(x) = \dfrac{x^2 - 2x - 3}{x-3}$

14. **(a)** $f(x) = \dfrac{x-4}{x^2 - 16}$

(b) $g(x) = \dfrac{x^3 - 1}{x-1}$

In problems 15 and 16, locate the horizontal asymptote of the graph of each function, and then display this asymptotic behavior graphically.

15. **(a)** $f(x) = \dfrac{-2x}{x+1}$

(b) $g(x) = \dfrac{1+3x}{2x-1}$

16. **(a)** $f(x) = \dfrac{x^2+1}{x^2-1}$

(b) $g(x) = \dfrac{2x^2 + x}{x^2 - 2x + 3}$

In problems 17–20, match the function to its hand-drawn graph in Figure 14. Locate all asymptotes.

17. $f(x) = \dfrac{x - 1}{x^2 + x}$ **18.** $g(x) = \dfrac{x^2}{x^2 - 4}$

19. $h(x) = \dfrac{2x}{x + 1}$ **20.** $f(x) = \dfrac{x^2 - 3x + 4}{x + 2}$

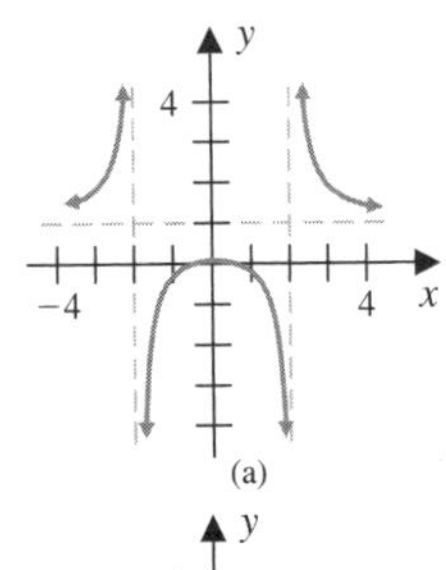

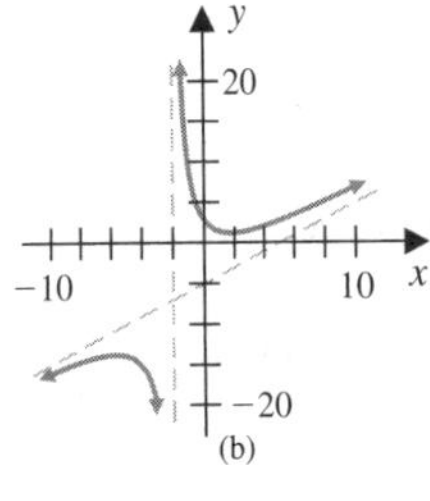

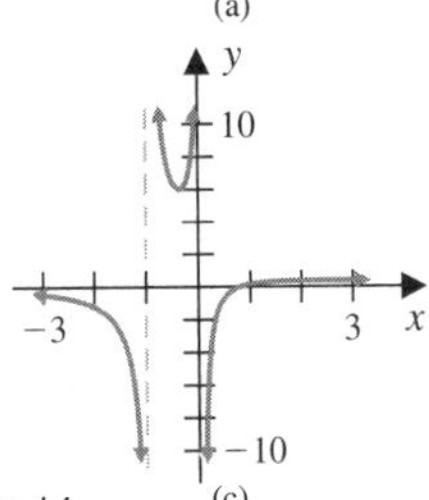

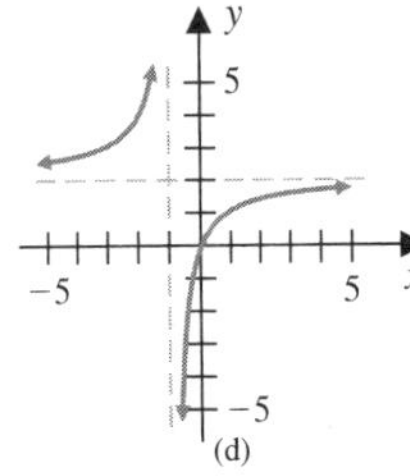

Figure 14

In problems 21–28, graph each function. Locate the asymptotes (if any) for each function.

21. $f(x) = \dfrac{2x}{x + 1}$

22. $g(x) = \dfrac{-3x + 1}{x + 4}$

23. $f(x) = \dfrac{3}{x^2 - 2x - 3}$

24. $h(x) = \dfrac{2}{x^2 - x - 2}$

25. $g(x) = \dfrac{-x}{x^2 - x - 6}$

26. $f(x) = \dfrac{-2x}{x^2 - 3x - 4}$

27. $f(x) = \dfrac{x^2 + 2}{x^2 - 4x + 3}$

28. $g(x) = \dfrac{x^2 - 4}{x^2 + 1}$

In problems 29–36, use the graph of an associated rational function to solve each inequality.

29. $2 - \dfrac{2}{x + 1} < 0$ (see problem 21)

30. $\dfrac{13}{x + 4} > 3$ (see problem 22)

31. $\dfrac{3}{x^2 - 2x - 3} \geq 0$ (see problem 23)

32. $\dfrac{2}{x^2 - x - 2} \leq 0$ (see problem 24)

33. $\dfrac{-x}{x^2 - x - 6} < 0$ (see problem 25)

34. $\dfrac{-2x}{x^2 - 3x - 4} > 0$ (see problem 26)

35. $\dfrac{x^2 + 2}{x^2 - 4x + 3} \leq 0$ (see problem 27)

36. $\dfrac{x^2 - 4}{x^2 + 1} \geq 0$ (see problem 28)

Applying the Concepts

37. **Business Sales Trend:** Suppose that the sale of S units of a certain product is related to the daily advertising expenditure x (in dollars) by the model

$$S(x) = \frac{5000x}{x + 100}$$

(a) What are the restrictions on x?
(b) Sketch the graph of S using the restrictions from part (a). Also, locate the asymptotes.
(c) What is the sales trend as the advertising expenditures increase?

38. **Government Securities:** Each weekday, national newspapers such as the *New York Times* and the *Wall Street Journal* publish a curve called a *yield curve* for the U.S. government securities. Suppose that the percentage yield p on government securities that mature after t years is modeled by the function

$$p = \frac{t + 0.1}{0.1t + 0.03}, \quad 0.25 \leq t \leq 30$$

(a) Graph this function and locate the asymptotes.
(b) Explain the implication of the horizontal asymptote on the graph of p. That is, analyze the trend of the yield as a function of maturity.

39. **Air Pollution:** Suppose that the cost C (in dollars) of removing x percent of air pollutants caused by automobile emissions in a certain town is given by the model

$$C = \frac{100{,}000x}{100 - 1.67x}, \quad 0 < x < 60$$

(a) Graph the function C.
(b) Use the graph to analyze the trend of the cost of pollution control by examining what happens as the air pollutants approach 60%.

40. Pain Relief: Researchers have determined that the percentage p of pain relief from x grams of aspirin is given by the model

$$p = \frac{100x^2}{x^2 + 0.02} \quad 0 < x < 1$$

(a) Graph p.

(b) Use the graph to approximate the value of x when the pain relief is very close to its maximum value of 100%.

(c) Explain the implication of the horizontal asymptote of the graph of p.

41. A closed box with a square base of x centimeters per side and a height of h centimeters has a volume of 500 cubic centimeters.

(a) Write a function for the surface area A of the box in terms of x.

(b) Graph A and to find the dimensions x and h that will minimize the area needed to manufacture the box. What is the minimum surface area? Round off the answers to two decimal places.

42. A city determined that the cost C (in millions of dollars) of operating a recycling center is given by the model

$$C(p) = \frac{1.1p}{90 - p}$$

where p is the percentage of the participants from the city residents.

(a) Graph C and locate the asymptotes of the graph of C.

(b) Explain the implication of the vertical asymptote on the graph of C as p approaches 90.

(c) Use the graph to analyze the trend of the cost if 80% of the residents are expected to participate.

Developing and Extending the Concepts

43. Let $f(x) = \frac{1}{x}$. Compute and simplify the expression

$$q = \frac{f(x) - f(2)}{x - 2}$$

Sketch the graph of q.

44. Let $f(x) = \frac{1}{x^2}$. Compute and simplify the expression.

$$q = \frac{f(x) - f(1)}{x - 1}$$

Sketch the graph of q.

In problems 45 and 46, find an equation of a rational function f that satisfies the following conditions.

45. f has real zeros at $-3, 1, 3$, a vertical asymptote at $x = 0$ and a horizontal asymptote at $y = 2$.

46. f has no real zeros, f has a vertical asymptote at $x = 3$ and it has a horizontal asymptote at $y = 4$.

47. Explain why the following statement is true: If the line with equation $x = k$ is a vertical asymptote of the graph of the rational function $y = f(x)$, then k is not in the domain of f.

48. Explain why the following statement is false: If the line with equation

$$y = k$$

is a horizontal asymptote of the graph of the rational function

$$y = f(x),$$

then k is not in the range of f.
Hint: Consider the function

$$f(x) = \frac{2x^2 + 1}{2x^2 - 3x}.$$

To solve problems 49 and 50, consider the rational function

$$f(x) = \frac{x^2 + 1}{x}$$

By division we can rewrite f as

$$f(x) = x + \frac{1}{x}$$

since $\frac{1}{x} \to 0$ as $x \to +\infty$ or $x \to -\infty$, this means that the graph of $y = f(x)$ eventually gets closer and closer to the line $y = x$ as $x \to +\infty$ or as $x \to -\infty$. The line $y = x$ is called an **oblique asymptote** of the graph of f.

In problems 49 and 50. Use the given graph of the rational function to find the equation of the oblique asymptotes.

49.

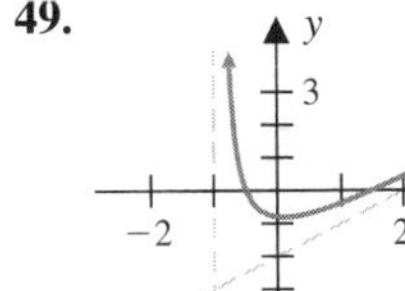

50.

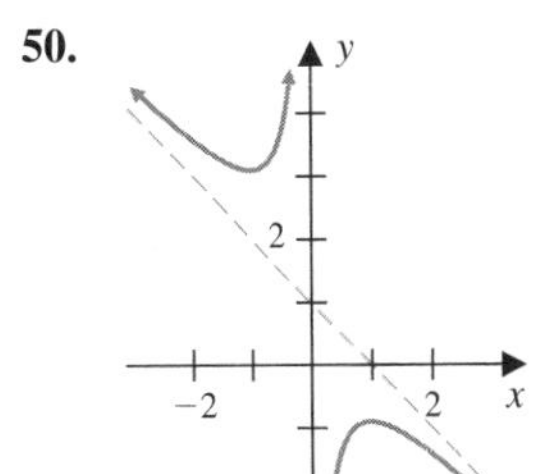

In problems 51 and 52, find the oblique asymptote of the graph of each function, and then demonstrate the asymptotic behavior by using a grapher to graph the function and its asymptote.

51. **(a)** $f(x) = \dfrac{-4x^2 + 1}{x + 2}$

(b) $g(x) = \dfrac{x^2 - 4x + 7}{x - 1}$

52. **(a)** $f(x) = \dfrac{3x^2 - 12x + 7}{x + 1}$

(b) $g(x) = \dfrac{x^3 - 2x^2}{x^2 - 1}$

Interlude: Progress Check 3.4

1. Give the rule of a rational function $r(x)$ such that the graph of $r(x)$ has one vertical asymptote, does not have a horizontal asymptote, has one x-intercept, and has a hole. Leave the numerator and denominator of $r(x)$ in factored form and show a graph of that function. (There are many possible correct answers here.)

2. Consider the rational function defined by $y = \dfrac{5x^2 - 15x - 50}{2x^2 - 8x - 10}$. Carefully sketch the graph of this function on the provided grid. On your graph, label all intercepts with exact coordinates. Indicate asymptotes with dashed lines.

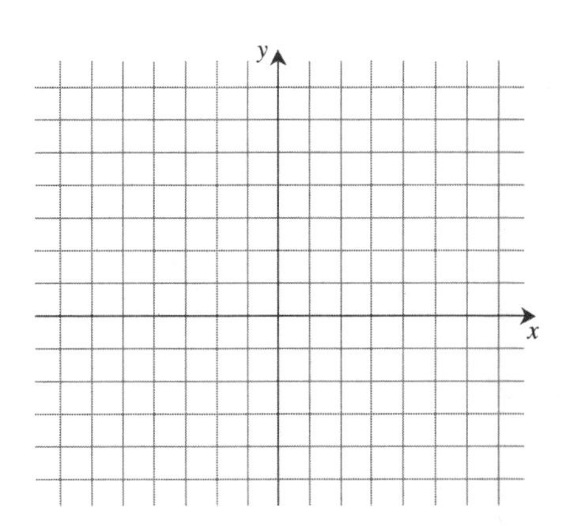

3. The cost per hour to operate a particular freight train that runs at an average speed of v miles per hour over the length of a trip is $300+\frac{v^2}{4}$ dollars. A 420-mile trip is being planned. Let $C(v)$ denote the total cost of the trip, in dollars.

(a) Give a formula for $C(v)$.

(b) Determine the average speed at which the train should be operated in order to minimize the total cost of the trip. Give an estimate accurate to three decimal places, labeled correctly. Show clearly how you arrived at your conclusion.

(c) What range of average speeds will guarantee a total trip cost of no more than $8000? Give all estimates accurate to three decimal places. Show clearly how you arrived at your conclusion.

CHAPTER 3 READINESS, PART 2

Provide complete solutions in the space provided.

A closed box with a square base is to have a volume of 24 cubic feet. The material for the sides, top, and base costs \$0.25, \$0.50, and \$1 per square foot, respectively.

1. Let x denote the length, in feet, of one of the edges of the base of the box, and let $C(x)$ denote the corresponding cost of the box, in dollars. Write a formula for $C(x)$.

2. Determine the dimensions of the box of least cost, and calculate the corresponding cost of that box. Demonstrate clearly how you arrived at your result.

A closed box with a square base is to have a volume of 24 cubic feet.

3. Let x denote the length, in feet, of one of the edges of the base of the box, and let $A(x)$ denote the corresponding amount of material needed to construct the box, in square feet. Write a formula for $A(x)$.

4. Determine the dimensions of the box that requires the minimum amount of material to construct, and calculate the corresponding amount of material. (Note: Compare this to your results from the problem on the previous page.)

Answers to Odd-Numbered Problems

PROBLEM SET 3.1

1. **(a)** B **(b)** C **(c)** D **(d)** A

3. Domain: $\mathbb{R}$;
Range: $[2, \infty)$;
Vertex: $(3, 2)$;
Axis of symmetry: $x = 3$

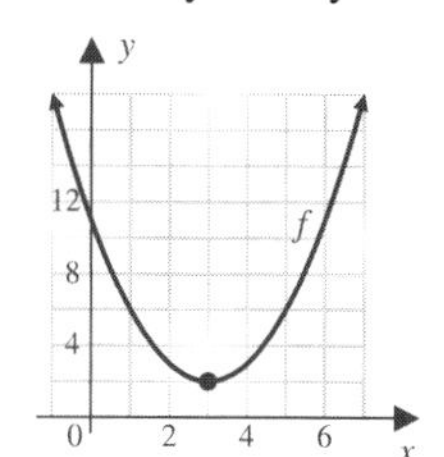

5. Domain: $\mathbb{R}$;
Range: $[-\infty, -3]$;
Vertex: $(-2, -3)$;
Axis of symmetry: $x = -2$

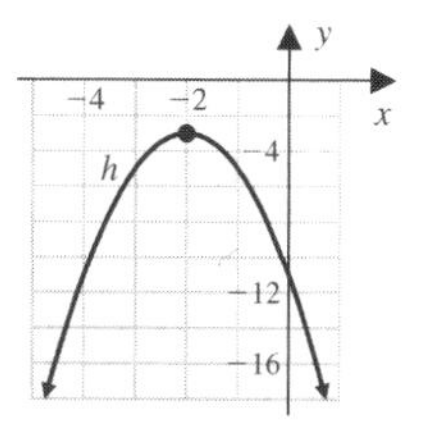

7. Domain: $\mathbb{R}$;
Range: $[1, \infty)$;
Vertex: $(-3, 1)$;
Axis of symmetry: $x = -3$

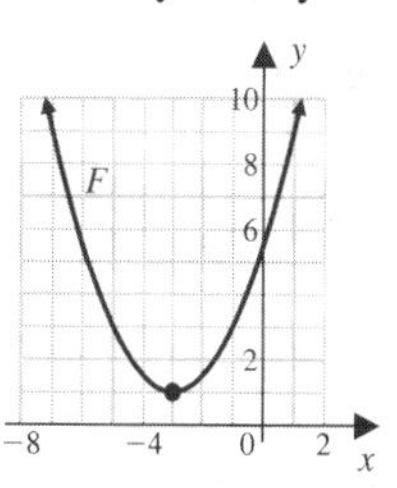

9. $f(x) = (x - 2)^2 - 9$;
Range: $[-9, \infty)$;
Vertex: $(2, -9)$;
Axis of symmetry: $x = 2$

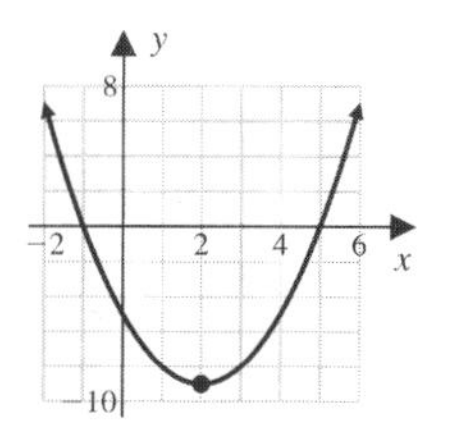

11. $f(x) = -(x + 3)^2 + 1$;
Range: $(-\infty, 1]$;
Vertex: $(-3, 1)$;
Axis of symmetry: $x = -3$

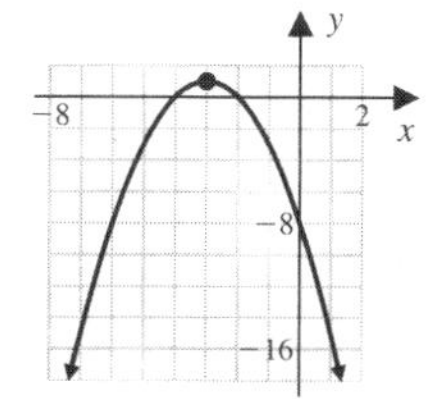

13. $f(x) = 3\left(x - \frac{5}{6}\right)^2 - \frac{277}{12}$;

Range: $\left[-\frac{277}{12}, \infty\right)$;

Vertex: $\left(\frac{5}{6}, -\frac{277}{12}\right)$;

Axis of symmetry: $x = \frac{5}{6}$

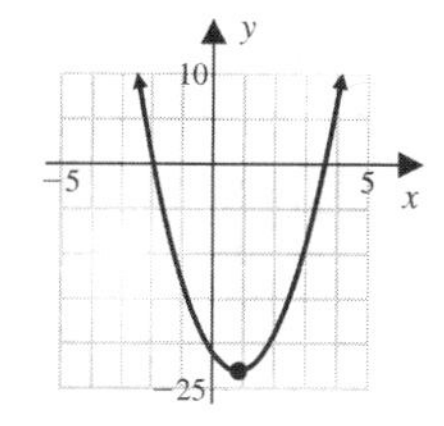

15. $f(x) = -5(x - 0.3)^2 + 4.45$;
Range: $(-\infty, 4.45]$;
Vertex: $(0.3, 4.45)$;
Axis of symmetry: $x = 0.3$

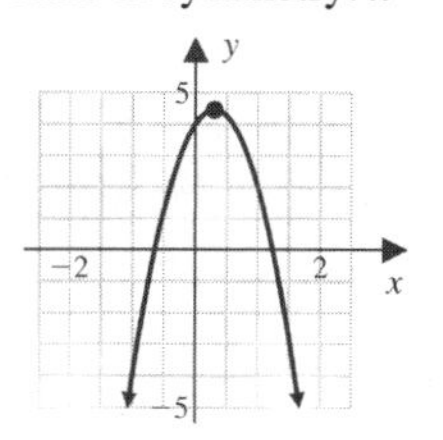

17. $f(x) = -\frac{1}{2}(x + 2)^2 + 3$;
Range: $(-\infty, 3]$;
Vertex: $(-2, 3)$;
Axis of symmetry: $x = -2$

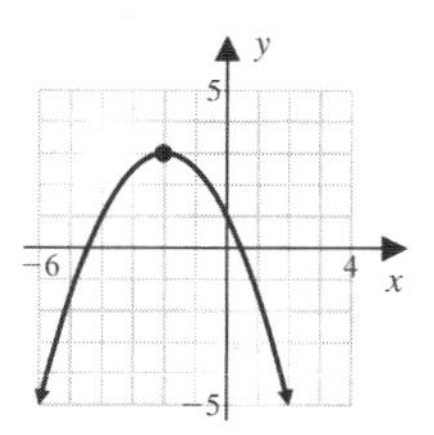

19. **(a)** $y = -x^2 + 4x$

(b) $y = \frac{2}{9}x^2 - 2$

21. $f(x) = \frac{1}{2}(x - 2)^2 - 2$

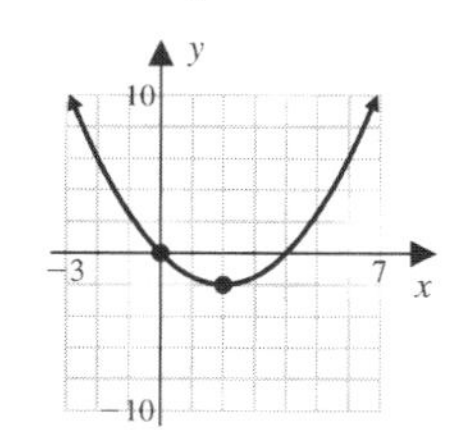

23. $f(x) = -x^2 + 1$

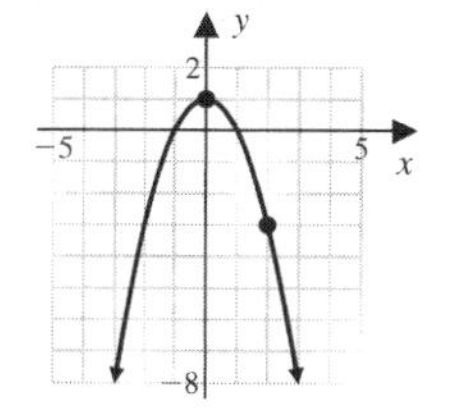

25. $(-\infty, -7)$ or $(2, \infty)$ **27.** $(-\infty, 1]$ or $\left[\frac{3}{2}, \infty\right)$ **29.** $\left(-\infty, -\frac{5}{2}\right]$ or $\left[\frac{4}{3}, \infty\right)$ **31.** Minimum $= -4$

33. Maximum $= -1$ **35.** Minimum $= -\frac{9}{8}$ **37.** $V = (-3, 4)$; $y = 4$; right **39.** $V = (-6, -3)$; $y = -3$; left

41. $V = (0, 3)$; $y = 3$; left **43.** $V = (-48, 2)$; $y = 2$; right **45. (a)** $P = 6y + y^2$ **(b)** -3 and 3; -9

47. (a) $P(5) = \$750$ thousand; $P(10) = \$1437.50$ thousand; $P(15) = \$2062.50$ thousand
(b) At 6250 employees, $P = \$4882.81$ (in thousands)
(c) As x increases, P increases to a maximum of \$4882.81 thousand, then decreases.

49. (a) $h(50) = 43.50$ ft; $h(100) = 63.50$ ft; $h(200) = 43.50$ ft
(b) Maximum height 66 ft (approx.); travels 125 ft (approx.) horizontally **(c)** 253.45 ft (approx.)

51. (a) 124.67 ft; 126.68 ft; 116.7 ft **(b)** Peak at approx. $x = 450.45$, $y = 126.76$ **(c)** 23.24 ft (approx.)

53. (a) $A = 500x - x^2$ **(b)** $0 \leq x \leq 500$ **(c)** 250 ft by 250 ft; 62,500 ft^2

55. 112.5 ft by 225 ft; maximum area is 25,312.5 ft^2

57. $y = 2$ ft, $x = 2$ ft

59. (b) $y = -7.355x^2 + 29.007x - 0.262$
(d) $x = 2.5, y \approx 26.3$
$x = 5, y \approx -39.1$
$x = 10, y \approx -445.7$

61. (b) $N = 3.471t^2 - 12.226t + 24.863$
(d) 2010; $t = 9, N \approx 196.0$
2015; $t = 14, N \approx 534.0$

63. (a) Two x intercepts **(b)** One x intercept **(c)** No x intercepts **65.** Two intercepts **67.** One intercept

PROBLEM SET 3.2

1. **(a)** Symmetric with respect to the y axis; resembles $y = x^2$ **(b)** Symmetric with respect to the origin; resembles $y = x^3$

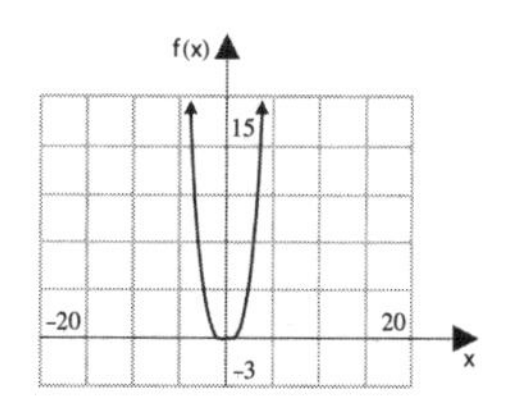

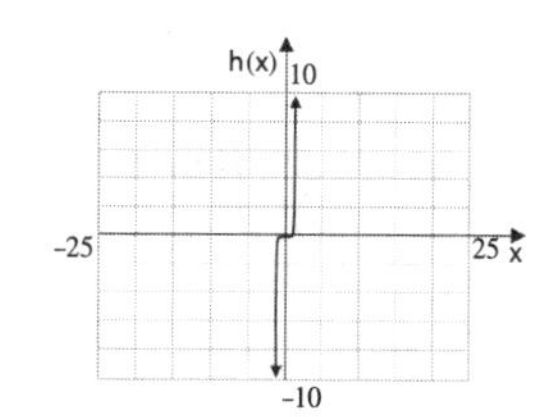

3.

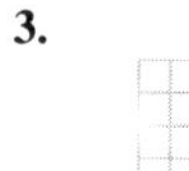

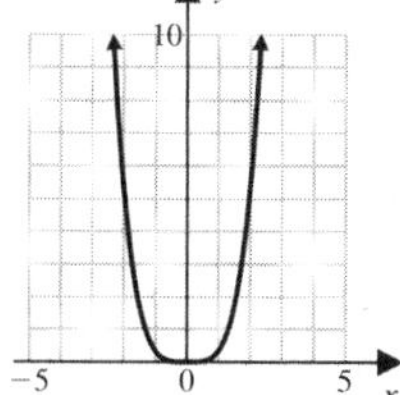

5.

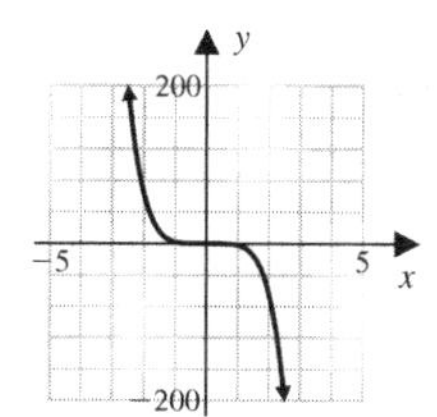

7.

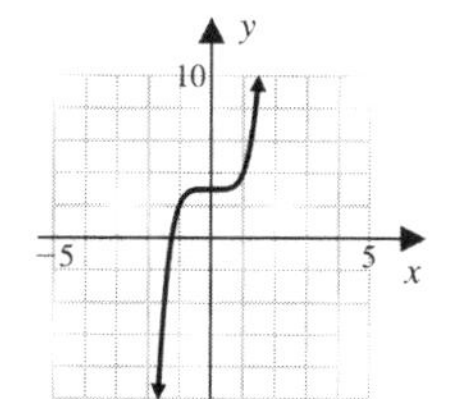

9.

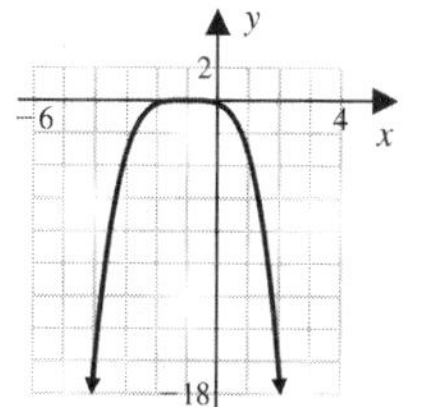

11.

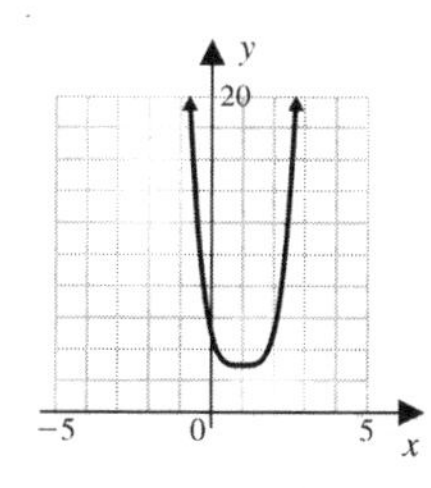

13.

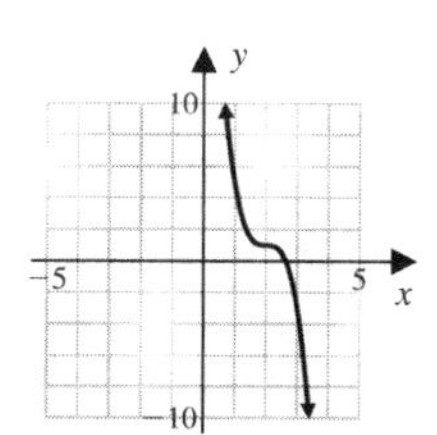

15. **(a)** $f(x) \to -\infty$ as $x \to +\infty$ **(b)** $f(x) \to -\infty$ as $x \to -\infty$ One turning point; a high point; continuous

17. **(a)** $f(x) \to +\infty$ as $x \to +\infty$ **(b)** $f(x) \to -\infty$ as $x \to -\infty$ No turning points; continuous

19. **(a)** $f(x) \to -\infty$ as $x \to +\infty$ **(b)** $f(x) \to +\infty$ as $x \to -\infty$ Two turning points; 1 low, 1 high point; continuous

21. **(a)** Model: $y = -x^3$; (A) **(b)** Model: $y = x^4$; (B)

23. **(a)** Does not show limit behavior as $x \to +\infty$ **(b)** Graph should be continuous.

25. Model: $y = x^4$
(a) $f(x) \to +\infty$ as $x \to +\infty$
(b) $f(x) \to +\infty$ as $x \to -\infty$
Three turning points;
1 peak, 2 valleys

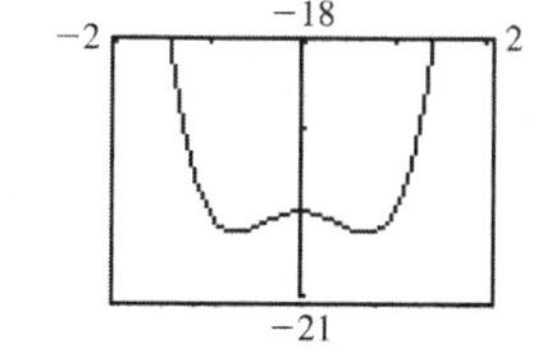

27. Model: $y = x^3$
(a) $f(x) \to +\infty$ as $x \to +\infty$
(b) $f(x) \to -\infty$ as $x \to -\infty$
Two turning points;
1 peak, 1 valley

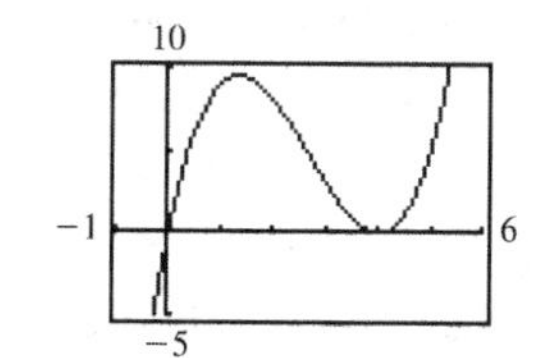

29. Model: $y = -x^4$
(a) $f(x) \to -\infty$ as $x \to +\infty$
(b) $f(x) \to -\infty$ as $x \to -\infty$
One turning point;
peak

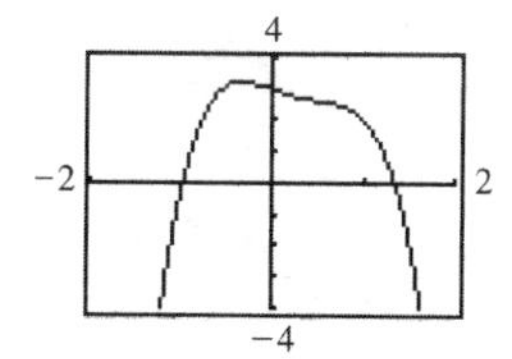

31. $-2, 0, 1$ zeros; $-2, 0, 1$; $(-2, 0)$ or $(1, \infty)$

33. $-3, 0, 4$; zeros $-3, 0, 4$; $(-\infty, -3)$ or $(0, 4)$

35. $1, \frac{5}{2}$; zeros $1, \frac{5}{2}$; $\left[\frac{5}{2}, \infty\right)$, and $x = 1$

37.

Real Zeros	Multiplicity	Graphical Behavior
-5	1 (odd)	crosses x axis at $x = -5$
-2	3 (odd)	crosses x axis at $x = -2$
5	2 (even)	touches x axis at $x = 5$ and bounces back

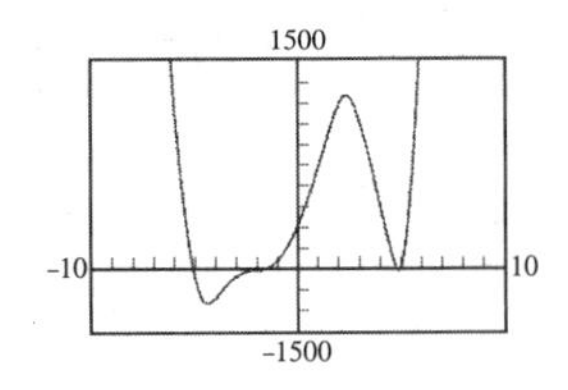

39.

Real Zeros	Multiplicity	Graphical Behavior
-3	2 (even)	touches x axis at $x = -3$ and bounces back
0	1 (odd)	crosses x axis at $x = 0$
4	3 (odd)	crosses x axis at $x = 4$

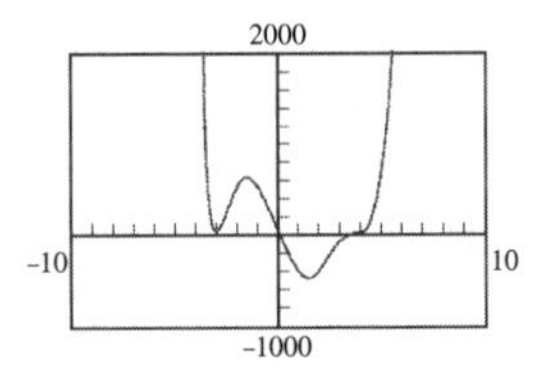

41. **(a)** 11:00 A.M. $f(3) = 7.2°F$
3:00 P.M. $f(7) = 26°F$
6:00 P.M. $f(10) = 12.8°F$

(b) $t = 2$, 10:00 P.M. and $t = 11$, 7:00 P.M.

(d) below zero between 8 P.M. and 10 P.M. and between 7 P.M. and 8 P.M. above zero between 10 P.M. and 7 P.M.

(c)

43. **(a)** $0 \le t \le 48$ **(c)** 3.82% at $t = 40$ **(d)** 6.80% at $t = 9$

45. **(a)** $V = x(20 - 2x)^2$

(b) $0 < x < 10$

(d) Maximum volume is 592.59 in.^{3} (approx.) when $x = 3.33$ in. (approx.)

47. **(a)** For $a > 0$ and $b > 0$: $f(x), g(x) \to +\infty$ as $x \to +\infty$; $f(x), g(x) \to +\infty$ as $x \to -\infty$.
For $a < 0$ and $b < 0$: $f(x), g(x) \to -\infty$ as $x \to +\infty$; $f(x), g(x) \to -\infty$ as $x \to -\infty$.

(b) For $a > 0$ and $b < 0$: $f(x) \to +\infty$ as $x \to +\infty$; $g(x) \to -\infty$ as $x \to +\infty$; $f(x) \to +\infty$ as $x \to -\infty$; $g(x) \to -\infty$ as $x \to -\infty$
For $a < 0$ and $b > 0$: $f(x) \to -\infty$ as $x \to +\infty$; $g(x) \to +\infty$; as $f(x) \to -\infty$ as $x \to -\infty$ $g(x) \to +\infty$ as $x \to -\infty$

49. **(a)** Concave down on $(-\infty, 0)$; concave up on $(0, \infty)$; point of inflection at $(0, 1)$

(b) Concave down on $(4, \infty)$; concave up on $(-\infty, 4)$; point of inflection at $(4, 1)$

51. $n > 1$

53.

55. They both approach $-\infty$ as $x \to -\infty$ and as $x \to +\infty$.
$R(x) \to 1$ as $x \to -\infty$ and as $x \to +\infty$.

57. **(b)** $y = 0.0018x^3 - 0.1795x^2 + 6.6212x - 43.8355$ **(d)** 40; 45; 55

PROBLEM SET 3.3

1. **(a)** -14 **(b)** 63 **(c)** 241 **3.** **(a)** Yes; $(x + 2)(x + 4)(x - 6)$ **(b)** Yes; $(x + 4)(x - 6)(x + 2)$ **(c)** No **(d)** Yes; $(x-6)(x+4)(x + 2)$ **5.** **(a)** $f(x) = (x - 1)(x - 2)(x - 3)$
7. **(a)** $f(x) = (x + 2)(2x^3 - 4x^2 + 8x - 13)$ **9.** $f(3) = 0$ **11.** $f(x) = -(x + 3)(x - 1)(x - 3)$
13. $-3, -2, 0, 2$; All are x intercepts. **15.** $-1, 2, -i, i$; x intercepts: $-1, 2$
17. $-2, 1, 2; f(x) = (x - 1)(x - 2)(x + 2)$ **19.** $1; h(x) = (x - 1)(5x^2 - 7x + 10)$
21. $-1, 1, 2; f(x) = (x + 1)(x - 1)^2(x - 2)^2$ **23.** -2 **25.** -5 and 5

27. $-2, 3, -\frac{1}{2}, \frac{1}{2}; f(x) = \frac{1}{4}(x + 2)(x - 3)(2x + 1)(2x - 1)$ **29.** $-1, \frac{2}{5}, \frac{1}{2}; f(x) = \frac{2}{3}(x + 1)\left(x - \frac{2}{5}\right)\left(x - \frac{1}{2}\right)$

31. $\frac{-1}{2}, -2\sqrt{3}, 2\sqrt{3}$ **33.** $-1, -\frac{1}{2}, 1, \frac{3}{2}; h(x) = 4(x + 1)\left(x + \frac{1}{2}\right)(x - 1)\left(x - \frac{3}{2}\right)$

35. $-1, 4; f(x) = (x + 1)(x - 4)(x^2 + 4)$ **37.** $-\frac{1}{2}; f(x) = \left(x + \frac{1}{2}\right)(x^2 - 4x + 5)$

39. $f(1) = -3, f(3) = 7$ **41.** $f(-3) = -42, f(-2) = 5$ **43.** $f(0.9) = 7.7401, f(2.1) = -7.5959$
45. $c = 0.3477$ **47.** $c = -1.1547$ **49.** $c = 4.7958$ **51.** $c = -1, 2; (2, \infty)$

53. $c = -1.4142; [-1.41, -1]$ or $[1, 4]$ **55.** **(a)** $V = 21\pi x^2 - \frac{\pi}{3}x^3$; **(b)** $0 < x \le 40$; **(c)** $x = 36$ ft.

57. **(a)** $V = 30\pi r^2 - \frac{\pi}{3}r^3$ **(b)** $0 < r \le 30$ **(c)** 6 ft. **59.** 1.83%; 4.13%; 10%

61. **(a)** 4200 ft 525 ft **(c)** $0 < x \le 525$ **(d)** Approx. 8.88 ft

63. False **65.** True **67.** False

PROBLEM SET 3.4

1. **(a)** Vertical asymptote:
$x = 0$;
Horizontal asymptote:
$y = 0$

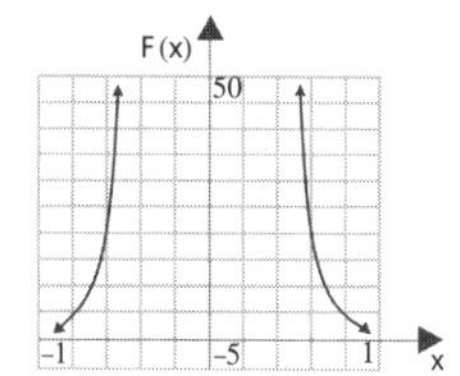

(b) Vertical asymptote:
$x = 0$;
Horizontal asymptote:
$y = 0$

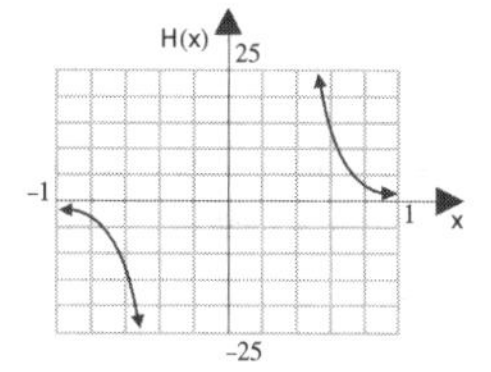

3. Vertical asymptote:
$x = 0$;
Horizontal asymptote: $y = -4$.
As $x \to 0$, $y \to +\infty$.
As $x \to +\infty$ and as
$x \to -\infty$, $y \to -4$.

5. Vertical asymptotes:
$x = -2$; and $x = 2$;
Horizontal asymptote:
$y = -0$.
As $x \to -2^-$, $y \to +\infty$;
as $x \to -2^+$, $y \to -\infty$;
As $x \to 2^-$, $y \to +\infty$;
as $x \to -2^+$, $y \to -\infty$;
As $x \to \infty$, $y \to 0$;
as $x \to -\infty$, $y \to 0$.

7. **(a)** Vertically stretch $1/x$ by 4,
shift to the left 3 units.
Vertical asymptote: $x = -3$;
Horizontal asymptote: $y = 0$

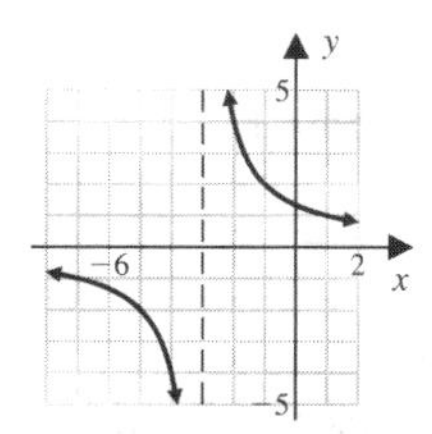

(b) Shift $1/x$ up 3 units.
Vertical asymptote: $x = 0$;
Horizontal asymptote: $y = 3$

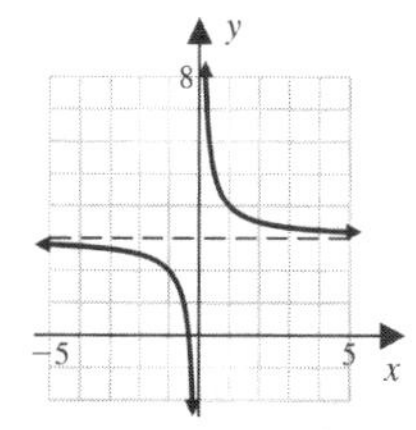

9. **(a)** Vertically stretch $1/x^2$ by 3,
reflect about the x axis,
and shift to left 1 unit.
Vertical asymptote: $x = -1$;
Horizontal asymptote: $y = 0$

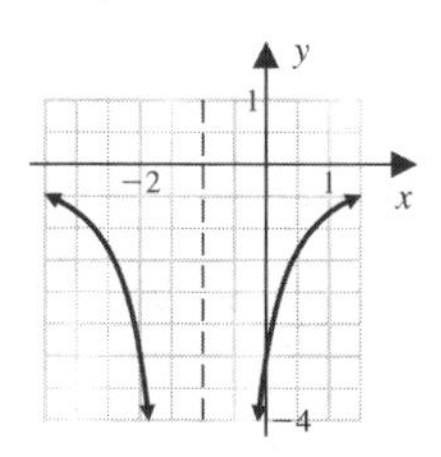

(b) Vertically stretch $1/x$ by 2, shift to the right 1 unit and up 3 units.
Vertical asymptote: $x = 1$;
Horizontal asymptote: $y = 3$

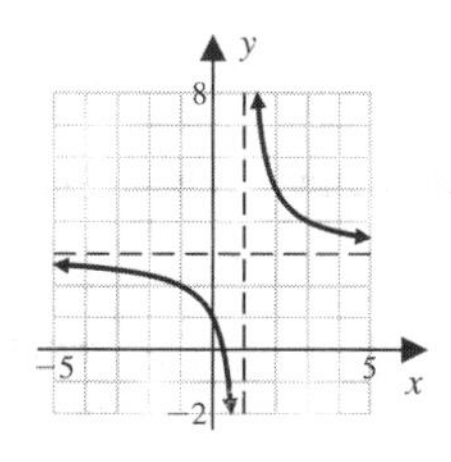

11. **(a)** Vertically stretch $1/x^2$ by 4, shift to the left 3 units, then down 1 unit.
Vertical asymptote: $x = -3$;
Horizontal asymptote: $y = -1$

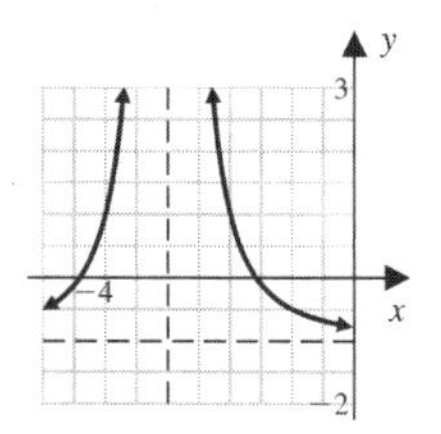

(b) Shift $1/x^2$ to the left 4 units,
then up 2 units.
Vertical asymptote: $x = -4$;
Horizontal asymptote: $y = 2$

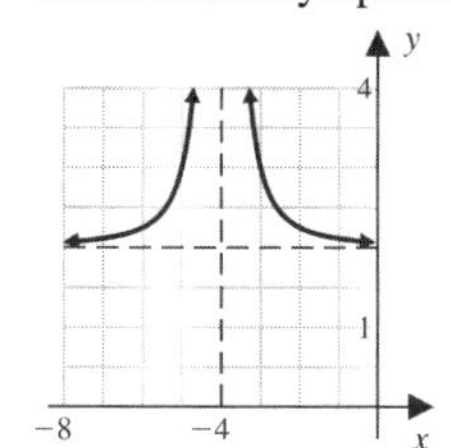

13. **(a)** No vertical asymptotes; hole at (1, 2)

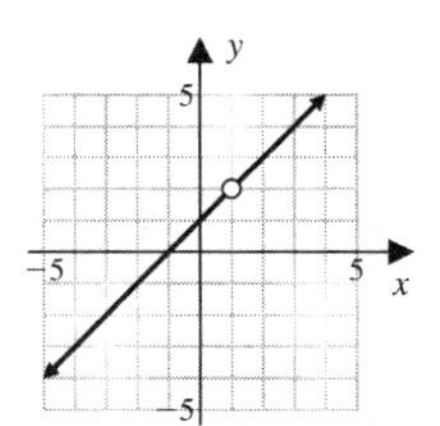

(b) No vertical asymptotes; hole at (3, 4)

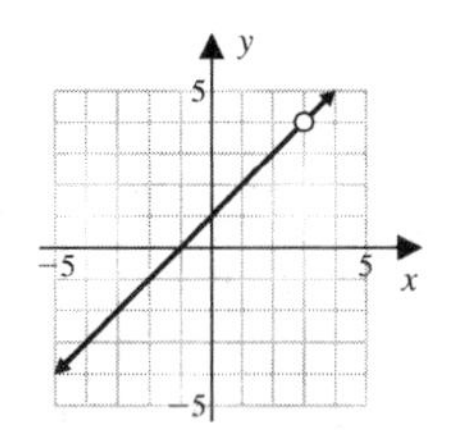

15. **(a)** Horizontal asymptote: $y = -2$

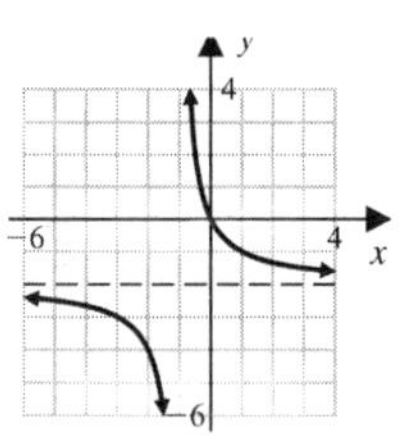

(b) Horizontal asymptote: $y = \frac{3}{2}$

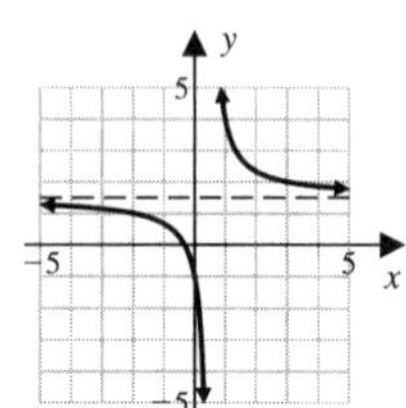

17. **(a)** C **(b)** Vertical asymptotes: $x = -1$ and $x = 0$; horizontal asymptote: $y = 0$

19. **(a)** d **(b)** Vertical asymptote: $x = -1$; horizontal asymptote: $y = 2$

21.

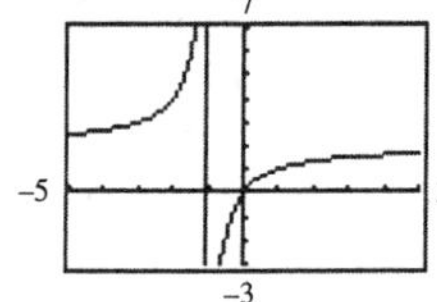

Vertical asymptote: $x = -1$ horizontal asymptote: $y = 2$

23.

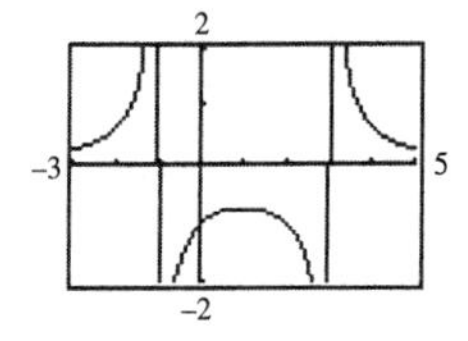

Vertical asymptotes: $x = -1, x = 3$ horizontal asymptote: $y = 0$

25.

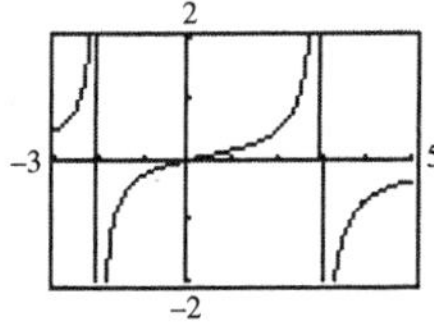

Vertical asymptote: $x = -2, x = 3$ horizontal asymptote: $y = 0$

27.

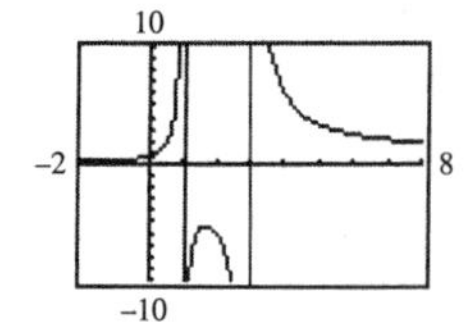

Vertical asymptotes: $x = 1, x = 3$ horizontal asymptote: $y = 1$

29. $(-1, 0)$ **31.** $(-\infty, -3)$ or $(1, \infty)$ **33.** $(-2, 0)$ or $(3, \infty)$ **35.** $[1, 3]$

37. **(a)** $x \leq 0$ **(b)** Horizontal asymptote: $S(x) = 5000$

(c) As expenditures increase, sales approach 5000 units. **39.** **(b)** As $x \to 60\%$, the cost approaches ∞.

41. **(a)** $A = 2x^2 + \frac{2000}{x}, x > 0$ **(b)** $x \approx 7.94$ cm

$h \approx 7.94$ cm

minimum $A \approx 377.98$ cm^2

43. $-\frac{1}{2x}, x \neq 2$ **45.** $f(x) = \frac{2(x-1)(x^2-9)}{x^3+4x}$

47. If k is in the domain of f, then $f(k)$ would have one defined value and $f(x)$ would not approach $+\infty$ or $-\infty$.

49. Oblique asymptote: Oblique asymptote: $y = \frac{1}{2}x - 1$

51. **(a)** Oblique asymptote $y = -4x + 8$

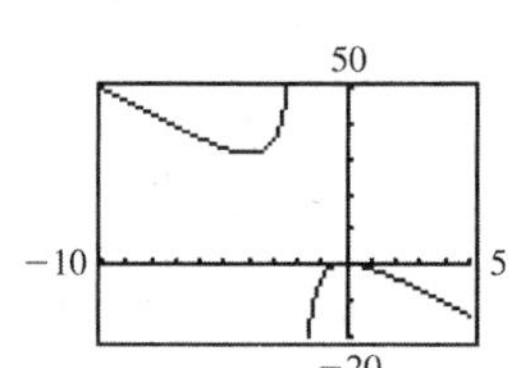

(b) Oblique asymptote $y = x - 3$

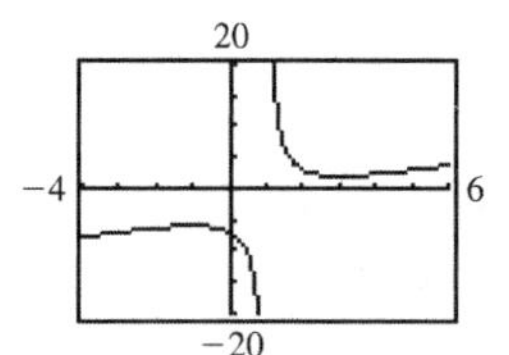

4 Exponential and Logarithmic Functions

CHAPTER CONTENTS

On December 26, 2003, there was an earthquake in Bam, Iran that caused considerable damage, and 20,000 people were killed from its effect. Its magnitude on the Richter scale was 6.5. Compare the intensity of this earthquake to the intensity of the Northern California earthquake in 1989, whose magnitude was 7.1. The solution of this problem is shown in Example 10 on p. 289.

Up to now we have studied algebraic functions formed by sums, differences, products, and quotients of polynomial functions. In this chapter, we introduce two types of functions that are closely related—*exponential functions* and the *inverses* of exponential functions—*logarithmic functions.* These functions belong to a classification of functions called *transcendental* functions.

We will examine their properties, graphs, and uses in a variety of applications. As we shall see, the exponential functions are especially suitable for developing mathematical models involving population growth or decline over time.

OBJECTIVES

1. Define Inverse Functions
2. Recognize the Symmetry of the Graphs of a Function and Its Inverse
3. Determine Whether a Function Has an Inverse
4. Find the Inverse of a Function
5. Solve Applied Problems

4.1 Inverse Functions

Our goal in this section is to develop the general notion of inverse functions. Later in this chapter we use the idea of inverse functions to introduce logarithmic functions.

Defining Inverse Functions

Let us begin by examining the composition of two functions

$$f(x) = 5x \quad \text{and} \quad g(x) = \frac{x}{5}$$

$$\begin{aligned} \text{to get } (f \circ g)(x) &= f[g(x)] \\ &= f\left(\frac{x}{5}\right) = 5 \cdot \frac{x}{5} \\ &= x \\ \text{and } (g \circ f)(x) &= g[f(x)] \\ &= g(5x) = \frac{5x}{5} = x \\ &= x \end{aligned}$$

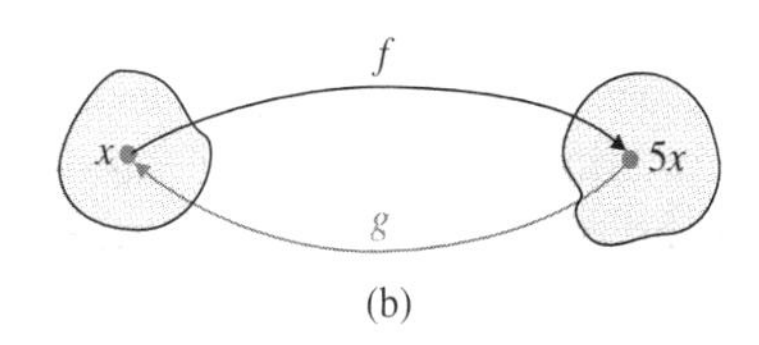

Figure 1

Thus, in this situation, for $f \circ g$, f "undoes" the output of function g, $x/5$, and produces a final output of x. The final output is the same as the original input (Figure 1a).

Similarly, for $g \circ f$, g "undoes" the output of function f and returns x as the final output (Figure 1b).

Two functions related in such a way are said to be *inverses* of each other.

> **Definition Inverse Functions**
>
> Two functions f and g are **inverses** of each other if and only if
>
> $$(f \circ g)(x) = f[g(x)] = x$$
>
> for every value of x in the domain of g, and
>
> $$(g \circ f)(x) = g[f(x)] = x$$
>
> for every value of x in the domain of f.

A function f is said to be **invertible** if such a function g exists.

EXAMPLE 1 Verifying that Two Functions Are Inverses of Each Other

Show that the functions $f(x) = 3x + 2$ and $g(x) = (1/3)x - 2/3$ are inverses of each other.

Not every function has an inverse.

Solution From the definition, we must verify that $f[g(x)] = x$ and $g[f(x)] = x$ We have

$$f[g(x)] = f\left(\frac{1}{3}x - \frac{2}{3}\right)$$
$$= 3\left(\frac{1}{3}x - \frac{2}{3}\right) + 2 = (x - 2) + 2 = x$$
$$g[f(x)] = g(3x + 2)$$
$$= \frac{1}{3}(3x + 2) - \frac{2}{3} = \left(x + \frac{2}{3}\right) - \frac{2}{3} = x$$

for all x. Thus, we conclude that f and g are indeed inverses of each other. ❖

Care must be taken not to confuse $y = f^{-1}(x)$, the inverse function of f, with $[f(x)]^{-1} = 1/f(x)$, the reciprocal of f(x).

When f and g are inverses of each other, we refer to g as the *inverse function of f,* and vice versa, and we write

$$g = f^{-1} \quad \text{or} \quad g^{-1} = f.$$

Thus, in Example 1, we could write either

$$f(x) = 3x + 2 \quad \text{and} \quad f^{-1}(x) = \frac{1}{3}x - \frac{2}{3}$$

or

$$g(x) = \frac{1}{3}x - \frac{2}{3} \quad \text{and} \quad g^{-1}(x) = 3x + 2.$$

Notice that the conditions stated in the definition of an inverse function can be restated as follows:

$$f^{-1}[f(x)] = x \quad \text{for every } x \text{ in the domain of } f$$
$$\text{and } f[f^{-1}(x)] = x \quad \text{for every } x \text{ in the domain of } f^{-1}$$

In what follows, we shall address the following three issues related to inverse functions:

1. How are the graphs of a function and its inverse related?
2. How can we determine whether a function has an inverse?
3. How do we find the inverse of a function if it exists?

Recognizing the Symmetry of the Graphs of a Function and Its Inverse

To understand the relationship between the graphs of a function and its inverse, let us consider the two functions in Example 1

$$f(x) = 3x + 2$$

and

$$f^{-1}(x) = \frac{1}{3}x - \frac{2}{3}.$$

Tables 1a and 1b are used to graph f and f^{-1} on the same coordinate system (Figure 2). The graph of the line $y = x$ is also shown in Figure 2.

TABLE 1

f:

x	y
0	2
1	5
−1	−1
−2	−4

(a)

f^{-1}:

x	y
2	0
5	1
−1	−1
−4	−2

(b)

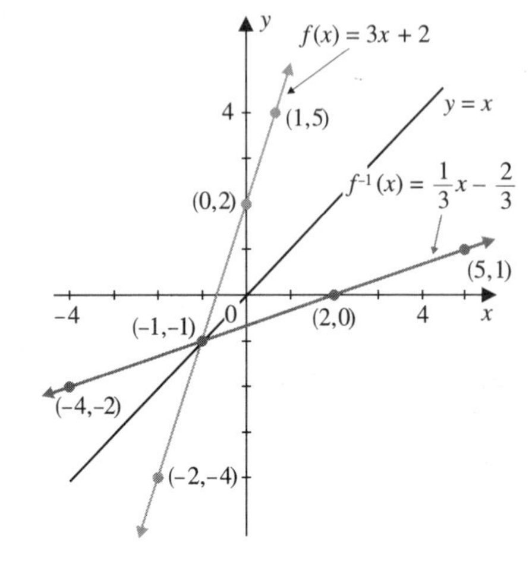

Figure 2

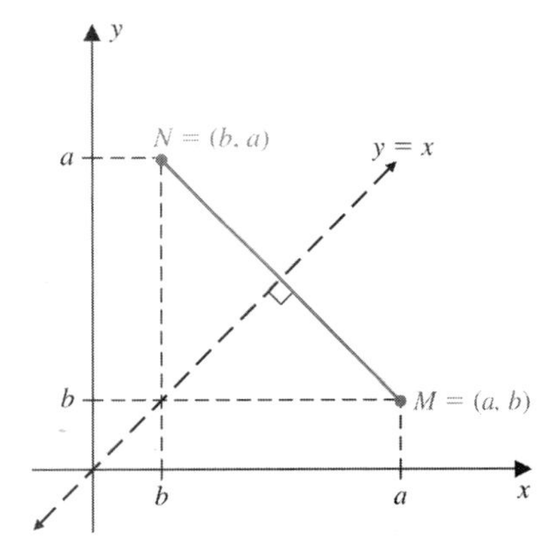

Figure 3

Observe from Table 1 the point (1, 5) on the graph of f, and the point (5, 1) on the graph of f^{-1}. Now notice the interchanging of x and y on the graphs of f and f^{-1}. That is, $f(1) = 5$ and $f^{-1}(5) = 1$. In other words, we get the point (5, 1) from the point (1, 5) by reflecting across the line $y = x$ (Figure 2). In general, as Figure 3 illustrates, the points $M = (a, b)$ and $N = (b, a)$ are symmetric across the line $y = x$.

This observation leads to the following result:

Property Graphs of f and f^{-1}

The graphs of f and f^{-1} are reflections of each other about the line $y = x$, as illustrated in Figure 4.

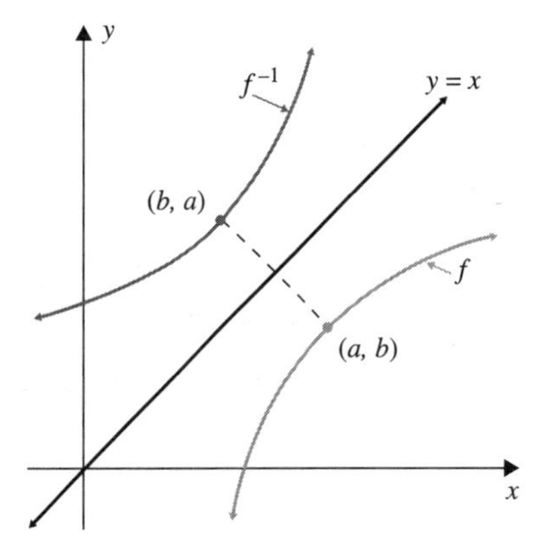

Figure 4

EXAMPLE 2 Graphing f^{-1} From the Graph of f

Use the graph of f given in Figure 5a to sketch the graph of f^{-1}. Identify the domain and range of f and f^{-1}.

Solution Because the graphs of f and f^{-1} are symmetric about the line $y = x$, the graph of f^{-1} is obtained by reflecting the graph of f about the line $y = x$ (Figure 5b).

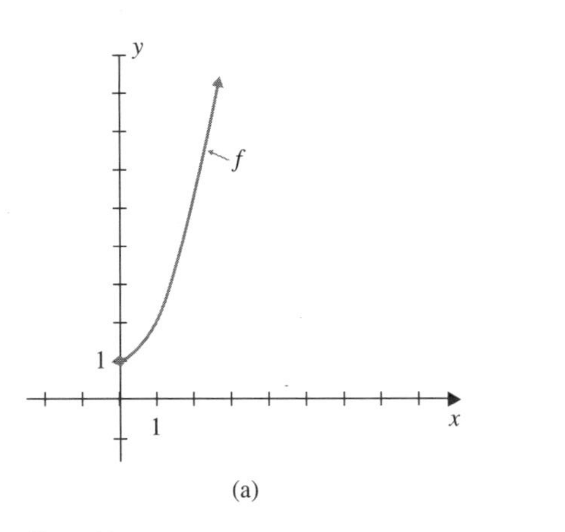

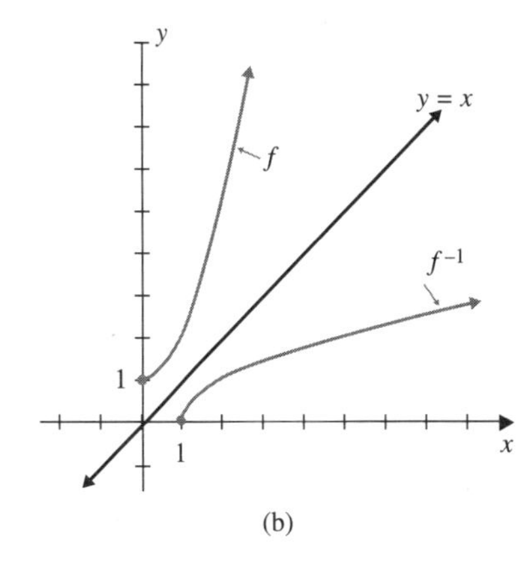

Figure 5

From the graphs in Figure 5, we observe that the domain of f is the interval $[0, \infty)$ and its range is the interval $[1, \infty)$. Also, we see that the domain of f^{-1} is the interval $[1, \infty)$ and its range is the interval $[0, \infty)$. In other words, the domain and range of f are, respectively, the range and domain of f^{-1}. ❖

This example illustrates the following property:

Property Domain and Range of f and f^{-1}

The domain of f is the same as the range of f^{-1}, and the range of f is the same as the domain of f^{-1}.

Determining Whether a Function Has an Inverse

There are two ways of determining whether a given function has an inverse:

(1) graphically and (2) analytically.

1. Determining Graphically Whether a Function Has an Inverse

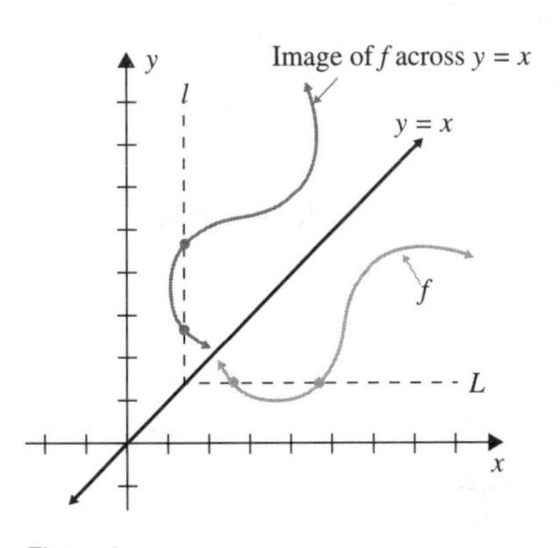

Figure 6

We can use the graph of a function to determine whether it has an inverse. Consider the graph of the function f in Figure 6. The reflected image of the graph of f about the line $y = x$ is not the graph of a function because we can draw a vertical line l that intersects the graph more than once. Thus, f cannot have an inverse. Notice that the horizontal line L, obtained by reflecting l across the line $y = x$, intersects the graph of f more than once.

This observation provides the basis for using the graph of a function to determine whether it has an inverse.

Property Horizontal-Line Test

A function f has an inverse if and only if no horizontal straight line intersects its graph more than once.

EXAMPLE 3 Determining Graphically Whether a Function Has an Inverse

Use the horizontal-line test to determine whether each function has an inverse.

(a) $f(x) = \sqrt{x}$ (b) $g(x) = x^2$

Solution The graphs of f and g are shown in Figures 7a and 7b, respectively.

(a) No horizontal line intersects the graph of f more than once (Figure 7a), so f has an inverse.
(b) Any horizontal line drawn above the x axis will intersect the graph of g twice (Figure 7b), so g does not have an inverse.

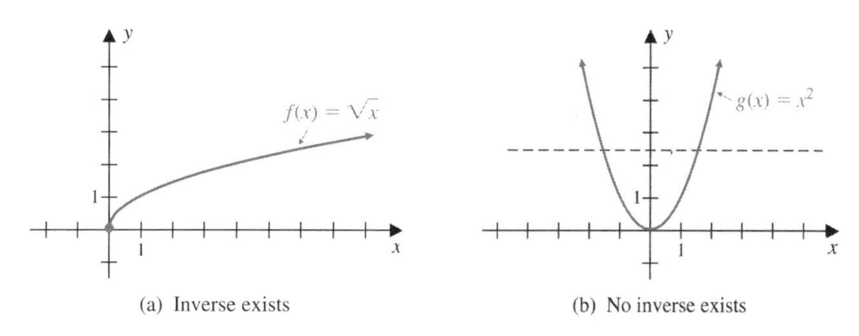

(a) Inverse exists (b) No inverse exists

Figure 7

2. Determining Analytically Whether a Function Has an Inverse

Let us review two characteristics of a function $y = f(x)$ that has an inverse:

(i) The graph of the function satisfies the vertical-line test. Thus for every input value x there is one and only one output value for y.
(ii) Because the function has an inverse, the graph must also satisfy the horizontal-line test. This means that every output value y has one and only one corresponding input value for x.

It follows that a function has an inverse if and only if each distinct input value always results in a distinct output value. A function that has this property is called a *one-to-one function.*

Definition One-to-One Function

A function f is said to be **one-to-one** if, whenever a and b are in the domain of f and $f(a) = f(b)$, it follows that $a = b$.

EXAMPLE 4 Determining One-to-One Functions

Use the definition of a one-to-one function to determine whether each function is one-to-one. Graph each function and use the graph to demonstrate the validity of the result.

(a) $f(x) = 3x - 2$
(b) $g(x) = 4x^2$

Solution (a) We assume that $f(a) = f(b)$ for some numbers a and b in the domain of f. According to the definition, we must show that $a = b$. We proceed as follows:

$f(a) = f(b)$	Assumption
$3a - 2 = 3b - 2$	Evaluate $f(a)$ and $f(b)$
$3a = 3b$	Add 2 to each side
$a = b$	Divide each side by 3

Hence, f is one-to-one. So f has an inverse. The horizontal-line test confirms that f is one-to-one (Figure 8a).

(b) Here we select $x = -1$ and $x = 1$, and observe that

$$g(-1) = 4(-1)^2 = 4 \quad \text{and} \quad g(1) = 4 \cdot 1^2 = 4.$$

Hence, $g(-1) = g(1)$ but $-1 \neq 1$, so g is not one-to-one. Thus g does not have an inverse. The horizontal-line test confirms that g is not one-to-one (Figure 8b).

The selection of specific numbers to show that a general property does not hold, as in Example 4b, illustrates a method of proof called proof by counterexample.

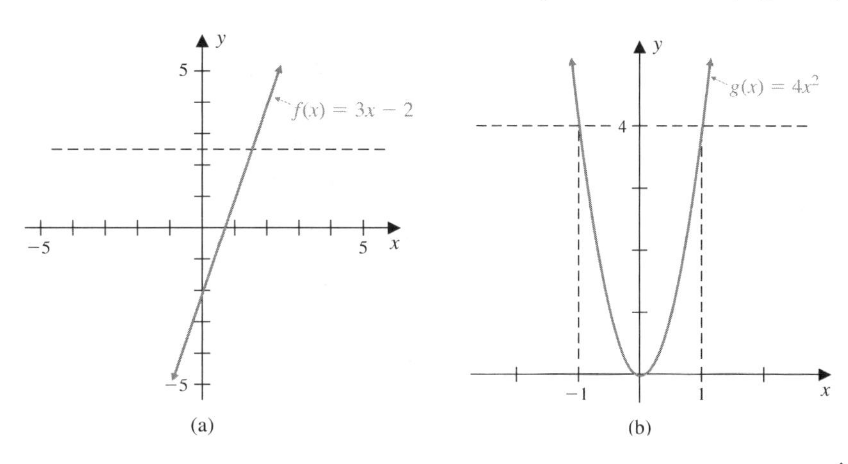

Figure 8 ❖

Note that functions that are either increasing or decreasing throughout their domains are one-to-one functions, and so they have inverses.

Finding the Inverse of a Function

We now turn our attention to answering the question:

If a function f has an inverse, how do we find the inverse function f^{-1}?

Earlier we discovered that whenever (a, b) is on the graph of f, then (b, a) is on the graph of f^{-1}. This means that f^{-1} is the function that can be formed by interchanging the roles of the domain and range values of f.

TABLE 2

Function f

Domain	Range
3	0
1	−4
5	2
7	11
23	−6

(a)

Function f^{-1}

Domain	Range
0	3
−4	1
2	5
11	7
−6	23

(b)

For instance, consider the function f defined by Table 2a. Since f is one-to-one, it has an inverse f^{-1}, which is formed by reversing the pairings shown in Table 2a to obtain the function f^{-1} defined by Table 2b.

If f is a function defined by the equation $y = f(x)$ and f has an inverse f^{-1}, then we have:

$y = f(x)$	y is expressed in terms of x
$f^{-1}(y) = f^{-1}[f(x)]$	Apply f^{-1} to each side
$f^{-1}(y) = x$	x is expressed in terms of y

This observation provides the basis for the following procedure, which enables us to find the equation for f^{-1} if f is defined by an equation.

Procedure Method for Finding f^{-1}

Step 1. Write the equation $y = f(x)$ that defines the one-to-one function f.

Step 2. Solve the equation $y = f(x)$ for x in terms of y, to obtain an equation for the inverse function $x = f^{-1}(y)$.

Step 3. Verify that the domain and range of f are, respectively, the same as the range and domain of f^{-1}.

EXAMPLE 5 Finding the Inverse of a Function

Let $f(x) = 2x - 3$.

(a) Use the above method to find $f^{-1}(x)$.

(b) Graph f and f^{-1} on the same coordinate system.

Solution The function

$$f(x) = 2x - 3$$

is a linear function. Since the slope is $m = 2$, the graph of f is increasing for all real numbers, so f is one-to-one and thus f^{-1} exists.

(a) To find f^{-1}, we proceed as follows:

Step 1. Let $y = f(x) = 2x - 3$.

Step 2. Solve the equation for x in terms of y as follows:

$$y = 2x - 3$$

$$x = \frac{y + 3}{2}$$

So

$$f^{-1}(y) = \frac{y + 3}{2}.$$

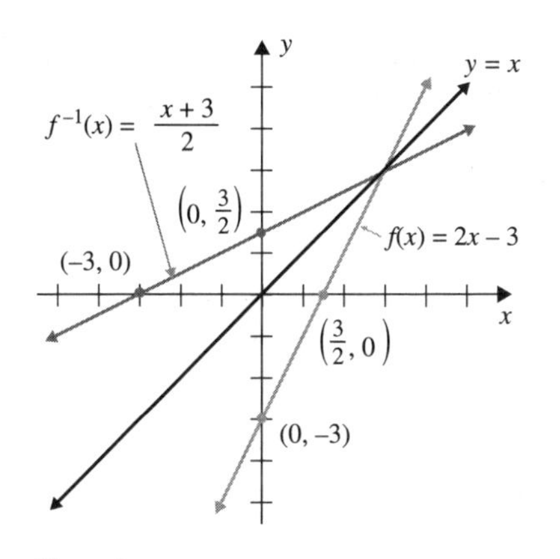

Figure 9

Following the convention of using x to represent the independent variable, we rewrite the latter equation as

$$f^{-1}(x) = \frac{x+3}{2}.$$

Step 3. The domain and range of f consist of all real numbers, which is the same for f^{-1}.

(b) The graphs of f and f^{-1} are shown in Figure 9. As expected, they are symmetric about the line $y = x$. ❖

Even though two functions have the same defining equation, one may have an inverse and the other may not. The next example illustrates the situation.

EXAMPLE 6 Comparing Two Functions with the Same Defining Equation

Let

$$f(x) = x^2 + 1$$

and

$$g(x) = x^2 + 1, \text{ where } x \geq 0.$$

(a) Use the graphs of f and g to explain why f does not have an inverse and g does have an inverse.

(b) Find $g^{-1}(x)$.

Solution (a) By inspecting the graph of f in Figure 10a, we see that it does not satisfy the horizontal-line test, so f does not have an inverse. On the other hand, the graph of g (Figure 10b) does satisfy the horizontal-line test, so g has an inverse function. Notice that both functions have the same defining equation, but g has a restricted domain that allows it to have an inverse.

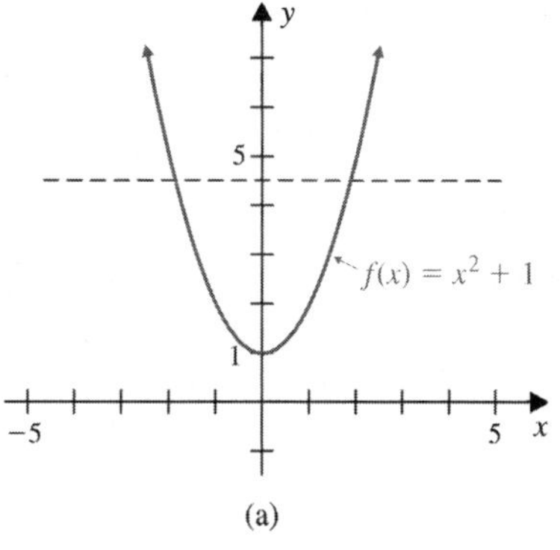

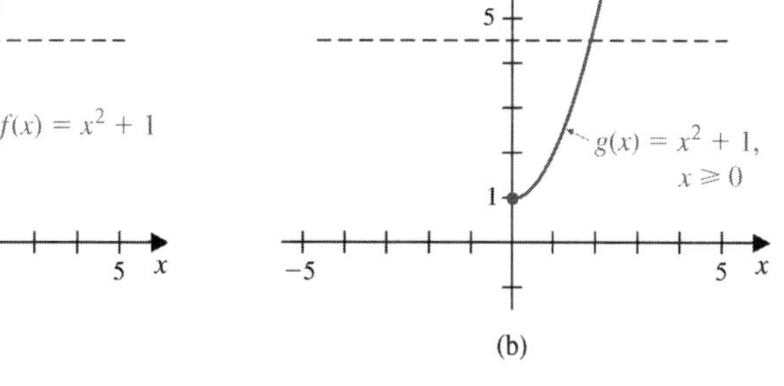

Figure 10

(b) We obtain an equation for the inverse function g^{-1} as follows:

Step 1. Let $y = x^2 + 1$, where $x \geq 0$

Step 2. Solve for x:

$$x^2 = y - 1$$

$$x = \sqrt{y-1} \quad x \geq 0 \text{ is a given restriction}$$

Thus $g^{-1}(y) = \sqrt{y-1}$.

By changing the variable we get $g^{-1}(x) = \sqrt{x-1}$

Step 3. The domain of g and the range of g^{-1} are the same, the interval $[0, \infty)$. Also, the range of g is the same as the domain of g^{-1}, the interval $[1, \infty)$. ❖

It should be noticed that at times we can find the inverse of the function

$$g(x) = x^2 + 1, x \geq 0$$

by inspecting the formation of its equation.

The function $g(x) = x^2 + 1$ is a composition of squaring x; and adding 1. The inverse function formed by "reversing" the composition of g by subtracting 1, then taking the square root to obtain

$$g^{-1}(x) = \sqrt{x-1}.$$

Solving Applied Problems

The next example shows an application of inverse functions.

EXAMPLE 7 Modeling an Insurance Premium Discount

Insurance companies reward students who have good grades and good driving records by discounting their automobile insurance premiums. Suppose that a student who qualifies is awarded a 20% discount in premiums.

(a) Find a function f that expresses the reduced premium R in terms of the original premium P.

(b) Find the inverse function f^{-1} of part (a) and interpret it in terms of the reduced and original premiums.

Solution (a) A 20% reduction in the original premium means the new premium is 80% of the original one. Therefore, the function f that expresses the reduced premium R in terms of the original premium P is given by

$$R = f(P) = 0.80P.$$

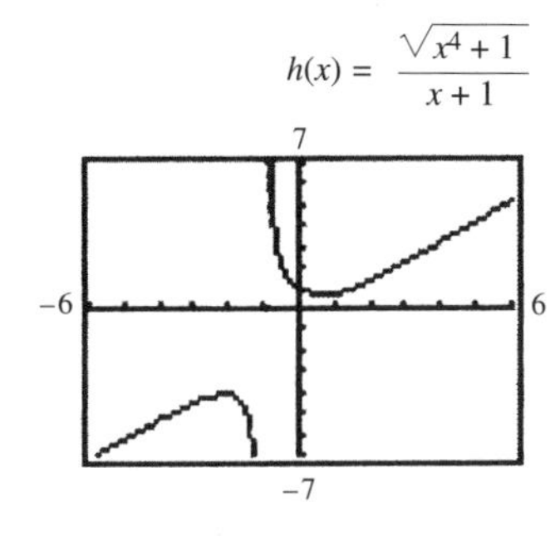

Figure 11

(b) The inverse function f^{-1} of $f(P) = 0.80P$ is obtained by solving the equation $R = 0.80P$ for P in terms of R to get

$$P = \frac{1}{0.80}R = 1.25R.$$

So

$$P = f^{-1}(R) = 1.25R.$$

This means that a 25% increase in the reduced premium is needed to obtain the original premium. ❖

Graphs provide us with an efficient way of determining whether a function has an inverse. If the function is difficult to graph by hand, then a grapher can help.

For instance, the graph of $h(x)\frac{\sqrt{x^4+1}}{x+1}$ (Figure 11) shows that the horizontal-line test fails. Therefore, h does not have an inverse.

PROBLEM SET 4.1

Mastering the Concepts

In problems 1–6, show that the functions f and g are inverses of each other by verifying that $f[g(x)] = x$ and $g[f(x)] = x$. Sketch the graphs of f and g on the same coordinate system to show that their graphs are symmetric about the line $y = x$.

1. $f(x) = 7x - 2$; $g(x) = \dfrac{x}{7} + \dfrac{2}{7}$

2. $f(x) = 1 - 5x$; $g(x) = \dfrac{1}{5} - \dfrac{x}{5}$

3. $f(x) = x^4$, where $x \geq 0$; $g(x) = \sqrt[4]{x}$

4. $f(x) = x^3$, $g(x) = \sqrt[4]{x}$

5. $f(x) = \dfrac{1}{x-2}$; $g(x) = \dfrac{1}{x} + 2$

6. $f(x) = \sqrt{x+3}$; $g(x) = x^2 - 3$, where $x \geq 0$

In problems 7 and 8, use the given graph of the one-to-one function f to do the following:

(i) Sketch the graph of f^{-1} by reflecting the graph of f about the line $y = x$.

(ii) Determine and compare the domains and ranges of f and f^{-1}.

7. **(a)**

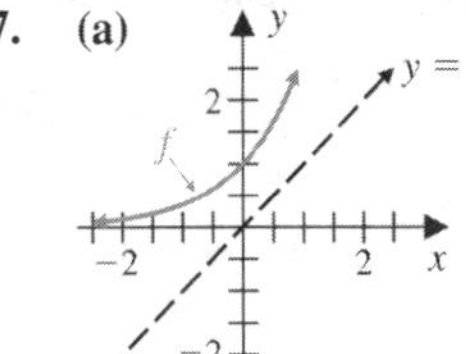

(b)

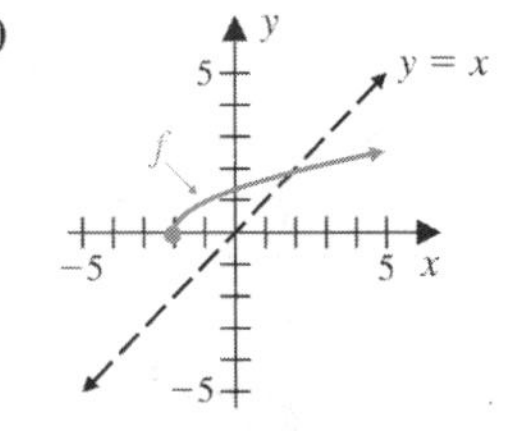

8. **(a)**

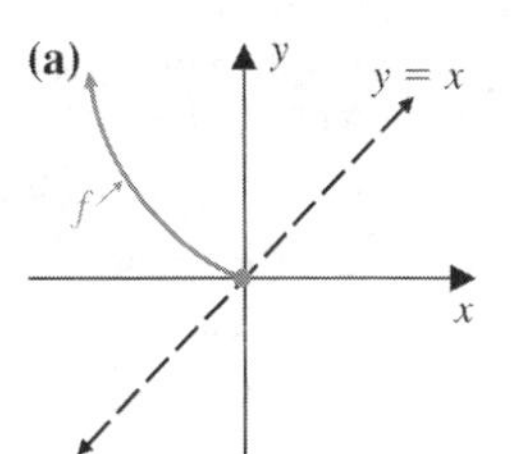

(b)

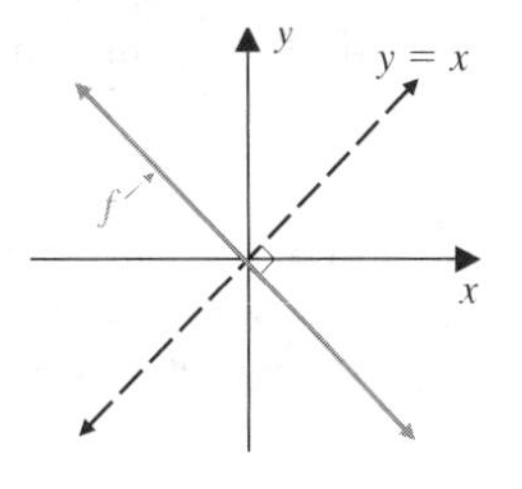

In problems 9 and 10, use the horizontal-line test to determine whether each of the functions whose graph is given has an inverse.

9. **(a)**

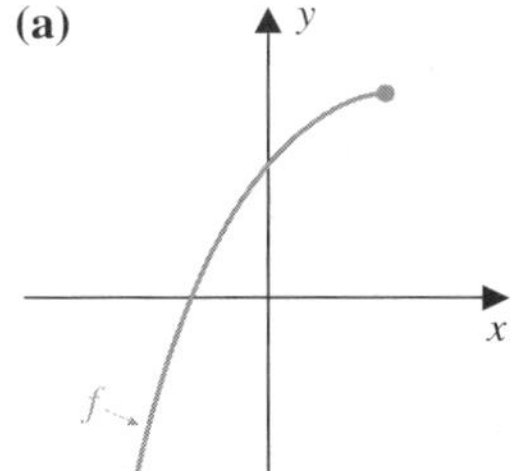

(b)

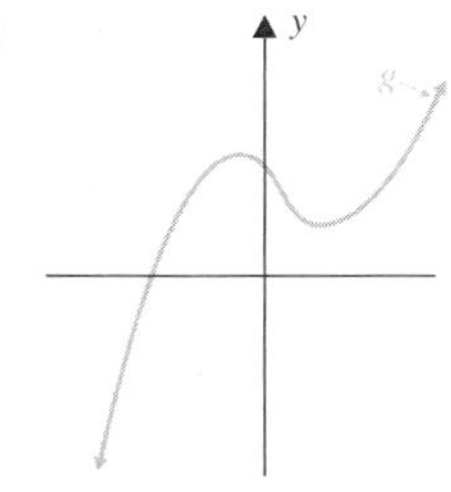

10. **(a)**

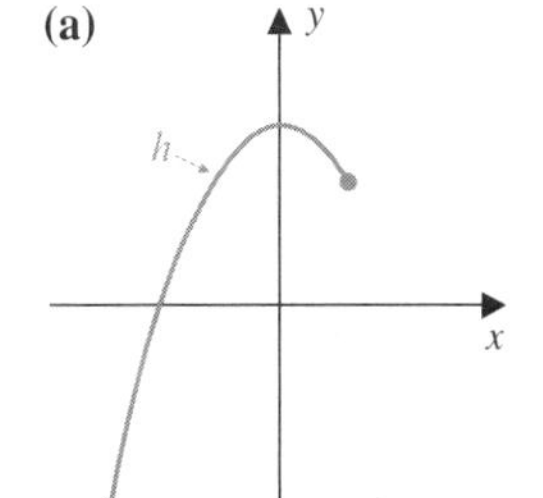

(b)

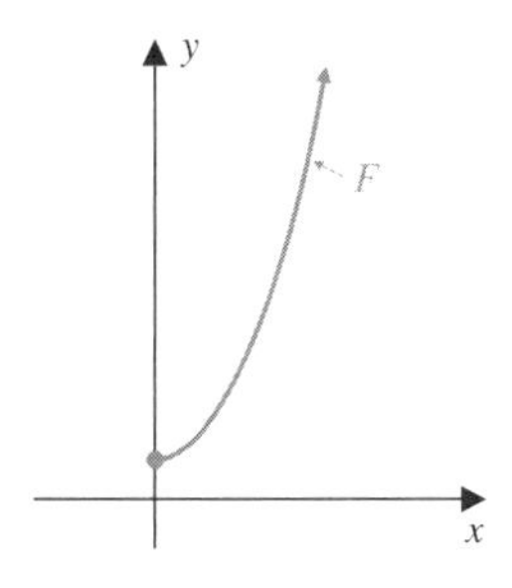

In problems 11–16, use the definition of a one-to-one function to determine whether f is one-to-one. Graph f to demonstrate the validity of the result.

11. $f(x) = -2x + 7$

12. $f(x) = -\dfrac{2}{3}x + 5$

13. $f(x) = 2x^2 + 1$

14. $f(x) = -|x|$

15. $f(x) = \sqrt{4 - x^2}$

16. $f(x) = \sqrt{5 + 3x}$

17. Write a table of values for f^{-1}, where f is given in Table 3. What are the domain and range of f? What are the domain and range of f^{-1}?

TABLE 3

x	1	2	3	4	5	6	7	8
$f(x)$	2	− 3	3	5	7	100	− 4	1

18. Determine whether each function has an inverse.

(a) $f(x)$ is the volume (in liters) of x kilograms of water.

(b) $f(n)$ is the number of students in a precalculus class whose birthday is on the nth day of the year.

(c) $f(t)$ is the number of customers in a department store at t minutes past 6:00 PM.

(d) $f(w)$ is the cost (in dollars) of express mailing a package that weighs w ounces.

In problems 19–32, sketch the graph of each function and use the graph to determine whether f^{-1} exists. If f^{-1} does not exist, explain why. If f^{-1} exists, use the procedure on page 248 to find $f^{-1}(x)$. Sketch the graphs of

$$f \text{ and } f^{-1}$$

on the same coordinate system.

19. $f(x) = 7x + 5$

20. $f(x) = 1 - 3x$

21. $f(x) = \dfrac{3}{x-1}$

22. $f(x) = \dfrac{2}{x+3}$

23. $f(x) = \sqrt{x-3}$

24. $f(x) = \sqrt{3-2x}$

25. $f(x) = x^3 - 8$

26. $f(x) = x^3 + 1$

27. $f(x) = |x + 4|$

28. $f(x) = -2x^2 + 5$

29. $f(x) = 2 - \sqrt[3]{x}$

30. $f(x) = \sqrt[3]{x+1} - 2$

31. $f(x) = x^2$, $x \geq 0$

32. $f(x) = \sqrt{1-x^2}$, $0 \leq x \leq 1$

In problems 33–36, the functions f and g have the same defining equation.

(a) Use the graphs of f and g to explain why f does not have an inverse but g does.

(b) Find $g^{-1}(x)$.

33. $f(x) = |x|$
$g(x) = |x|, x \leq 0$

34. $f(x) = -x^2 + 4;$
$g(x) = -x^2 + 4, x \geq 0$

35. $f(x) = \sqrt{1-x^2};$
$g(x) = \sqrt{1-x^2}, \ 0 \leq x \leq 1$

36. $f(x) = \sqrt{4-x^2};$
$g(x) = \sqrt{4-x^2}, \ -2 \leq x \leq 0$

37. Figure 12 shows the graph of a function f. Use symmetry and transformations to sketch the graph of each function.

(a) $y = f^{-1}(x)$ **(b)** $y = f^{-1}(x) + 1$

(c) $y = f^{-1}(x - 1)$ **(d)** $y = f^{-1}(-x)$

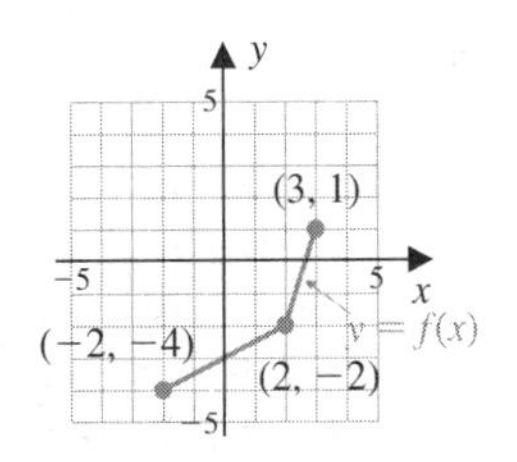

Figure 12

38. Figure 13 shows the graph of a function f. Use symmetry and transformations to sketch the graph of each function.

(a) $y = f^{-1}(x)$ **(b)** $y = f^{-1}(x - 2)$

(c) $y = f^{-1}(x - 2)$ **(d)** $y = -f^{-1}(x)$

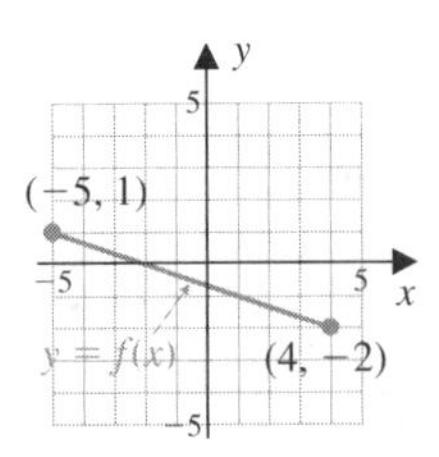

Figure 13

In problems 39–42, use a grapher to sketch the graph of each function. Use the horizontal-line test to determine whether the function has an inverse.

39. $f(x) = \dfrac{x+2}{\sqrt[3]{x}+1}$

40. $f(x) = \dfrac{3x}{1-\sqrt{x}}$

41. $f(x) = \dfrac{x^3 - |x|}{x}$

42. $f(x) = \dfrac{x^3 - 4x}{x^3 - 4}$

Applying the Concepts

43. Wages: To avoid layoffs, workers at a certain company accept a 15% reduction in wages.

(a) Find a function f that expresses the reduced wages W in terms of the original wages P.

(b) Find the inverse function of part (a) and interpret it in terms of the reduced and original wages.

44. Currency Exchange: At one time, \$100 U.S. was worth 116 Euro:

(a) Write a function f that expresses the number of dollars $f(c)$ in terms of the number of Euro c.

(b) Write a function g that expresses the number of Euro $g(d)$ in terms of the number of U.S. dollars d.

(c) What is the relationship between f and g? Explain.

45. Physics—Temperature Conversion: The function that converts degrees Celsius (C) to degrees Fahrenheit (F) is defined by the equation

$$F = f(C) = \frac{9}{5}C + 32.$$

(a) Find f^{-1} and interpret the results in terms of Fahrenheit and Celsius readings.

(b) Graph f and f^{-1} on the same coordinate system.

(c) Find the Celsius measurements corresponding to 85°F and 113°F.

46. Business—Cost Function: A lumber yard will deliver wood for \$6 per board foot plus a fixed delivery charge of \$30.

(a) Write the equation that expresses the cost function C (in dollars) in terms of x board feet of lumber delivered; that is, find $C = f(x)$.
(b) Find f^{-1} and interpret the result in terms of cost and board feet.
(c) Graph f and f^{-1} on the same coordinate system.
(d) Find the number of board feet delivered if the cost is \$270.

47. Sales Commission: A salesperson makes a commission of 40% of total sales plus a salary of \$25,000 per year.

(a) Determine the function that describes the total annual income y (in dollars) in terms of the total sales x (in dollars).
(b) Find the inverse of this function and graph it.
(c) Interpret the inverse function in terms of sales and income.

48. Dress Sizes: A size 4 dress in the United States corresponds to a European size 32 dress, and a size 12 dress in the United States corresponds to a European size 48. Assume that the relationship between the sizes in the United States and Europe is linear.

(a) Express the European dress size D as a function f of the dress size x in the United States.
(b) Use the result in part (a) to find the European dress sizes that correspond to sizes 6, 8, and 10 in the United States.
(c) Explain why f has an inverse. Find f^{-1}, graph it, and interpret it in terms of dress sizes.
(d) Use the inverse function to find dress sizes in the United States to correspond to sizes 40, 52, and 60 in Europe.

Developing and Extending the Concepts

49. Use the horizontal-line test to determine whether each function is one-to-one. If the function is one-to-one, find its inverse. Determine the domains and ranges of f and f^{-1}.

(a) $$f(x) = \begin{cases} \frac{1}{2}x - 4 & \text{if } x < 0 \\ x - 4 & \text{if } x \geq 0 \end{cases}$$

(b) $$f(x) = \begin{cases} x^3 & \text{if } x < 0 \\ -\sqrt{x} & \text{if } x \geq 0 \end{cases}$$

50. Graph

$$f(x) = \frac{x+1}{x}$$

and

$$g(x) = \frac{1}{1+x}.$$

Do these functions appear to be inverses of each other? Confirm your conclusion algebraically.

51. Explain why $(f^{-1})^{-1} = f$.

52. Give examples of functions f and g to confirm that the inverse of the quotient function $(f/g)^{-1}$ does not necessarily equal the quotient function g/f. That is, give examples of functions f and g to show that $(f/g)^{-1} \neq g/f$.

53. Suppose $f(x) = 2x + 4$ and $g(x) = 3x - 1$.
(a) Determine f^{-1}, g^{-1}, and $(f \circ g)^{-1}$.
(b) Show that $(f \circ g)^{-1} \neq f^{-1} \circ g^{-1}$.
(c) Show that $(f \circ g)^{-1} = g^{-1} \circ f^{-1}$.
(d) Generalize your results to express $(g \circ f)^{-1}$ in terms of g^{-1} and f^{-1} for any two invertible functions f and g. Give examples to support your assertion.

54. Graph each function and then use symmetry with respect to the line $y = x$ to graph f^{-1} without first finding the equation for f^{-1}. Then verify the result by finding the equation for f^{-1} and graphing it.
(a) $f(x) = 3x + 2$
(b) $f(x) = 2x^3 - 1$
(c) $f(x) = 1 - 7x$
(d) $f(x) = \sqrt[3]{x+1}$

55. Match each function f with its inverse g. Then use a grapher to graph the function and its inverse on the same coordinate system.

(a) $f(x) = 5x^3 + 10$
(b) $f(x) = (5x + 10)^3$
(c) $f(x) = 5(x + 10)^3$
(d) $f(x) = (5x)^3 + 10$

(A) $g(x) = \dfrac{\sqrt[3]{x} - 10}{5}$
(B) $g(x) = \sqrt[3]{\dfrac{x - 10}{5}}$
(C) $g(x) = \sqrt[3]{\dfrac{x}{5}} - 10$

OBJECTIVES

1. Define Exponential Functions
2. Graph Exponential Functions
3. Solve Applied Problems

4.2 Exponential Functions

As we will see, exponential functions are especially useful in modeling the growth or decline of such varied phenomena as population, radioactivity, and investment yields.

Defining Exponential Functions

Recall how we define b^x, where b is a positive constant, for all rational values x. For instance,

$$3^2 = 9, \quad 4^{3/2} = \left(\sqrt{4}\right)^3 = 2^3 = 8, \quad \text{and} \quad 7^{1/3} = \sqrt[3]{7}.$$

It is possible to extend the notion of exponents to include all irrational numbers. However, a thorough explanation depends on concepts studied in calculus. For now, it is enough to know that if r is a rational number approximately equal to an irrational number x, then b^r is approximately equal to b^x. The better that r approximates x, the better that b^r approximates b^x. For instance, Table 1 shows a pattern of improving approximations of 4^π rounded to four decimal places. As the value of the exponent gets closer and closer to π, the value of 4^π gets closer and closer to a real number whose approximate decimal value is 77.8802.

With this background we are ready to define exponential functions.

TABLE 1
Approximations of 4^π

Approximation of π		Approximation of 4^π	
Improved accuracy ↓	3.14	77.7085	↓ Improved accuracy
	3.142	77.9242	
	3.1416	77.8810	
	3.14159	77.8799	
	⋮	⋮	

In the definition, $b \neq 1$ *because* $f(x) = 1^x = 1$, *which is considered to be a constant function rather than an exponential function.*

Definition Exponential Function—Base b

A function of the form

$$f(x) = b^x \text{ where } b > 0 \text{ and } b \neq 1$$

is called an **exponential function of base b.** The independent variable x, which represents any real number is called the **power** or **exponent.**

Thus

$$f(x) = 3^x \text{ is an exponential function with base 3}$$

and

$$g(x) = 5^{-x} = \left(\frac{1}{5}\right)^x \text{ is an exponential function with base } \frac{1}{5}.$$

Among the exponential functions, there is one that is encountered frequently in mathematics. This function is referred to as the *natural exponential function.*

Its base is denoted by the symbol e in honor of the Swiss mathematician Leonard Euler (1707–1783). The value of e can be approximated by evaluating the expression

$$\left(1 + \frac{1}{x}\right)^x$$

for larger and larger values of x.

Table 2 shows some of the calculations for specific values of x.

TABLE 2

x	1	10	100	1,000	10,000	100,000	1,000,000
$\left(1 + \frac{1}{x}\right)^x$	2	2.59374	2.70481	2.71692	2.71815	2.71827	2.71828

This leads us to the following definition:

Definition Natural Exponential Function

Let x be a real number, the function f defined by

$$f(x) = e^x$$

is called the **natural exponential function** with base e.

This function is used extensively in formulas that model real-world phenomena. To 12 decimal places.

$$e = 2.718281828459$$

EXAMPLE 1 Evaluating Exponential Functions

Given the functions

$$f(x) = 3^x, \; g(x) = e^x, \; h(x) = 2^{-x}, \text{ and } k(x) = e^{-x}$$

find each of the following values. Round off the answers to two decimal places.

(a) $f(\sqrt{2})$ (b) $g\left(\frac{3}{2}\right)$ (c) $h(2.7)$ (d) $k(\pi)$

Solution (a) $f\left(\sqrt{2}\right) = 3^{\sqrt{2}} = 4.73\left(\text{approx.}\right)$ (b) $g\left(\frac{3}{2}\right) = e^{\frac{3}{2}} = 4.48\ \left(\text{approx.}\right)$

(c) $h(2.7) = 2^{-2.7} = 0.15$ (approx.) (d) $k(\pi) = e^{-\pi} = 0.04$ (approx.) 

Graphing Exponential Functions

To examine the common features of exponential functions, we use the point-plotting method to graph two such functions—one with a base greater than 1 and another with a base less then 1.

EXAMPLE 2 Graphing Exponential Functions

Sketch the graph of each exponential function. Determine the domain, range, and any asymptotes.

(a) $f(x) = 2^x$ (b) $g(x) = \left(\frac{1}{2}\right)^x$

Since $2^x = 0$ has no solution, there is no x intercept; that is, the graph does not intersect the x axis.

Solution (a) Table 3 lists some ordered pairs of numbers that satisfy $f(x) = 2^x$. After plotting these points and then connecting them with a smooth continuous curve, we obtain the graph (Figure 1a). Since 2^x is defined for any real number x, the domain of f consists of all real numbers $\mathbb{R}$. From the graph, we see that the range consists of all positive real numbers. Because of the limit behavior

$$f(x) \to 0 \text{ as } x \to -\infty$$

the x axis is a horizontal asymptote.

TABLE 3

x	$f(x) = 2^x$	$g(x) = \left(\frac{1}{2}\right)^x$
-3	1/8	8
-2	1/4	4
-1	1/2	2
0	1	1
1	2	1/2
2	4	1/4
3	8	1/8

(a)

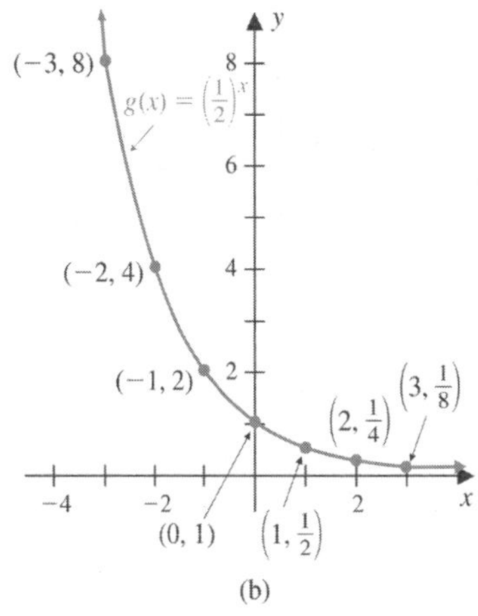

(b)

Figure 1

(b) Table 3 lists some ordered pairs of numbers that satisfy the equation $g(x) = (1/2)^x$. After plotting these points and then connecting them with a smooth continuous curve, we obtain the graph (Figure 1b). From the graph, we see that the domain consists of all real numbers $\mathbb{R}$ and the range consists of all positive real numbers. Since

$$g(x) \to 0 \text{ as } x \to +\infty$$

the x axis is a horizontal asymptote.

As illustrated in Figure 2a, graphs of functions of the form $f(x) = b^x$, where $b > 1$, resemble the graph of $y = 2^x$.

Similarly, if $0 < b < 1$, graphs of functions of the form $f(x) = b^x$ have the same general shape and characteristics as the graph of $y = (1/2)^x$ (Figure 2b). Notice the similarities between the graphs in Figure 2a and in Figure 2b.

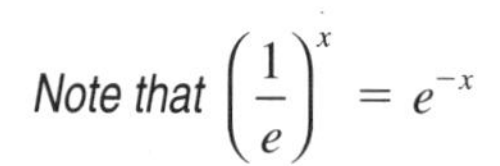

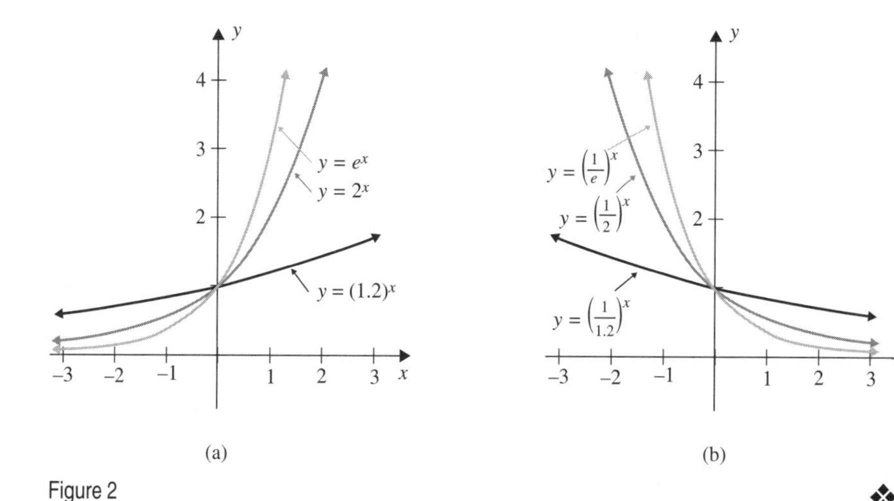

Figure 2

❖

In each case the graphs contain the point (0, 1). The functions are increasing in Figure 2a and decreasing in Figure 2b. Also, all functions have the x axis as a horizontal asympote.

These observations lead us to the general properties given in Table 4.

TABLE 4

General Graphs and Properties of $f(x) = b^x$, $b > 0$, $b \neq 1$

Properties	General Graphs
1. The domain includes all real numbers.	$f(x) = b^x$, $b > 1$
2. The range is the interval $(0, \infty)$.	$f(x) = b^x$, $0 < b < 1$
3. The function increases everywhere if $b > 1$ and decreases everywhere if $0 < b < 1$.	
4. The x axis is a horizontal asymptote.	
5. The function is one-to-one. That is, $b^u = b^v$ if and only if $u = v$.	
6. The graph contains the point (0, 1).	

Known graphs of exponential functions, along with transformations, can be used to obtain graphs of other exponential functions.

EXAMPLE 3 Using Transformations to Graph

(a) Describe a sequence of transformations that will transform the graph of

$$y = \left(\frac{1}{3}\right)^x \text{ into the graph of } G(x) = -\left(\frac{1}{3}\right)^x + 1.$$

Sketch the graph of G. Determine the domain and range of G, and identify any horizontal asymptote.

(b) Sketch the graph of

$$f(x) = -e^{x+3}$$

by using a sequence of transformations of the graph of $y = e^x$ and find the horizontal asymptote.

Solution (a) As displayed in Figure 3a, the graph of G can be obtained by performing the following sequence of transformations:

(i) Reflect the graph of $y = (1/3)^x$ across the x axis to get the graph of

$$g(x) = -\left(\frac{1}{3}\right)^x.$$

(ii) Vertically shift the graph of g up 1 unit to get the graph of G.

From the graph, we see that the domain of G consists of all real numbers $\mathbb{R}$. By reading the graph of G, we find that its range is the interval $(-\infty, 1)$. Because of the 1 unit vertical shift upward, the horizontal asymptote of the graph of g (the x axis) moves 1 unit above the x axis to become the line $y = 1$, which is the horizontal asymptote of the graph of G, that is,

$$G(x) \to 1 \text{ as } x \to +\infty.$$

(b) To graph $f(x) = -e^{x+3}$, we first reflect the graph of $y = e^x$ about the x axis to get the graph of $h(x) = -e^x$. Then the graph of $f(x) = -e^{x+3} = h(x + 3)$ is obtained by shifting the graph of h to the left 3 units (Figure 3b). The horizontal asymptote of $y = e^x$ is not affected by these transformations; it is the x axis for f as well.

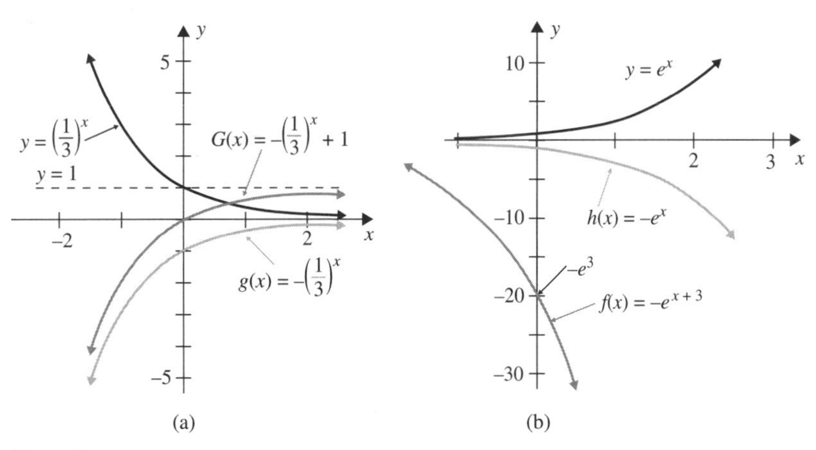

Figure 3

Property 5 in Table 4 indicates that $f(x) = b^x$, where $b > 0$ and $b \neq 1$, so that

$$b^u = b^v, \text{ if and only if } u = v$$

This property is used to solve certain types of exponential equations, as the next example illustrates.

EXAMPLE 4 Using the One-to-One Property

Solve each of the given equations by using the one-to-one property.

(a) $3^{2x+1} = 27$ (b) $4^{x-0.5} = 8$

Solution (a) First we express each side of the equation $3^{2x+1} = 27$ in terms of base 3 to get

$$3^{2x+1} = 3^3.$$

Using the one-to-one property, we equate the exponents to get

$$2x + 1 = 3$$

so that

$$x = 1.$$

(b) In this situation, we look for a common base by proceeding as follows:

$4^{x-0.5} = 8$	Given
$(2^2)^{x-0.5} = 2^3$	Express each side in base 2
$2^{2x-1} = 2^3$	Equate exponents
$2x - 1 = 3$	Solve
$x = 2$	

❖

Solving Applied Problems

Exponential functions are used to model real-world phenomena in a variety of applications, such as *compound interest, population growth and decay,* and *radio-isotope dating.*

Modeling Compound Interest

Suppose \$3,000 is deposited in a saving account that pays a 3.6% annual interest rate, but that the interest is compounded or credited each quarter of the year.

The **interest rate per period** is the annual rate 3.6% = 0.036 divided by the number of compounding periods per year, 4. That is, the interest rate per period for this situation is given by $\dfrac{0.036}{4} = 0.009.$

If this money is left in the account for 3 years, then there are a total of 3(4) = 12 quarters, and the accumulated amount due at the end of 3 years is given by

$$S = 3000(1 + 0.009)^{12} = 3340.53 \text{ (approx.)}.$$

Thus an investment of \$3000 that pays 3.6% annual interest compounded quarterly accumulates \$3340.53 (including the original \$3000) over 3 years.

This illustration leads to the following formula:

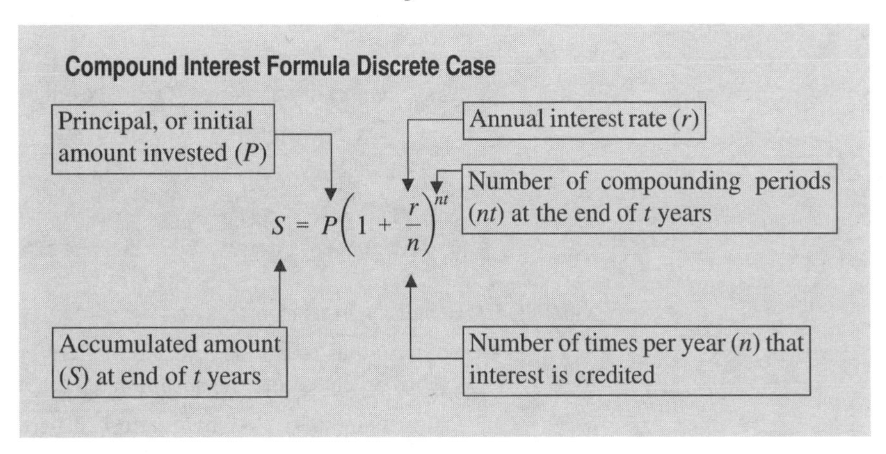

The accumulated amount S includes the principal P.

This formula is applicable when the interest is compounded at *discrete* intervals of time such as yearly, quarterly, monthly, weekly, daily, and so forth.

Now let us examine what happens to the formula

$$S = P\left(1 + \frac{r}{n}\right)^{nt}$$

if we allow the number of annual compounding periods n to increase without bound.

First we rewrite the formula as follows:

$$S = P\left[1 + \frac{1}{\left(\frac{n}{r}\right)}\right]^{\left(\frac{n}{r}\right)rt}.$$

Next, we replace $\frac{n}{r}$ with x to get

$$S = P\left[\left(1 + \frac{1}{x}\right)\right]^{xrt} = P\left[\left(1 + \frac{1}{x}\right)^{x}\right]^{rt}.$$

By allowing $n \to +\infty$, correspondingly, $\frac{n}{r} = x \to +\infty$, so that

$$\left(1 + \frac{1}{x}\right)^{x} \to e \text{ and, in turn,}$$

$$\left[\left(1 + \frac{1}{x}\right)^{x}\right]^{rt} \to e^{rt}$$

Thus if we allow $n \to +\infty$, it turns out that correspondingly,

$$P\left(1 + \frac{r}{n}\right)^{nt} \to Pe^{rt}.$$

In other words, if the interest is compounded without interruption, the accumulated amount is found by using the following formula:

Compounding Interest Formula Continuous Case

If P is invested at an annual rate r compounded *continuously,* then the accumulated amount S in the account at the end of t years is given by

$$S = Pe^{rt}.$$

EXAMPLE 5 Comparing Investments

Suppose that \$5,000 is invested in a credit union paying an annual interest rate of 4% for 3 years. Find the accumulated amount if the interest is compounded:

(a) semiannually (b) quarterly (c) weekly

(d) daily (e) hourly (f) continuously

Solution Under the given conditions, $P = 5000$, $r = 0.04$, and $t = 3$, so that if n is the number of compounding periods per year, then the accumulated amount S is given by

$$S = 5000\left(1 + \frac{0.04}{n}\right)^{3n}$$

for the discrete situations in parts (a) through (e), whereas,

$$S = 5000e^{(0.04)(3)}$$

for the continuous situation in part (*f*). ❖

The accumulated amount S for each situation is computed in Table 5.

TABLE 5

Compounding Period	n	Accumulated Amount S	
(a) Semiannually	2	\$5,630.81	Discrete case; use $S = 5000\left(1 + \frac{0.04}{n}\right)^{3n}$
(b) Quarterly	4	\$5,634.13	
(c) Weekly	52	\$5,637.22	
(d) Daily	365	\$5,637.45	
(e) Hourly	$(24)(365) = 8760$	\$5,637.48	
(f) Continuously	———	\$5,637.48}	Continuous case; use $S = 5000e^{(0.04)(3)}$

Notice that the largest gains appear in going from semiannually to quarterly to weekly. Subsequently, the changes in the value of S lessen as n increases. In fact, S is tending toward \$5,637.48 as n gets larger and larger.

Modeling Growth and Decay

Base e is used in mathematical modeling of phenomena that exhibit continuous (uninterrupted) growth or decay as follows:

Modeling Continuous Growth or Decay

An exponential function of the form

$$P = P_0e^{kt} \text{ where } k > 0$$

is used to model continuous **growth.** Its general graph is shown in Figure 4a.

An exponential function of the form

$$P = P_0e^{-kt} \text{ where } k > 0$$

is used to model continuous **decay.** Its general graph is shown in Figure 4b.

In either case, P_0 is the initial quantity, t is the time elapsed, and k is the rate of growth or decay.

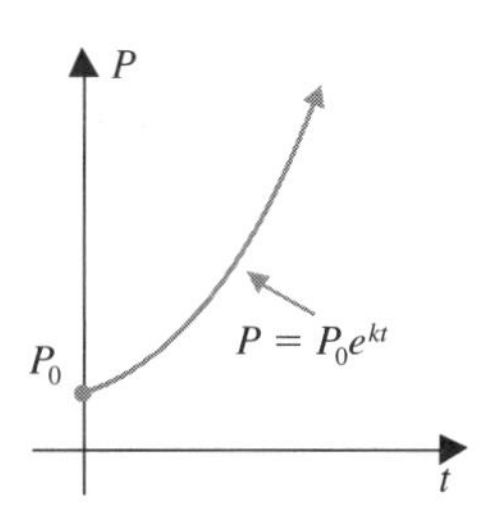

(a) Exponential continuous growth

(b) Exponential continuous decay

Figure 4

We say that P is growing or decaying *exponentially* at a *continuous* rate of k.

For example, suppose that the initial population of a certain city is 1.013 million, and the population P (in millions) grows exponentially at a continuous rate of about 0.58% per year. The growth model for this population t years later is given by the function

$$P(t) = 1.013e^{0.0058t}.$$

If we replace t by 1, 2, 3, and so on, we observe that the corresponding output values increase exponentially as the input values increase.

For the decay situation, the outputs decrease as the input values increase, as illustrated in the next example.

TABLE 6

Percentage of C-14 Present After t Years

t *Years*	P	Percentage of C-14 Present
0	1.00	100
1000	0.89	89
2000	0.79	79
5000	0.55	55
6000	0.48	48
10,000	0.30	30

Modeling Radiocarbon Dating

Radioactive carbon-14 (C-14) is found in all living things. Once an organism dies, the C-14 begins to decay radioactively. The less C-14 found in a specimen, the older the specimen. Archaeologists are able to date objects such as relics and fossils by determining the remaining amounts of C-14. The process is called *radiocarbon dating.*

EXAMPLE 6 Modeling Radioactive Decay of Carbon-14

The percentage P (as a decimal) of C-14 found in a specimen t years after it begins to decay is modeled by the exponential function

$$P = e^{-0.000121t}.$$

Use the model to determine the percentage P of C-14 still present in an organism t years after it begins to decay if $t = 0$; 1000; 2000; 5000; 6000; 10,000. Round off the answers to two decimal places.

Solution We substitute each value of t into the function $P = e^{-0.000121t}$ and then use a calculator to evaluate. The results are summarized in Table 6.

If the fossil has half of its C-14 remaining, it follows that $P = 0.50$ in the model in Example 6. So the solution of the equation

$$0.50 = e^{-0.000121t}$$

or, equivalently,
$$0 = e^{-0.000121t} - 0.50$$

will give us the number of years of decay it took to reach this state.

The solution of this latter equation is the same as the x intercept of the graph of the function

$$f(x) = e^{-0.000121x} - 0.50.$$

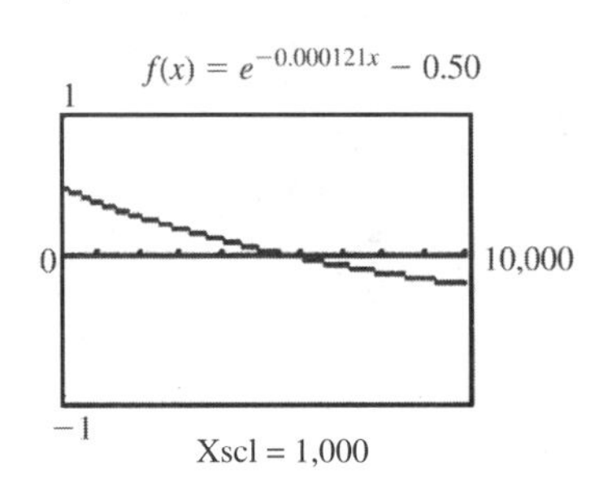

Figure 5

By using a grapher to graph f (Figure 5), along with the ZERO -finding feature we find that x is approximately 5730 (to the nearest 10 years). Thus, it took about 5730 years to lose half the C-14 content. ❖

The number of years that it takes for half the amount of a radioactive substance to decay is called its **half-life.** So in this situation we found that the half-life of C-14 is about 5730 years.

So far we have considered polynomial best fit regression models. Now we extend the idea to exponential regression models created by the least-squares method.

EXAMPLE 7 Modeling Population Growth

The projected population growth in the United States for the years from 2000 through 2100 as reported by the U.S. Census Bureau is given in Table 7 (see Example in Section 3.1).

TABLE 7

Year	t	U.S. Population P (nearest thousand)
2000	0	275,306
2010	10	299,862
2020	20	324,927
2030	30	351,070
2040	40	377,350
2050	50	403,687
2060	60	432,011
2070	70	463,639
2080	80	497,830
2090	90	533,605
2100	100	570,954

(Source: U.S. Census Bureau, January 13, 2000)

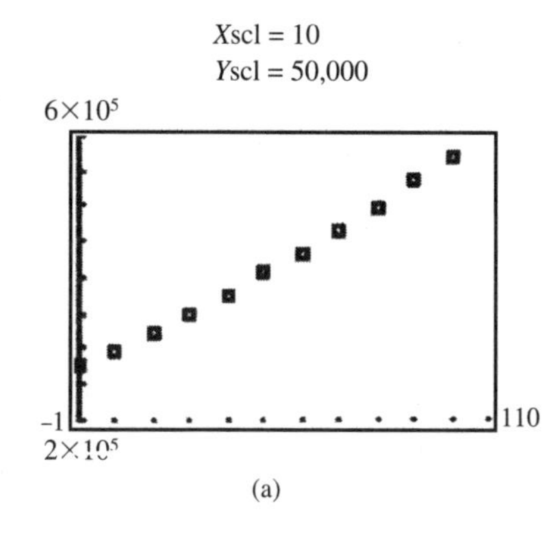

(a)

(a) Plot a scattergram of the data. Locate the t values on the horizontal axis and the P values on the vertical axis.

(b) Use a grapher and the exponential regression feature to determine the exponential model of the form

$$P = a \cdot b^t$$

that best fits the data. Round a and b to four decimal places.

(c) Sketch the graph of the function in part (b).

(d) Assuming this trend will continue, predict the population of the United States in the years 2015, 2075, and 2110. Round off the answers to the nearest thousand.

Solution (a) The scattergram for the given data is shown in Figure 6a, where P is used for the vertical axis and t is used for the horizontal axis.

(b) By using a grapher we find the exponential regression model to be

$$P = 279992.9292 \cdot 1.0072^t.$$

(c) The graph of the exponential regression equation in part (b) is shown in Figure 6b along with the scattergram of the given data.

(d) To predict the population for the years 2015, 2075, and 2110, we substitute $t =$ 15, 75, and 110, respectively, into the regression equation to get the population approximations:

311,805,000 for the year 2015
479,541,000 for the year 2075
and 616,419,000 for the year 2110 ❖

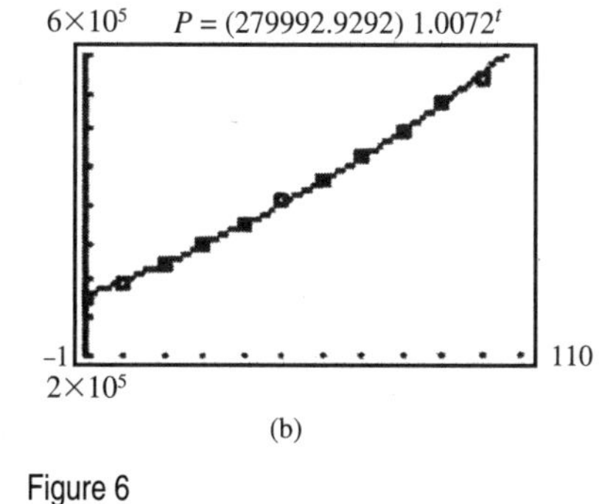

(b)

Figure 6

PROBLEM SET 4.2

In problems 1–4, use a calculator to approximate each expression, rounded to four decimal places.

1. **(a)** 2^{π}
(b) $2^{-\sqrt{2}}$
(c) $1000(5^{-\pi})$
(d) $100(1.02)^{-8}$

2. **(a)** π^3
(b) $5^{-\sqrt{7}}$
(c) $2^{\sqrt[3]{5}}$
(d) $10^x(3.14)^{\frac{1}{3}}$

3. **(a)** $e^{3.4}$
(b) $e^{-\pi}$
(c) $e^{\sqrt{2}}$
(d) $1 - 4e^{-0.53}$

4. **(a)** $e^{-2.5}$
(b) $e^{-\sqrt{3}}$
(c) $e^{\sqrt{2.5}}$
(d) $3(1 + 2e^{0.9})$

In problems 5 and 6, let $f(x) = 4^{-\left(\frac{x}{2}\right)}$ and $g(x) = \left(\frac{2}{3}\right)^x$.

Find each value. Round off each answer to two decimal places.

5. **(a)** $f(-1)$
(b) $f(\sqrt{3})$
(c) $g(-2)$
(d) $g(0.75)$

6. **(a)** $f(0.3)$
(b) $f(\sqrt{5})$
(c) $g(-\sqrt{7})$
(d) $g(1.31)$

In problems 7 and 8, graph the functions f, g, and h on the same coordinate system. Describe the relationship among f, g, and h.

7. $f(x) = e^x$, $g(x) = 2 + e^x$, $h(x) = -2 + e^x$

8. $f(x) = 3^x$, $g(x) = 3^x - 1$, $h(x) = 3^x + 2$

In problems 9 and 10, each curve is the graph of an exponential function $y = b^x$ or b^{-x} that contains the given point. Find the value of b in each case.

9.

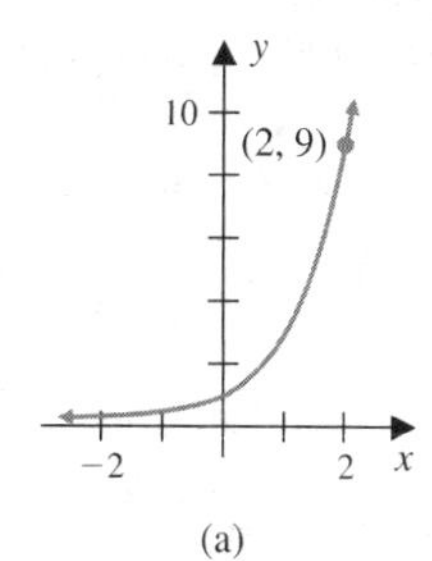

(a)

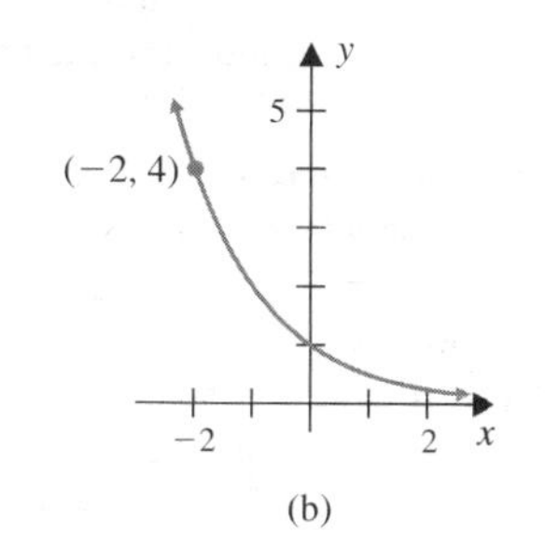

(b)

10.

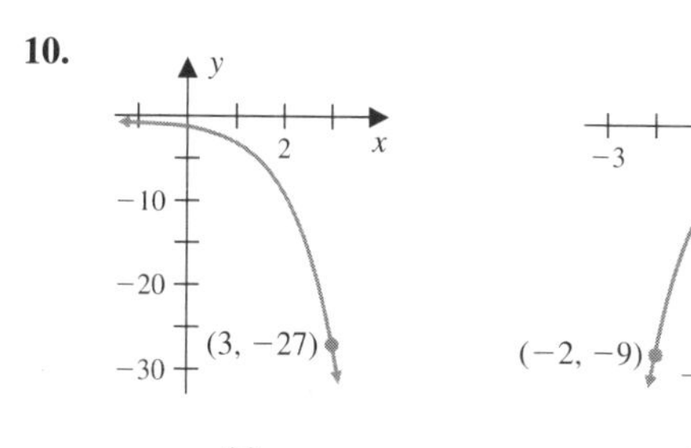

(a) (b)

In problems 11 and 12, suppose that each curve is the graph of an exponential function of the form

$$y = A \cdot b^x$$

that contains the given pair of points. Find the values of A and b

11. **(a)** $P_1 = (0, 3)$ and $P_2 = (1, 9)$
(b) $P_1 = (0, 1)$ and $P_2 = (-2, 4)$

12. **(a)** $P_1 = (0, 2)$ and $P_2 = (1, 6)$
(b) $P_1 = (0, -2)$ and $P_2 = (1, -4)$

13. Each graph below is a transformation of the graph of the function $y = e^x$. Match each function with its graph.
(a) $f(x) = e^x + 1$ **(b)** $f(x) = e^{x+1}$
(c) $f(x) = -e^x$ **(d)** $f(x) = e^x - 1$

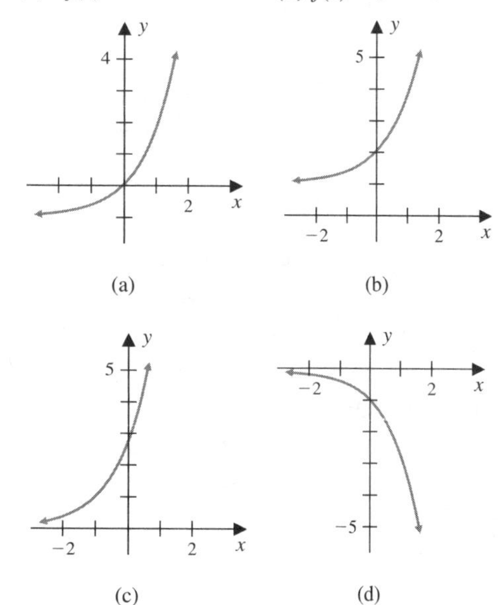

14. Each graph below is a transformation of the graph of the function $y = (0.4)^x$. Match each given function with its graph.

(a) $g(x) = (0.4)^x - 1$ **(b)** $f(x) = (0.4)^{x-1}$
(c) $h(x) = -(0.4)^x$ **(d)** $g(x) = (0.4)^{x+1}$

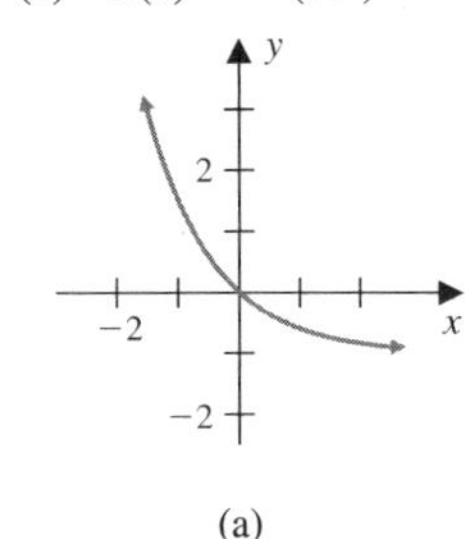

(a)

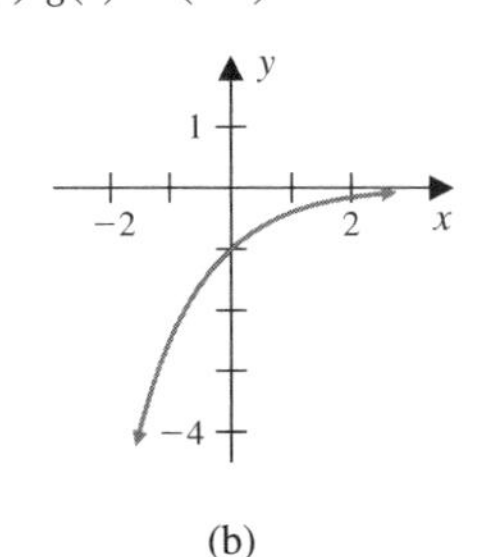

(b)

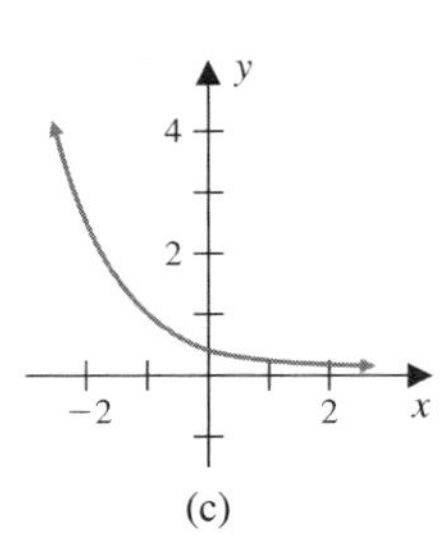

(c)

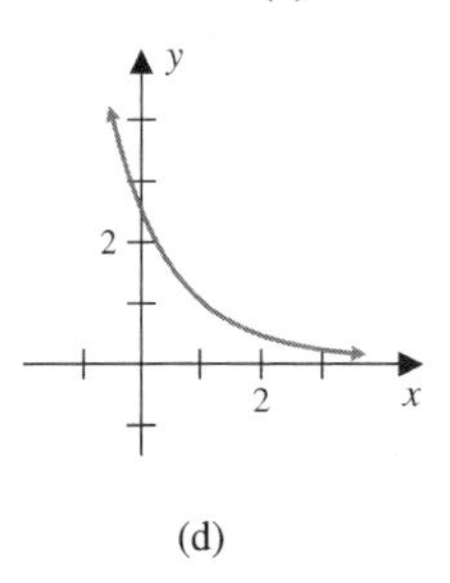

(d)

In problems 15–20, solve each equation for x.

15. $2^{3x} = 2^{4x-2}$ **16.** $5^{2-3x} = 5^{5x-6}$
17. $3^{x^2} = 3^{5x+6}$ **18.** $4^{x-1} = 8^{2x}$
19. $25^{3x} = (125)^{x-1}$ **20.** $9^{x-1} = 27^{2x}$

In problems 21–28, describe a sequence of transformations that will transform the graph of a function of the form

$$y = b^x \text{ or } y = e^x$$

to the graph of the given function. Sketch the graph in each case. Determine the domain and range, and indicate whether the function increases or decreases. Also identify any horizontal asymptote.

21. $f(x) = 5^{x+1}$ **22.** $h(x) = 4^{x-2}$
23. $f(x) = 2^x + 3$ **24.** $g(x) = 4^x - 1$
25. $f(x) = -2e^x$ **26.** $h(x) = 1 + e^x$
27. $f(x) = e^{-x} - 2$ **28.** $f(x) = 2 - 3e^{-x}$

Applying the Concepts

In problems 29–48 round off the answers to two decimal places.

29. Compound Interest: If \$5000 is invested in a certificate of deposit for 5 years at an annual interest rate of 5.75%, how much money will be in the account if the interest rate is compounded:

(a) Semiannually? **(b)** Quarterly?
(c) Monthly? **(d)** Weekly?

30. Compound Interest: Suppose that \$10,450 is invested in a money market account paying an annual interest rate of 4.25% compounded monthly.

(a) Express the accumulated amount S as a function of elapsed time t in years.

(b) After 3 years, how will the earned interest in this investment compare to the earned interest if the money is invested in a bank that pays an annual interest rate of 4.015% compounded daily?

31. Cumulative Earnings: Suppose that \$5000 is invested in a money market account paying an annual interest rate of 5.75% compounded continuously.

(a) Express the accumulated amount S as a function of the number of elapsed years t.

(b) What is the accumulated amount after 4 years?

(c) Will this investment earn more or less than another investment of \$5000 that pays an annual interest rate of 6% compounded semiannually? Explain the difference.

32. Cumulative Interest Trend: The accumulated amount S of an investment of \$10,000 paying continuously compounded interest over t years is given by

$$S = 10{,}000e^{0.0625t}.$$

(a) What is the annual interest rate being paid on this investment?

(b) What is the accumulated amount after 5 years?

(c) Suppose \$10,000 is invested in an account that pays 6.25% compounded daily. How much will accumulate after 5 years? Compare this investment to the one in part (b).

33. Population Growth: In 1999, the world population reached about 6 billion people, and it was growing at a continuous rate of about 1.3% per year.

(a) Find the exponential growth model for the world population P (in billions) t years after 1999.

(b) Use the model to predict the world population in the year 2009.

34. Population Growth: According to the U.S. Census Bureau, the population of the United States in 1999 was approximately 273 million people, and it was growing at a continuous rate of about 0.85% per year.

(a) Find the exponential growth model that expresses the population P (in millions) t years after 1999.

(b) Predict the population in the year 2050.

35. Bacteria Growth: A biologist studying bacteria in a swimming pool determines that the initial population of bacteria is 25,000 per cubic inch, and it is growing at a continuous rate given by the model

$$P = 25{,}000e^{0.1t}$$

where P represents the number present after t days.

(a) What is the continuous growth rate of the bacteria?

(b) Predict the number of bacteria per cubic inch after 5 days.

36. **Bacteria Growth:** The growth rate of bacteria in foods is used to determine the safe shelf-life of various products. Once the bacteria count reaches a certain level, these products are no longer safe to eat. If the daily continuous growth rate of bacteria in a certain brand of cheese is 5% of its population each day, how many times the current count level will the bacteria count be after 140 days?

37. **Car Trade-in Value:** Suppose that the *trade-in value* V (in dollars) of an automobile that is t years old is given by the formula

$$V = P(1 - r)^t$$

where r is the annual percentage of the depreciation and P is the original price of the car (in dollars). A certain make of car is purchased for \$19,545, and it depreciates in value by 20% each year.

(a) Write an equation that expresses the trade-in value V as a function of t.

(b) Find the trade-in value after 1 year, after 3 years, and after 5 years. Round off answers to the nearest dollar.

38. **Real Estate Appreciation:** Because of inflation, the value of a home often increases with time. State and local agencies use the formula

$$S = C(1 + r)^t$$

to assess the value S (in dollars) of a home for property tax purposes, where C dollars is the cost of the home when new, t is the age of the home (in years), and r is the annual rate of inflation. Assume that the inflation rate is 3.1% and that a home originally cost \$135,000.

(a) Write an equation that expresses the assessed value S as a function of t.

(b) If the inflation rate remains constant at 3.1%, what is the assessed value after 1 year? 5 years? 10 years? 15 years?

39. **Recycling:** In the state of Michigan, stores charge an additional 10¢ deposit for each bottle of beverage sold. Customers return these bottles for refunds, and then the stores return them to the beverage companies for recycling. Suppose that 65% of all bottles distributed will be recycled every year. If a company distributed 2.5 million bottles in 1 year, the number N (in millions) of recycled containers still in use after t years is given by the model

$$N = 2.5(0.65)^t.$$

How many recycled bottles (in millions) are still in use after 1 year? 3 years? 5 years? Round off the answers to two decimal places.

40. **Electric Circuit:** The electric current I (in amperes) flowing in a series circuit having inductance L henrys, resistance R ohms, and electromotive force E volts (Figure 7) is given by the model

$$I = \frac{E}{R}\left(1 - e^{-Rt/L}\right)$$

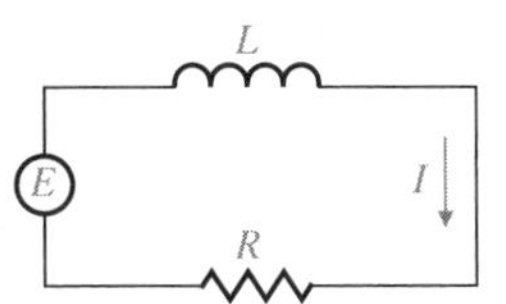

Figure 7

where t is the time (in seconds) after the current begins to flow.

(a) If $E = 7.5$ volts, $R = 3$ ohms, and $L = 0.83$ henry, express I as a function of t.

(b) Find the amount of current that flows in the circuit after 0.3 second; 0.5 second; and 1 second.

41. **Environmental Science:** According to the Bouguer–Lambert law, the percentage P of light that penetrates ordinary seawater to a depth of d feet is given by the function

$$P = e^{-0.044d}.$$

What is the percentage of light penetration in seawater to a depth of 4 feet? 6 feet?

42. **Physics—Law of Cooling:** Suppose that a heated object is placed in cooler surroundings.

Newton's law of cooling relates the temperature T of the heated object to the temperature T_0 of the cooler surroundings. It states that after t units of time,

$$T = T_0 + Ce^{-kt}$$

where k and C are positive constants associated with the cooling object.

Suppose that a baked cake is removed from an oven and cooled to a room temperature of 70°F. The temperature T of the cake after t minutes is given by

$$T = 70 + 280e^{-0.14t}.$$

(a) What was the temperature of the cake when it was first removed from the oven?

(b) What is the temperature of the cake after 5 minutes? 10 minutes? 20 minutes?

43. **Radioactive Decay:** Radioactive elements decay exponentially. The decay model for a specific radioactive element is

$$A = A_0e^{-0.04463t}$$

where A is the amount present after t days and A_0 is the amount present initially. Assume there is a block of 50 grams of the element at the start. How much of the element remains after
10 days?
15 days?
30 days?

44. Financing Retirement: A method for financing a retirement is to invest an amount of money at a given rate compounded annually and then withdraw a fixed amount of money from it each year. The maximum amount of withdrawal W possible if an investment P is to last n years is given by the model

$$W = (W - Pr)(1 + r)^n.$$

If \$400,000 is invested in an account that pays 5% with the intention of making withdrawals for 15 years, how much can be withdrawn each year?

45. Population Trend: Ecologists have determined that the population N of bears in a certain protected forest area since 1996 ($t = 0$) is represented by the data in the following table.

Year	t	Population N
1996	0	225
1997	1	229
1998	2	234
1999	3	239
2000	4	244
2001	5	248
2002	6	253
2003	7	259
2004	8	264
2005	9	272

(a) Construct a scattergram that exhibits the given data. Locate the t values on the horizontal axis and the N values on the vertical axis.
(b) Use the exponential regression feature on a grapher to determine the model $N = a \cdot b^t$ that best fits the data. Round off a and b to three decimal places.
(c) Sketch the graph of the function N in part (b).
(d) Assuming that the trend of this model will continue, predict the number of bears that will inhabit the region in the year 2016.

46. Profit Trend: Data in the following table shows the annual profit P of a company (in thousands of dollars) after t years.

Year t	Profit P
0	150
1	210
2	348
3	490
4	660
5	872
6	1400

(a) Construct a scattergram that exhibits the given data. Locate the t values on the horizontal axis and the P values on the vertical axis.
(b) Use the exponential regression feature on a grapher to determine the model $P = a \cdot b^t$ that best fits the data. Round off a and b to three decimal places.
(c) Sketch the graph of the function P in part (b).
(d) Assuming the trend of this model will continue, predict the profit after 10 years, to the nearest thousand dollars.

47. Inflation: The following data represent the amount A that \$1 will be worth t years from now, assuming that inflation runs steadily at 4% per year.

Year t	Amount A
0	\$1
1	\$0.96
2	\$0.92
3	\$0.88
4	\$0.85
5	\$0.82
6	\$0.78
7	\$0.75
8	\$0.72

(a) Construct a scattergram that exhibits the given data. Locate the t values on the horizontal axis and the A values on the vertical axis.
(b) Determine how closely the exponential model $A = (0.96)^t$ best fits the given data by comparing the function values (rounded to two decimal places) to the table values.

(c) Sketch the graph of the function A in part (b).
(d) Assuming that the trend of this model continues, predict how much \$1 now will be worth in 17 years. Round off to two decimal places.

48. Depreciation: Data in the following table show the depreciation of a car that was purchased for \$24,000 in 2005.

Year	2005	2006	2007	2008	2009
t	0	1	2	3	4
Value N	24,000	20,300	16,820	14,400	11,920

(a) Plot the scattergram of the data. Locate the t values on the horizontal axis and the N values on the vertical axis.
(b) Determine how closely the exponential function

$$N = 24{,}000(0.84)^t$$

best fits the given data by comparing the function values (rounded to the nearest dollar) to the table values.
(c) Sketch the graph of the function N in part (b).
(d) Assuming this trend continues, what will be the value of the car after 7 years and after 9 years? Round off the answers to the nearest dollar.

Developing and Extending the Concepts

49. Consider the exponential function

$$f(x) = b^x.$$

(a) Why does the definition exclude the value $b = 1$? How about $f(x) = 1^x$? Is this an exponential function?
(b) Why do we have the restriction $b > 0$? Is there any problem in defining and graphing the function $f(x) = (-3)^x$? Explain.

50. The *hyperbolic sine function,* denoted by *sinh,* and the *hyperbolic cosine function,* denoted by *cosh,* are defined as follows:

$$\sinh x = \frac{e^x - e^{-x}}{2} \quad \text{and} \quad \cosh x = \frac{e^x + e^{-x}}{2}$$

These functions are used in engineering. Show that $(\cosh x)^2 - (\sinh x)^2 = 1$ for every number x.

51. Let $f(x) = e^x$.
(a) Find the values of x such that

$$f(x^2) = f(7x - 12).$$

(b) Find and simplify

$$(f(x) + f(-x))^2 - (f(x) - f(-x))^2.$$

52. Civil Engineering: When a flexible cord or cable hangs freely from its ends, it forms an arc called a *catenary.* Its equation has the form

$$y = \frac{k}{2}\left(e^{cx} + e^{-cx}\right)$$

for appropriate choices of the constants c and k. The famous Gateway Arch in St. Louis, Missouri, has the shape of an *inverted catenary* (Figure 8). Suppose the arch is placed in a coordinate system so that the x axis is ground level and the y axis is the axis of symmetry of the arch. Assume its equation is given by

$$y = -63.85(e^{(x/127.7)} + e^{(-x/127.7)}) + 757.70$$

where x is in feet.

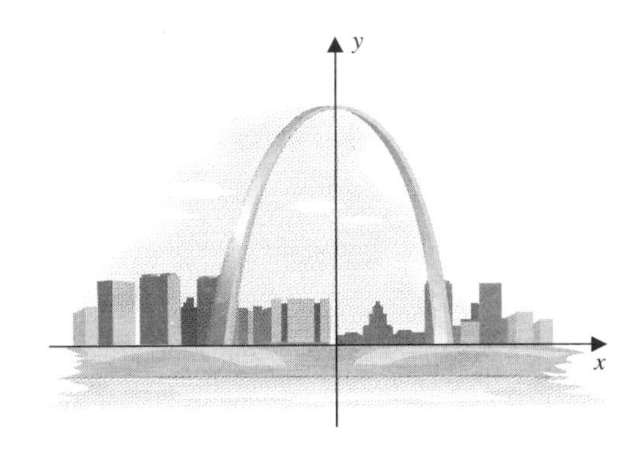

Figure 8

(a) Use a grapher to sketch the graph of this equation, and compare the shape of the curve to Figure 8.
(b) Locate the highest point (*apex*) of the arch.
(c) Find the distance between the bases at ground level.

53. Find the real zeros of the function f if
(a) $f(x) = x^3e^x + 8x^2e^x - 29xe^x + 44e^x$
(b) $f(x) = 5x^3e^{-2x} - 20x^2e^{-2x} + 5xe^{-2x} + 30e^{-2x}$

54. Which of the given functions are equal to each other? Justify your assertion algebraically and illustrate the validity of your conclusion graphically.
(i) $f(x) = 2^{-(x-4)}$
(ii) $g(x) = (0.5)^{x-4}$
(iii) $h(x) = (0.25)^{2x-8}$

1 : 375
2 : 281.25
3 : 210.9
4 : 158.2
5 : 118.67
6 . 88.9
7 . 66.7
8 . 50.1
9 . 37.5
10 . 28.2
11 . 21.1
12 . 15.8
13 . 11.9
14 . 8.9

OBJECTIVES

1. Convert Exponentials to Logarithmic Forms and Vice Versa
2. Evaluate Logarithms Base 10 and Base e
3. Use the Properties of Logarithms
4. Use the Change-of-Base Formula
5. Solve Applied Problems

4.3 Properties of Logarithmic Functions

In Section 4.1, we learned that the graph of the inverse function f^{-1} is a reflection of the graph of f across the line $y = x$. In Section 4.2, we indicated that exponential functions are one-to-one, so they have inverses. Consequently, if we graph the function $f(x) = b^x$ and reflect its graph across the line $y = x$, the result is the graph of f^{-1}. This new function is given the name **logarithmic function with base b,** and it is written as $f(x) = \log_b x$.

For example, as Figure 1 shows, the graph of $f^{-1}(x) = \log_2 x$ is the reflection of the graph of $f(x) = 2^x$ across the line $y = x$. Since logarithms, in a sense, evolve as inverses of exponential functions, the algebra of logarithms is derived from the algebra of exponents, as we shall see.

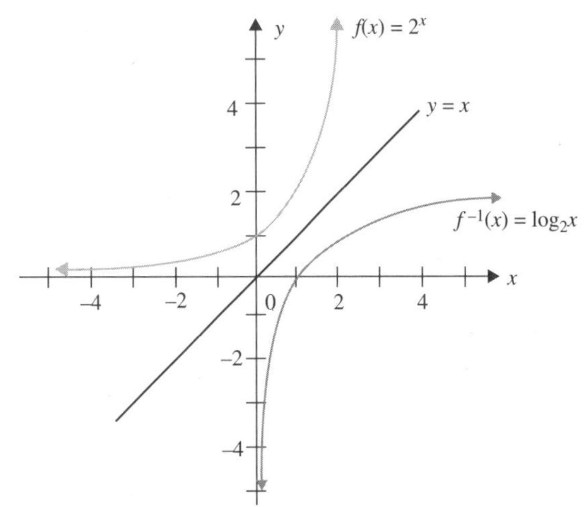

Figure 1

Converting Exponentials to Logarithmic Forms and Vice Versa

A logarithm is actually an exponent. For instance, consider the expression

$$49 = 7^2.$$

We refer to

2 as the *logarithm* of 49 with base 7,

and we write

$$\log_7 49 = 2.$$

The logarithm value 2 is the exponent to which 7 is raised to get 49.

In general, we have the following definition:

Definition Logarithm

Let $b > 0$ and $b \neq 1$. The **logarithm of x with base b,** which is represented by y, is defined by

$$y = \log_b x \qquad \text{if and only if} \qquad x = b^y$$

for every $x > 0$ and for every real number y.

In words, this definition says that

> the logarithm y is the exponent to which b is raised to get x.

The two equations in the above definition are equivalent and as such can be used interchangeably. The first equation is in *logarithmic form* and the second is in *exponential form.* The diagram below is helpful when changing from one form to the other:

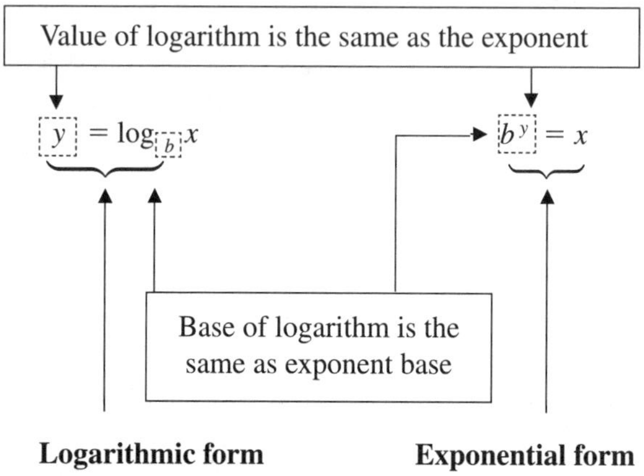

Value of logarithm is the same as the exponent

$y = \log_b x$ $\quad\quad$ $b^y = x$

Base of logarithm is the same as exponent base

Logarithmic form $\quad\quad$ **Exponential form**

TABLE 1

x	$\log_2 x$
2^1	1
2^2	2
2^3	3
2^{-1}	-1
2^{-2}	-2
2^{-3}	-3

It is important to observe that $\log_b x$ is an exponent. For instance, the numbers in the right column of Table 1 are the logarithms (base 2) of the numbers in the left column.

EXAMPLE 1 Converting Exponentials to Logarithmic

Convert each exponential form equation to logarithmic form.

(a) $5^2 = 25$ $\quad$ (b) $\sqrt{49} = 7$ $\quad$ (c) $e^{-1} = \frac{1}{e}$

Solution The equivalencies are listed in the following table.

Exponential Form	Logarithmic Form
(a) $5^2 = 25$	$\log_5 25 = 2$
(b) $\sqrt{49} = 49^{\frac{1}{2}} = 7$	$\log_{49} 7 = \frac{1}{2}$
(c) $e^{-1} = \frac{1}{e}$	$\log_e \frac{1}{e} = -1$

❖

EXAMPLE 2 Converting Logarithmic to Exponentials

Convert each logarithmic form to an equivalent exponential form.

(a) $\log_2 16 = 4$ $\quad$ (b) $\log_9 3 = \frac{1}{2}$ $\quad$ (c) $\log_5 \frac{1}{25} = -2$

(d) $\log_{10} 17 = t$ $\quad$ (e) $\log_e(2x - 5) = y$

Solution The following table lists the equivalency of each logarithmic form.

Logarithmic Form	Exponential Form
(a) $\log_2 16 = 4$	$2^4 = 16$
(b) $\log_9 3 = \frac{1}{2}$	$9^{\frac{1}{2}} = 3$
(c) $\log_5 \frac{1}{25} = -2$	$5^{-2} = \frac{1}{25}$
(d) $\log_{10} 17 = t$	$10^t = 17$
(e) $\log_e(2x - 5) = y$	$e^y = 2x - 5$

❖

At times, we can find the numerical value of a logarithm by converting to exponential form and then using the one-to-one property of exponents, as the next example illustrates.

EXAMPLE 3 **Evaluating Logarithms**

Evaluate each logarithm.

(a) $\log_4 64$ (b) $\log_3 \frac{1}{9}$ (c) $\log_e e^4$ (d) $\log_{\frac{1}{4}} 16$

Solution In each case we let u equal the given expression, and write the logarithmic equation in its equivalent exponential form and then solve the resulting equation for u as shown in the following table.

Logarithmic Form	Exponential Form	Result
(a) $\log_4 64 = u$	$4^u = 64$ $4^u = 4^3$	$u = 3$ or $\log_4 64 = 3$
(b) $\log_3 \frac{1}{9} = u$	$3^u = \frac{1}{9}$ $3^u = 3^{-2}$	$u = -2$ or $\log_3 \frac{1}{9} = -2$
(c) $\log_e e^4 = u$	$e^u = e^4$	$u = 4$ or $\log_e e^4 = 4$
(d) $\log_{\frac{1}{4}} 16 = u$	$\left(\frac{1}{4}\right)^u = 16$ $4^{-u} = 4^2$	$u = -2$ or $\log_{\frac{1}{4}} 16 = -2$

We use the equivalency between the logarithmic and exponential forms to solve certain equations involving logarithms, as the next example shows.

EXAMPLE 4 **Solving Logarithmic Equations**

Solve each logarithmic equation for x.

(a) $\log_3 x = 4$ (b) $\log_5 \frac{1}{125} = x$

(c) $\log_x 8 = 3$ (d) $3 + \log_e 2x = t$

Solution First we rewrite the given equation in exponential form and then solve the resulting equation.

(a) $\log_3 x = 4$ is equivalent to $3^4 = x$, so $x = 81$.

(b) $\log_5 \dfrac{1}{125} = x$ is equivalent to

$$5^x = \frac{1}{125}$$

$$5^x = 5^{-3} \qquad \text{Rewrite } \frac{1}{125} = 5^{-3}$$

$$x = -3 \qquad \text{One-to-one property}$$

(c) $\log_x 8 = 3$ is equivalent to

$$x^3 = 8$$

$$x^3 = 2^3 \qquad \text{Rewrite } 8 = 2^3$$

$$x = 2 \qquad \text{One-to-one property}$$

(d) $3 + \log_e 2x = t$ is equivalent to $\log_e 2x = t - 3$

$$2x = e^{t-3} \qquad \text{Convert to exponential form}$$

$$x = \frac{1}{2}e^{t-3} \qquad \text{Divide by 2}$$

❖

Evaluating Logarithms—Base 10 and Base *e*

The base of a logarithmic function can be any positive number except 1. However, the two bases that are most widely used are 10 and *e*.

A logarithm with base 10 is called a **common logarithm.** Its value at *x* is denoted by **log *x*,** that is,

$$\log x = \log_{10} x.$$

A logarithm with base *e* is called a **natural logarithm,** and its value at *x* is denoted by **ln *x*,** that is,

$$\ln x = \log_e x.$$

Since $b^1 = b$, it follows that $\log_b b = 1$, and so

$$\log 10 = 1 \text{ and } \ln e = 1.$$

Sometimes we can use the definition of logarithms to evaluate common and natural logarithms easily. For instance:

$$\log 1000 = 3 \qquad \text{since } 10^3 = 1000$$

$$\log 0.0001 = -4 \qquad \text{since } 10^{-4} = 0.0001$$

$$\ln e^5 = 5 \qquad \text{since } e^5 = e^5$$

$$\ln \frac{1}{e^2} = -2 \qquad \text{since } e^{-2} = \frac{1}{e^2}$$

Scientific calculators have the following function keys:

log for base 10 and an "ellen" **ln** key for the base e

to approximate values of logarithms. Table 2 lists some calculator approximations of logarithms (rounded to four decimal places) along with the equivalent exponential forms.

TABLE 2
Examples—Calculator Values of Logarithms

Approximate Logarithm Value	Exponential Form
$\log 4819 = 3.6830$	$10^{3.6830} = 4819$
$\log 0.897 = -0.0472$	$10^{-0.0472} = 0.897$
$\ln 45 = 3.8067$	$e^{3.8067} = 45$
$\ln 0.057 = -2.8647$	$e^{-2.8647} = 0.057$

Next we consider the problem of finding the value of x when the value of $\log x$ or $\ln x$ is given.

EXAMPLE 5 Solving Logarithmic Equations

Use a calculator to approximate the value of x to four decimal places.

(a) $\log x = 0.7235$
(b) $\ln x = 1.4674$
(c) $\log(x - 1) = -1.7$

Solution Using a calculator and rounding the results to four decimal places, we get:

(a) $\log x = 0.7235$ is equivalent to $x = 10^{0.7235} = 5.2905$.
(b) $\ln x = 1.4674$ is equivalent to $x = e^{1.4674} = 4.3379$.
(c) $\log(x - 1) = -1.7$ is equivalent to $x - 1 = 10^{-1.7}$, so $x = 1 + 10^{-1.7}$ or $x = 1.0200$. ❖

Using the Properties of Logarithms

Since logarithms are exponents, we can use the properties of exponents to derive some basic properties of logarithms.

For instance, notice that if $y = \log_b N$, then $N = b^y$. Substituting for y in the latter equation, we get

$$N = b^{\log_b N}$$

Basic Properties of Logarithms

Suppose that M, N, and b are positive real numbers, where, $b \neq 1$, and r is any real number. Then we have:

(i) The Product Rule:

$$\log_b MN = \log_b M + \log_b N$$

(ii) The Quotient Rule:

$$\log_b \frac{M}{N} = \log_b M - \log_b N$$

(iii) The Logarithm of a Power:

$$\log_b N^r = r \log_b N$$

These properties can be described in words as follows:

(i) The logarithm of a product of numbers is the sum of logarithms of the numbers.
(ii) The logarithm of a quotient of numbers is the difference of the logarithms of the numbers.
(iii) The logarithm of a power of a number is the exponent times the logarithm of the number.

The proofs of these properties are based on properties of exponents. For instance, a proof of Property (i) follows.

Since

$$M = b^{\log_b M} \text{ and } N = b^{\log_b N},$$

then

$$MN = b^{\log_b M} \cdot b^{\log_b N}$$
$$= b^{\log_b M + \log_b N}$$

By converting the equation to logarithmic form, we get

$$\log_b MN = \log_b M + \log_b N.$$

The basic properties of logarithms for the special cases of the common and natural logarithms are stated in Table 3.

TABLE 3
Basic Properties of Common and Natural Logarithms

Common Logarithms *Base 10*	Natural Logarithms *Base e*
(i) $\log MN = \log M + \log N$	(i) $\ln MN = \ln M + \ln N$
(ii) $\log \frac{M}{N} = \log M - \log N$	(ii) $\ln \frac{M}{N} = \ln M - \ln N$
(iii) $\log N^r = r \log N$	(iii) $\ln N^r = r \ln N$

In Property (iii), if the value of N is the same as the base, then we get $\log_b b^r = r \log_b b = r \cdot 1 = r$, so

$$\log_b b^r = r.$$

Specifically for $b = 10$ and $b = e$, we have

$$\log 10^r = r \text{ and } \ln e^r = r.$$

The following identities follow directly from the definition of logarithms and their properties ($b > 0$, $b \neq 1$, and $x > 0$).

Summary of Logarithmic Identities

1. $\log_b 1 = 0$ because $b^0 = 1$
2. $\log_b b = 1$ because $b^1 = b$
3. $\log_b b^x = x$ because $b^x = b^x$
4. $b^{\log_b x} = x$ because if $y = \log_b x$ then $b^y = x$

The next two examples illustrate how the properties of logarithms are used to manipulate logarithmic expressions algebraically.

EXAMPLE 6 **Using the Properties of Logarithms**

Write each expression as a sum or difference of multiples of logarithms.

(a) $\log_8 \dfrac{x}{5}$ (b) $\log(x^7 y^{11})$ (c) $\ln\left[\dfrac{(y+7)^3}{\sqrt{y}}\right]$

Solution We assume that all numbers whose logarithms are taken are positive.

(a) $\log_8 \dfrac{x}{5} = \log_8 x - \log_8 5$ Property (ii)

(b) $\log(x^7 y^{11}) = \log x^7 + \log y^{11}$ Property (i)

$= 7 \log x + 11 \log y$ Property (iii)

(c) $\ln\left[\dfrac{(y+7)^3}{\sqrt{y}}\right] = \ln(y+7)^3 - \ln \sqrt{y}$ Property (ii)

$= \ln(y+7)^3 - \ln y^{1/2}$ $\sqrt{y} = y^{1/2}$

$= 3 \ln(y+7) - \dfrac{1}{2} \ln y$ Property (iii) ❖

EXAMPLE 7 **Combining Logarithmic Expressions**

Write each expression as a single logarithm.

(a) $2 \ln x - 3 \ln y$, where $x > 0$ and $y > 0$

(b) $\log(c^2 - cd) - \log(2c - 2d)$, where $c > d$.

Solution (a) $2 \ln x - 3 \ln y = \ln x^2 - \ln y^3$ Property (iii)

$$= \ln\left(\frac{x^2}{y^3}\right)$$ Property (ii)

(b) $$\log(c^2 - cd) - \log(2c - 2d) = \log\left[\frac{c^2 - cd}{2c - 2d}\right]$$ Property (ii)

$$= \log\left[\frac{c(c-d)}{2(c-d)}\right] = \log\left(\frac{c}{2}\right)$$

❖

Notice that, although the properties of the logarithms tell us how to compute the logarithm of a product or a quotient, there is no corresponding property for the logarithm of a sum or difference. That is,

$$\log_b(M + N) \neq \log_b M + \log_b N$$

and

$$\log_b(M - N) \neq \log_b M - \log_b N$$

In calculus, we encounter expressions involving exponential and logarithmic functions with base e that have to be simplified by using the properties and identities for logarithms.

EXAMPLE 8 Simplifying Expressions Involving Base *e*

Simplify each expression

(a) $e^{3 \ln(2x+1)}$ (b) $e^{4-2 \ln x}$

Solution (a) $e^{3 \ln(2x+1)} = e^{\ln(2x+1)^3}$ Property (iii)

$= (2x + 1)^3$ $e^{\ln u} = u$ Identity 4

(b) $e^{4-2 \ln x} = e^4 \cdot e^{-2 \ln x}$ $e^{u+v} = e^u \cdot e^v$

$= e^4 \cdot e^{\ln x^{-2}}$ Property (iii)

$= e^4 \cdot x^{-2}$ $e^{\ln u} = u$ Identity 4

$= \dfrac{e^4}{x^2}$ $a^{-n} = \dfrac{1}{a^n}$

Using the Change-of-Base Formula

To determine logarithms with bases other than 10 or e, such as $\log_3 7$, we use the following *change-of-base formula:*

Change-of-Base Formula

$$\log_b x = \frac{\log_a x}{\log_a b}$$

where a and b are positive real numbers different from 1, and x is positive.

We can verify the formula as follows:

Let $y = \log_b x$

$$\begin{aligned}
b^y &= x && \text{Exponential form} \\
\log_a b^y &= \log_a x && \text{Take the logarithm of each side to base } a \\
y \log_a b &= \log_a x && \text{Property (iii)} \\
y &= \frac{\log_a x}{\log_a b} && \text{Solve for } y \\
\log_b x &= \frac{\log_a x}{\log_a b} && y = \log_b x
\end{aligned}$$

In practice, we usually choose either e or 10 for the new base, so that a calculator can be used to evaluate the necessary logarithm. Thus,

$$\log_b x = \frac{\log x}{\log b} \quad \text{or, equivalently,} \quad \log_b x = \frac{\ln x}{\ln b}.$$

EXAMPLE 9 **Using the Change-of-Base Formula**

Use the change-of-base formula to calculate each expression. Round off the answers to three decimal places.

(a) $\log_3 5$ (b) $\log_{\sqrt{2}} \sqrt{11}$

Solution Rounding off to three decimal places, we get:

(a) $\log_3 5 = \dfrac{\log 5}{\log 3} = 1.465$ or $\log_3 5 = \dfrac{\ln 5}{\ln 3} = 1.465$

(b) $\log_{\sqrt{2}} \sqrt{11} = \dfrac{\log \sqrt{11}}{\log \sqrt{2}} = 3.459$ or $\log_{\sqrt{2}} \sqrt{11} = \dfrac{\ln \sqrt{11}}{\ln \sqrt{2}} = 3.459$ ❖

Solving Applied Problems

Applications as diverse as measuring the strength of an earthquake, determining the acidity of a liquid, and modeling the profit from a business make use of logarithmic functions.

EXAMPLE 10 **Measuring the Strength of an Earthquake**

The **Richter scale** provides us with a way of grading the strength of an earthquake on a scale from 1 to 10—the larger the number, the more severe the earthquake. For a given earthquake, suppose that the largest seismic wave recorded on a seismograph is I and the smallest seismic wave recorded for the area is I_0. Then the **magnitude** M of the earthquake is given by

$$M = \log \frac{I}{I_0}$$

where I is called the **amplitude** of the earthquake, I_0 is called the **zero-level amplitude,** and the ratio I/I_0 is called the **intensity** of the earthquake.

(a) The 1989 Northern California earthquake had a magnitude of 7.1. Find the intensity of the earthquake and express its amplitude in terms of the zero-level amplitude.

(b) The 2003 earthquake in Bam, Iran had a magnitude of 6.5. Compare the intensity of this earthquake to the intensity of the 1989 Northern California earthquake.

Solution (a) Substituting $M = 7.1$ into the formula $M = \log\frac{I}{I_0}$ we get

$$7.1 = \log\frac{I}{I_0}$$

Converting to exponential form, we find that the intensity is given by

$$\frac{I}{I_0} = 10^{7.1} \text{ so that } I = 10^{7.1}\, I_0.$$

That is, the amplitude of the earthquake was $10^{7.1}$ (approx. 12,589,254) times the zero-level amplitude.

(b) For the Bam, Iran earthquake, we have

$$6.5 = \log\frac{I}{I_0}$$

so the intensity is $\frac{I}{I_0} = 10^{6.5}$ (approx. 3,162,278).

Since $10^{7.1} = (10^{6.5})(10^{0.6})$ it follows that the intensity of the California earthquake was $10^{0.6}$ or approximately 4 times the intensity of the Bam, Iran earthquake.

PROBLEM SET 4.3

Mastering the Concepts

In problems 1 and 2, write each exponential equation in logarithmic form.

1. **(a)** $5^3 = 125$
(b) $32^{1/5} = 2$
(c) $17 = e^t$
(d) $b^x = 13z + 1$

2. **(a)** $\left(\frac{1}{2}\right)^{-3} = 8$
(b) $9^{3/2} = 27$
(c) $10^w = x - 3$
(d) $a^8 = y - 5$

In problems 3 and 4, write each logarithmic equation in exponential form.

3. **(a)** $\log_9 81 = 2$
(b) $\log 0.0001 = -4$
(c) $\log_c 9 = w$
(d) $\ln\frac{1}{2} = -1 - 3x$

4. **(a)** $\log_{36} 216 = \frac{3}{2}$
(b) $\ln s = t$
(c) $\log x = \sqrt{2}$
(d) $\log_b 7 = -3 + 8x$

In problems 5–8, find the exact value of each logarithm.

5. **(a)** $\log_2 \frac{1}{8}$
(b) $\ln e^4$
(c) $\log \sqrt[3]{10}$
(d) $\log_8 4$

6. **(a)** $\log_7 7$
(b) $\log_5 \frac{1}{625}$
(c) $\log \sqrt{10}$
(d) $\ln \sqrt[3]{e}$

7. **(a)** $\log_8 1$
(b) $\log_{1/2} 4$
(c) $\log_{10} 10^3$
(d) $\log_8 2$

8. **(a)** $\log_{0.5} 0.5$
(b) $\ln e^7$
(c) $10^{\log 3}$
(d) $e^{-\ln 4}$

In problems 9–14, solve each logarithmic equation for x by using its exponential form.

9. **(a)** $\log_6 x = 2$
(b) $\log_x \frac{1}{4} = -\frac{1}{2}$

10. **(a)** $\log_{27} x = \frac{1}{3}$
(b) $\log_x 4 = \frac{2}{3}$

11. **(a)** $\log_2 16 = 3x - 5$
(b) $\log_3(x + 1) = 2$

12. **(a)** $\log_5(5x - 1) = -2$
(b) $\log_5(x^2 - 4x) = 1$

13. **(a)** $y = 3 - \log x$
(b) $y = 8 + \ln(2x + 1)$

14. **(a)** $y = \ln(x - 1)$
(b) $y = 7 + \log(4 - x)$

In problems 15 and 16, use a calculator to approximate the value of each logarithm to four decimal places. Express the result in the equivalent exponential form.

15. **(a)** $\log 39.2$
(b) $\ln 961$
(c) $\log\frac{3}{4}$
(d) $\ln 0.23$
(e) $\ln 10$

16. **(a)** $\ln 39.2$
(b) $\log 961$
(c) $\ln\frac{3}{4}$
(d) $\log 0.23$
(e) $\log e$

In problems 17–20, use a calculator to approximate the value of x to four decimal places.

17. **(a)** $\ln x = 2.37$
(b) $\log x = 0.4137$
(c) $\ln x = -2.5$
(d) $\ln(2x - 1) = -3.8$

18. **(a)** $\log x = 2.37$
(b) $\ln x = 0.4137$
(c) $\log x = -2.5$
(d) $\log(2x - 1) = -3.8$

19. **(a)** $x = \frac{\log 3}{\log 1.71}$
(b) $x = \frac{\ln 3}{\ln 1.52}$

20. **(a)** $x = \frac{\ln 0.2}{-0.03}$
(b) $x = \frac{\log 0.7}{-0.05}$

In problems 21–26, write each expression as a sum or difference of multiples of logarithms.

21. **(a)** $\log_3 x(x + 1)$
(b) $\log_3 \frac{18}{x + 2}$

22. **(a)** $\log_b \frac{x^9}{y^7}$
(b) $\log_b x^2y^3$

23. **(a)** $\log_b(x + 3)^4$
(b) $\log x\sqrt{2x + 1}$

24. **(a)** $\log \sqrt[6]{\frac{x^5}{y^2}}$
(b) $\ln \frac{x^5 y^2}{\sqrt[5]{z}}$

25. **(a)** $\ln[y(3x + 1)^{2/3}]$
(b) $\ln(x^2 + 7x)$

26. **(a)** $\log_5 \frac{x - 2}{x^2 - 4}$
(b) $\log \frac{5x(x^2 + 1)^2}{(x + 1)\sqrt{7x + 3}}$

In problems 27–30, use the properties of logarithms to write each expression as a single logarithm.

27. **(a)** $\log_5 \frac{5}{7} + \log_5 \frac{40}{25}$

(b) $\log_2 \frac{32}{11} + \log_2 \frac{121}{16} - \log_2 \frac{4}{5}$

28. **(a)** $\log(a^2 - ab) - \log(7a - 7b)$

(b) $\log\left(a + \frac{a}{b}\right) - \log\left(c + \frac{c}{b}\right)$

29. **(a)** $\ln(x^2 - 9) - \ln(x^2 - 6x + 9)$

(b) $\ln(3x^2 + 7x + 4) - \ln(3x^2 - 5x - 12)$

30. **(a)** $\log\left(\frac{1}{4} - \frac{1}{x^2}\right) - \log\left(\frac{1}{2} - \frac{1}{x}\right)$

(b) $3\ln(x + 2) - \frac{1}{3}\ln x - \frac{1}{3}\ln(1 - 2x)$

In problems 31–34, simplify each expression.

31. **(a)** $e^{\ln 5}$
(b) $e^{-2\ln 3}$

32. **(a)** $e^{3+4\ln x}$
(b) $e^{\ln(1/x)}$

33. **(a)** $e^{-7-\ln x}$
(b) $\ln e^{x^2-4}$

34. **(a)** $e^{(\ln x^2)-1}$
(b) $e^{\ln x - 3\ln y}$

In problems 35 and 36, compute the value to three decimal places after using the change-of-base formula to write each logarithm as a quotient of:

(i) Common logarithms **(ii)** Natural logarithms

35. **(a)** $\log_2 5$
(b) $\log_7 2.89$
(c) $\log_4 \frac{3}{23}$
(d) $\log_3 e$

36. **(a)** $\log_8 13$
(b) $\log_4 \frac{1}{17}$
(c) $\log_3 0.46$
(d) $\log_{11} 7$

Applying the Concepts

37. Installment Loan Payment: An equation related to a loan of P dollars to be paid off in n equal monthly installment payments of A dollars with an interest rate of r percent per period is given by

$$\ln A - \ln P = \ln r + n\ln(1 + r) - \ln[(1 + r)^n - 1]$$

where r = (Annual loan rate)/12.

(a) Solve this equation for A in terms of P, r, and n.

(b) Express A as a function of n if \$18,000 is borrowed to purchase a boat at an annual rate of 10%. What are the monthly payments if the loan is to be paid off in 3 years? 4 years? 5 years?

38. Depreciation: Used-car dealers determine the value of a car by applying the formula

$$\log(1 - r) = \frac{1}{t}(\log w - \log p)$$

where p (in dollars) is the purchase price of a car when it was new, and w (in dollars) is its value t years later at an annual rate of depreciation r.

(a) Solve this equation for r in terms of t, w, and p.

(b) Express the annual rate of depreciation r as a function of t if a new car was purchased for \$18,300 and sold t years later for \$7500. What is the rate of depreciation if the car was sold after 3 years? 3 years and 6 months? 4 years? Round off the answers to two decimal places.

39. Population Growth—Fruit Flies: The fruit fly *Drosophila melanogaster* is used in some genetic studies in laboratories. The number N of fruit flies in a colony after t days of breeding is given by the equation

$$\ln(230 - N) - \ln N = \ln 6.931 - 0.1702t.$$

(a) Express N as a function of t.

(b) How many fruit flies are in the colony initially? After 10 days? After 16 days? After 30 days?

40. Physics—Speed of a Rocket: Suppose a rocket with mass M_0 (in kilograms) is moving in free space at a speed of V_0 (in kilometers per second). When the rocket is fired, its speed V is given by

$$V = V_0 + V_1(\ln M_0 - \ln M)$$

where M is the mass of the propellant that has been burned off and V_1 is the speed of the exhaust gases produced by the burn. Solve the equation for M in terms of V, V_0, V_1, and M_0.

41. Magnitude of an Earthquake: The amplitude of the earthquake in Japan in May 1983 was $10^{7.7}$ times the zero-level amplitude. Find the magnitude of the earthquake on the Richter scale.

42. Earthquake Comparisons: The 1988 Armenian earthquake had a magnitude of 6.8 on the Richter scale, and the 1985 earthquake in Mexico City had a magnitude of 7.8. How many times more powerful was the intensity of the Mexico City earthquake than the Armenian earthquake?

*In problems 43 and 44, the concentration of hydrogen ions in a substance is denoted by $[H^+]$, measured in moles per liter. The **pH** of a substance is defined by the logarithmic function*

$$\text{pH} = -\log\,[\text{H}^+].$$

This function is used to measure the *acidity* of the substance. The pH of neutral distilled water is 7. A substance with a pH of less than 7 is known as an *acid,* whereas a substance with a pH of more than 7 is called a *base.*

43. **pH Scale:** Find the pH of each substance and determine whether it is an acid or a base. Round off the answers to one decimal place.
   (a) Beer: $[H^+] = 3.16 \times 10^{-3}$ mole per liter
   (b) Eggs: $[H^+] = 1.6 \times 10^{-8}$ mole per liter
   (c) Milk: $[H^+] = 4 \times 10^{-7}$ mole per liter

44. **pH Scale:**
   (a) Express the hydrogen ion concentration $[H^+]$ as a function of pH.
   (b) Use the function in part (a) to find the hydrogen ion concentration (in moles per liter) of: calcium hydroxide, pH = 13.2; vinegar, pH = 3.1; and tomatoes, pH = 4.2. Round off the answers to three decimal places.

In problems 45 and 46, the loudness L of a sound is measured by

$$L = 10 \log \frac{I}{I_0}$$

Here, L is the number of *decibels*, I is the intensity of the sound (in watts per square meter), and I_0 is the smallest intensity that can be heard. Suppose the intensity I_0 is 10^{-12} watt per square meter.

45. **Sound Intensity:**
   (a) Find the number of decibels of the noise of a truck passing a pedestrian at the side of the road if the sound intensity of the truck is 10^{-3} watt per square meter.
   (b) Find the number of decibels of sound from a jet plane with sound intensity $8.3 * 10^2$ watts per square meter. Round off the answer to two decimal places.

46. **Sound Intensity:**
   (a) The threshold of pain from loud sound is considered to be about 120 decibels. Find the intensity of such a sound in watts per square meter.
   (b) Find the intensity of a whisper in watts per square meter if the whisper is 25 decibels.

Developing and Extending the Concepts

47. Use $M = 10{,}000$, $N = 10$, $b = 10$, and $p = 3$ to show that each statement is false.
   (a) $\log_b \dfrac{M}{N} = \dfrac{\log_b M}{\log_b N}$
   (b) $\dfrac{\log_b M}{\log_b N} = \log_b M - \log_b N$
   (c) $\log_b M \cdot \log_b N = \log_b M + \log_b N$
   (d) $\log_b MN = \log_b M \cdot \log_b N$
   (e) $\log_b M^p = (\log_b M)^p$
   (f) $(\log_b M)^p = p \log_b M$

48. Show that: $\ln\left(\dfrac{\sqrt{3}+\sqrt{2}}{\sqrt{3}-\sqrt{2}}\right) = 2\ln\left(\sqrt{3}+\sqrt{2}\right)$.

In problems 49 and 50, find values for a and b that satisfy the given equation.

49. $\log(a + b) = \log a + \log b$
50. $\log(a - b) = \log a - \log b$

In problems 51 and 52, use the change-of-base formula to prove each equation.

51. $\log_b a = \dfrac{1}{\log_a b}$

52. $\dfrac{\log_b x}{\log_{ab} x} = 1 + \log_b a$

53. Find the domain and the x intercepts of the graph of
$$y = -1 + \ln(x - 2).$$
Also, find the y intercept (if any).

54. Criticize the statement:

$$1 < 2$$

$$\frac{1}{8} < \frac{2}{8} \qquad \text{Divide by 8}$$

$$\frac{1}{8} < \frac{1}{4} \qquad \text{Reduce}$$

$$\left(\frac{1}{2}\right)^3 < \left(\frac{1}{2}\right)^2$$

$$\log\left(\frac{1}{2}\right)^3 < \log\left(\frac{1}{2}\right)^2$$

$$3\log\frac{1}{2} < 2\log\frac{1}{2} \qquad \text{Divide by } \log\frac{1}{2}$$

$$3 < 2$$

55. Solve for x.
$$\log_3 x + \frac{1}{\log_x 3} = 4$$

56. Solve for x.
$$3 + \log_3(2x + 7) + \log_3(2x - 1) = 5 + \log_3 x$$

Interlude: Progress Check 4.3, Part 1

1. Convert the given exponential functions to the form $P(t) = Ae^{rt}$. Round constants to four decimal places.

(a) $P(t) = 34(1.56)^t$

(b) $P(t) = 99(0.68)^t$

(c) $P(t) = 150(4)^{\frac{t}{8}}$

2. Fill in the blanks.

(a) If $\log\left(\frac{A}{\alpha}\right) = 3$ and $\log\left(\frac{B}{\alpha}\right) = 5$, then the value of B is ____________ times the value of A.

(b) If $\log_2(3x) = 9$ and $\log_2(3y) = 4$, then the value of x is ____________ times the value of y.

3. Fill in the blanks. Give results as percentages, accurate to two decimal places.

(a) An incremental growth rate of 8.9% per year corresponds to a continuous growth rate of ____________ % per year.

(b) An incremental decay rate of 13.5% per hour corresponds to a continuous decay rate of ____________ % per hour.

(c) A continuous growth rate of 21% per second corresponds to an incremental growth rate of ____________ % per second.

(d) A continuous decay rate of 5.8% per minute corresponds to an incremental decay rate of ____________ % per minute.

4. In an experiment, the volume of liquid in a beaker decreases by 40% over the course of ten minutes. Assume that the volume of liquid in the beaker is decreasing exponentially.

(a) Determine the incremental percent per minute rate of decrease in volume. Give as a percentage, accurate to two decimal places.

(b) Determine the continuous percent per minute rate of decrease in volume. Give as a percentage, accurate to two decimal places.

Interlude: Progress Check 4.3, Part 2

1. Assume that x and y are positive numbers. Completely simplify $\dfrac{\ln e^{-2x}}{10e^{2\ln x - 3\ln y}}$.

2. The *doubling function* $D(r) = \dfrac{\ln 2}{\ln(1+r)}$ gives the number of years it takes to double your money when it is invested at annual interest rate r (expressed as a decimal), with interest compounded annually.

(a) Find the time it takes to double your money at each of the following annual interest rates: 4%, 6%, 8%.

(b) Round the answers in part a.) to the nearest year and compare them with these numbers: $\dfrac{72}{4}, \dfrac{72}{6}, \dfrac{72}{8}$. Use this evidence to state a "rule of thumb" for determining approximate doubling time, without using the function D. (This rule of thumb, which has long been used by bankers, is called the **rule of 72.**)

3. For some airplanes, the minimum runway length (in thousands of feet) required for takeoff is given by $L(x) = 3 \log^{x}$, where x is the weight of the airplane, in pounds.

(a) What is the heaviest such an airplane that is able to take off using a 15,000-foot runway? Give your result to the nearest pound.

(b) Let w denote the weight of such an airplane, in thousands of pounds. Let R be a function that gives the minimum runway length for an airplane of weight w. Write an equation for R in terms of $\log w$.

5. One way of describing the brightness of a star is by a magnitude, m, given by

$$m = 6 - 2.5 \log\left(\frac{l}{l_0}\right)$$

where l is the light flux (apparent brightness) of the star and l_0 is the light flux of the dimmest stars still visible to the naked eye. Which is brighter: a magnitude 1 star, or a magnitude 6 star? How many times brighter than the dimmer star is the brighter star?

OBJECTIVES

1. Define Logarithmic Functions
2. Graphing Logarithmic Functions by Converting to Exponential Form
3. Transforming Graphs of Logarithmic Functions
5. Solve Applied Problems

4.4 Graphing Logarithmic Functions

As a follow up to our coverage of the algebra of logarithms in Section 4.3, we shall now study logarithmic functions in more detail. Our interest is focused on graphing these functions and exploring their applications.

Defining Logarithmic Functions

We now define a new class of functions called *logarithmic functions,* which are inverses of exponential functions, as follows:

Definition Logarithmic Function

The inverse of the exponential function $f(x) = b^x$ is the **logarithmic function** defined by $g(x) = \log_b x$, and vice versa.

Table 1 lists examples of exponential functions along with their inverses.

TABLE 1
Examples—Inverses of Exponential Functions

Exponential Function	Inverse Function
$f(x) = 4^x$	$f^{-1}(x) = \log_4 x$
$f(x) = (1/3)^x$	$f^{-1}(x) = \log_{1/3} x$
$f(x) = 10^x$	$f^{-1}(x) = \log x$
$f(x) = e^x$	$f^{-1}(x) = \ln x$

Since $g(x) = \log_b x$ is the inverse of $f(x) = b^x$, it follows that the domain of f, which includes all real numbers, is the range of g.

By examining the definition of the logarithmic function, we observe that if

$$y = g(x) = \log_b x, \quad \text{then} \quad x = b^y \quad \text{where} \quad b > 0 \quad \text{and} \quad b \neq 1.$$

Since b^y is always positive under these conditions, it follows that x is always positive. Thus, in the real number system, we can only evaluate the logarithm of a positive number.

Table 2 lists some logarithmic functions with their domains.

TABLE 2

Function	Condition	Domain
1. $f(x) = \log_3(x - 2)$	$x - 2 > 0$	$x > 2$ or $(2, \infty)$
2. $f(x) = \log_5(4 - 7x)$	$4 - 7x > 0$	$x < \frac{4}{7}$ or $\left(-\infty, \frac{4}{7}\right)$

EXAMPLE 1 **Finding the Domain**

Find the domain of

$$f(x) = \ln (x^2 - x - 6).$$

Solution To find the domain of f we use the fact that $\ln (x^2 - x - 6)$ is defined only if

$$x^2 - x - 6 = (x - 3)(x + 2) > 0.$$

Solving this inequality, we get

$$x < -2 \quad \text{or} \quad x > 3.$$

Thus the domain of f consists of all real numbers in the intervals

$$(-\infty, -2) \quad \text{and} \quad (3, \infty).$$

Graphing Logarithmic Functions by Converting to Exponential Form

One approach to graphing a logarithmic function $y = \log_b x$ is to first convert its equation to the equivalent exponential form $x = b^y$. Then use the *point-plotting method* to sketch the graph of the equation $x = b^y$.

For example, to graph the logarithmic function

$$y = f(x) = \log_2 x.$$

First, we rewrite f in the form

$$x = 2^y$$

and then select values of y and their corresponding x values as shown in the table of values.

Then we graph the latter form by using point-plotting (Figure 1).

$x = 2^y$	$y = \log_2 x$
$2^{-2} = \frac{1}{4}$	-2
$2^{-1} = \frac{1}{2}$	-1
$2^0 = 1$	0
$2^1 = 2$	1
$2^2 = 4$	2

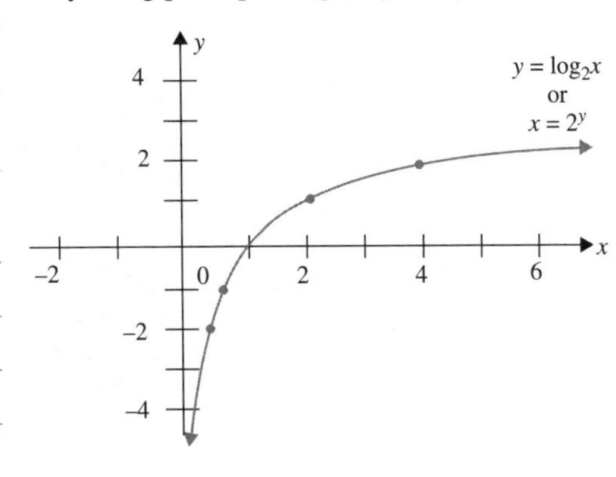

Figure 1

An alternate approach is to graph the function $y = 2^x$ and reflect the graph across the line $y = x$, as shown in Figure 2. Note that the horizontal asymptote (the x axis) for $y = 2^x$ converts to a vertical asymptote (the y axis) for $y = \log_2 x$.

That is, $y = \log_2 x \to -\infty$ as $x \to 0^+$.

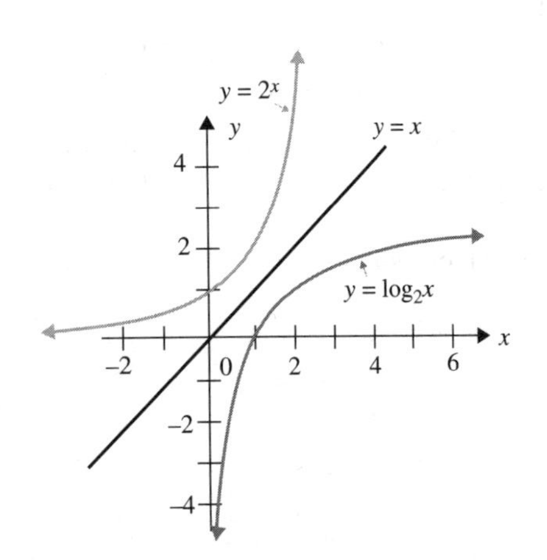

Figure 2

Note that the technique of interchanging the variables in $y = 2^x$ to obtain $f^{-1}(x) = \log_2 x$ causes the domain and range to be interchanged. That is, the domain of $y = 2^x$, $\mathbb{R}$, is the range of $y = \log_2 x$ and the range of $y = 2^x$, all positive real numbers, is the domain of $y = \log_2 x$.

EXAMPLE 2 **Graphing Logarithmic Functions Using Exponential Forms**

Graph each function by using its exponential form, and determine the domain and range.

(a) $f(x) = \log_4 x$ (b) $g(x) = \log_{1/4} x$

Solution (a) To graph f we first convert $y = f(x) = \log_4 x$ to its equivalent form, $x = 4^y$. Next, we form a table of values for $x = 4^y$ by substituting values for y and then finding the corresponding x values (Table 3).

Finally, we plot the points and connect them with a smooth curve to get the graph of f (Figure 3a).

The domain is the interval $(0, \infty)$ and the range includes all real numbers $\mathbb{R}$.

(b) Similarly, $y = g(x) = \log_{1/4} x$ is converted to the equivalent form $x = (1/4)^y$. Then this equation is graphed by point-plotting as in part (a) (Table 3, Figure 3b).

The domain is the interval $(0, \infty)$ and the range includes all real numbers $\mathbb{R}$.

TABLE 3

$x = 4^x$	$x = \left(\frac{1}{4}\right)^y$	y
1	1	0
4	$\frac{1}{4}$	1
16	$\frac{1}{16}$	2

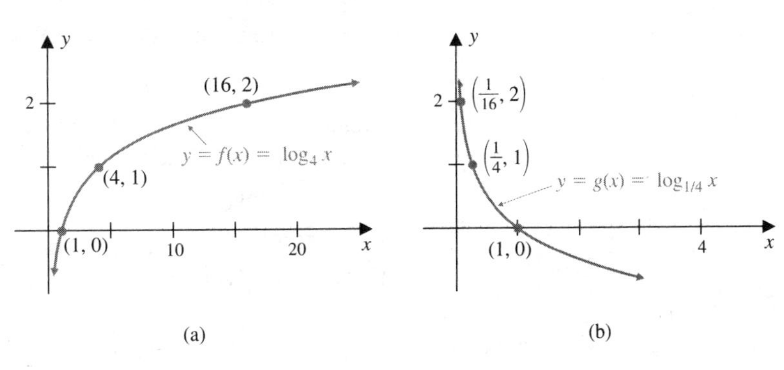

Figure 3

❖

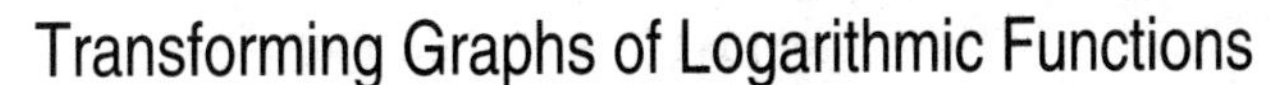

Transforming Graphs of Logarithmic Functions

The graphs in Figure 3 above serve as prototypes for the general graphs and properties of the logarithmic functions $f(x) = \log_b x$ for base $b > 1$ and for base $0 < b < 1$, respectively (Table 4).

TABLE 4

General Graphs and Properties of $f(x) = \log_b x$

Properties	General Graphs
1. The domain is the interval $(0, \infty)$.	$f(x) = \log_b x$, $b > 1$
2. The range includes all real numbers.	
3. The function increases on interval $(0, \infty)$ if $b > 1$ and decreases on interval $(0, \infty)$ if $0 < b < 1$.	
4. The y axis is a vertical asymptote.	
5. The function is one-to-one.	$f(x) = \log_b x$, $0 < b < 1$
6. The graph contains the point $(1, 0)$.	
7. The x intercept of the graph is 1. There is no y intercept.	
8. The graph is smooth and continuous.	

Examples of specific graphs for $b > 1$ and $0 < b < 1$ are shown in Figures 4a and 4b, respectively. We can use transformations of graphs of the functions

$$y = \log_b x$$

to graph other logarithmic functions.

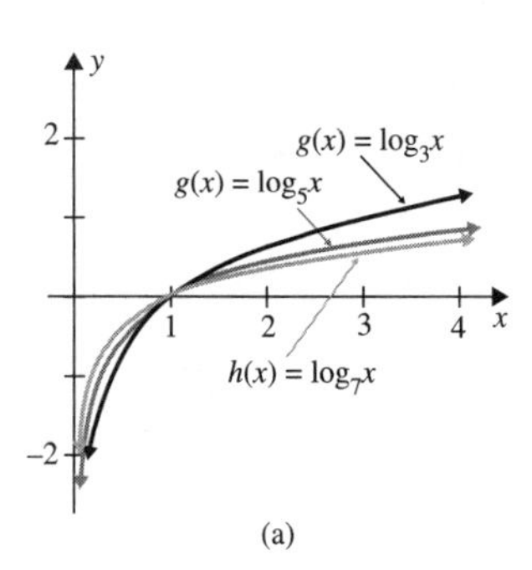

(a)

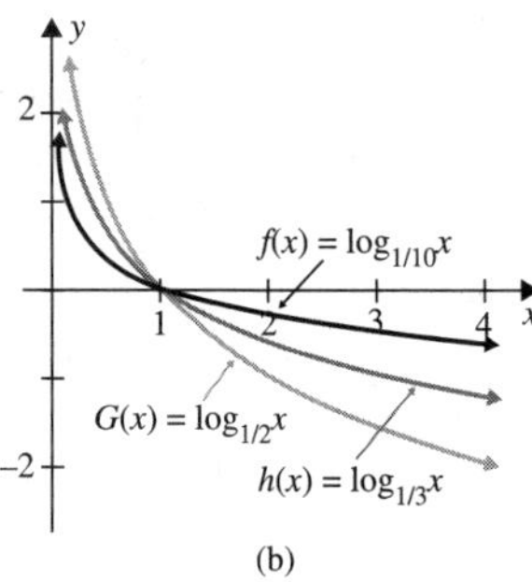

(b)

Figure 4

EXAMPLE 3 **Transforming Graphs of Logarithmic Functions**

Use a transformation of the graph of

$$y = \log_3 x$$

to sketch the graph of each function. Also, find the domain and range of each function.

(a) $f(x) = -\log_3 x$

(b) $g(x) = \log_3(-x)$

(c) $h(x) = \log_3(x - 1)$

Solution (a) We can graph the function

$$f(x) = -\log_3 x$$

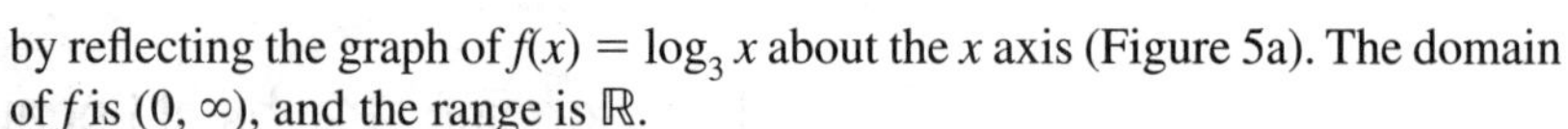

by reflecting the graph of $f(x) = \log_3 x$ about the x axis (Figure 5a). The domain of f is $(0, \infty)$, and the range is $\mathbb{R}$.

(b) To obtain the graph of

$$g(x) = \log_3(-x)$$

we reflect the graph of $g(x) = \log_3 x$ about the y axis (Figure 5b). The domain of g is $(-\infty, 0)$ and the range is $\mathbb{R}$.

(c) The graph of h is obtained by shifting the graph $y = \log_3 x$ one unit to the right (Figure 5c). The domain of h is $(1, \infty)$ and the graph shows that the range consists of all real numbers $\mathbb{R}$. Note that the line $x = 1$ is a vertical asymptote.

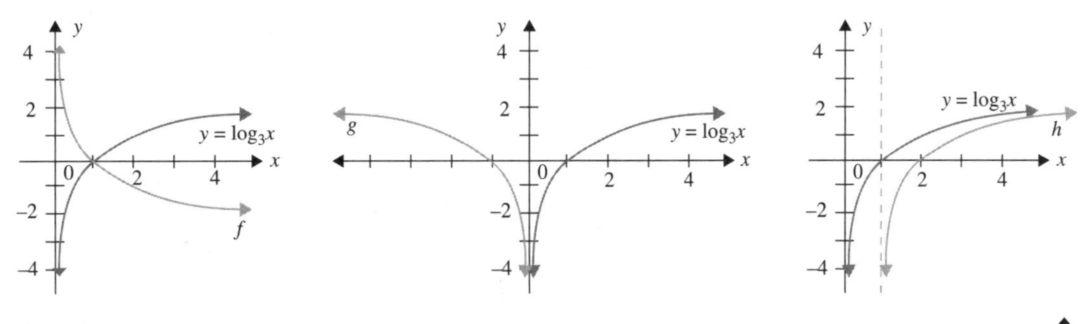

Figure 5

❖

Graphers can be helpful in displaying the domain and the vertical asymptotes of more complicated logarithmic functions. For instance, Figure 6 shows the graph of the function

$$f(x) = \ln(x^2 - x - 6)$$

in Example 1 on page 300. The graph of f indicates the domain consists of all real numbers in the intervals $(-\infty, -2)$ or $(3, \infty)$. Also the equations of the vertical asymptotes are

$$x = -2 \text{ and } x = 3.$$

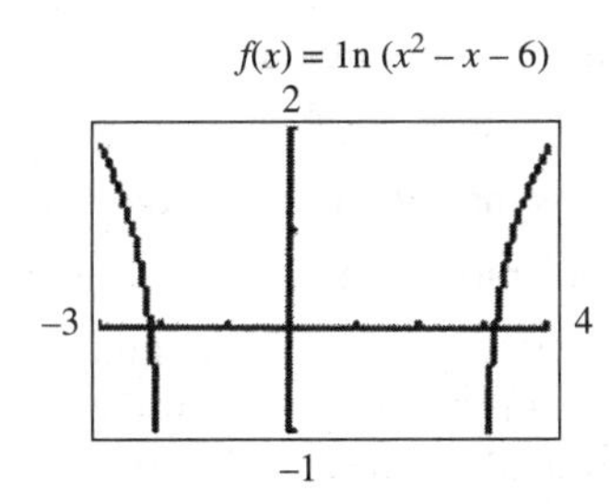

Figure 6

Graphers, along with the change of base formula, can be used to graph logarithmic functions with bases different than 10 or e. The next example illustrates the process.

EXAMPLE 4 **Using a Change-of-Base to Graph**

Use the change-of-base formula and a grapher to sketch the graph of $f(x) = \log_8 x$. Also determine the domain and the vertical asymptote.

Solution First we rewrite the function in terms of base 10 or base e. We'll select base 10. Thus

$$f(x) = \log_8 x = \frac{\log x}{\log 8}.$$

Next we use a grapher to graph $f(x) = (\log x)/(\log 8)$. A viewing window of the graph is shown in Figure 7. Since we can evaluate logarithms only for positive real numbers, it follows that the domain includes all values x in the interval $(0, \infty)$. The y axis is a vertical asymptote.

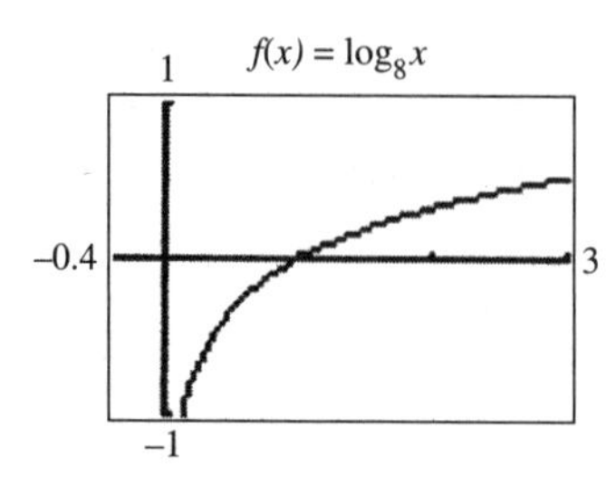

Figure 7

Solving Applied Problems

At times, logarithmic functions are used for best-fitting regression models, as the next example illustrates.

EXAMPLE 5 Modeling a Pricing Trend

A manufacturer charges different prices P (in dollars) for each calculator according to the number Q of calculators purchased each year by a school, as specified in the table.

Price P (in dollars) per Calculator	Quantity Q Purchased
120	95
110	160
100	270
90	285
80	305
70	320

(a) Plot a scattergram of this data with the price P as the horizontal axis and the quantity Q as the vertical axis.

(b) Use a grapher and the logarithmic regression feature to determine the logarithmic model of the form

$$Q = a + b\ln P$$

that best fits the data. Use four decimal places for a and b.

(c) Sketch the graph of the function in part (b).

(d) Use this pricing schedule to find the number of calculators needed to be purchased for the school year if the price of each calculator is to be \$115, \$95, or \$75. Round off the answers to the nearest integer.

Solution (a) The scattergram for the given data is shown in Figure 8, where P is used for the horizontal axis and Q is used for the vertical axis.

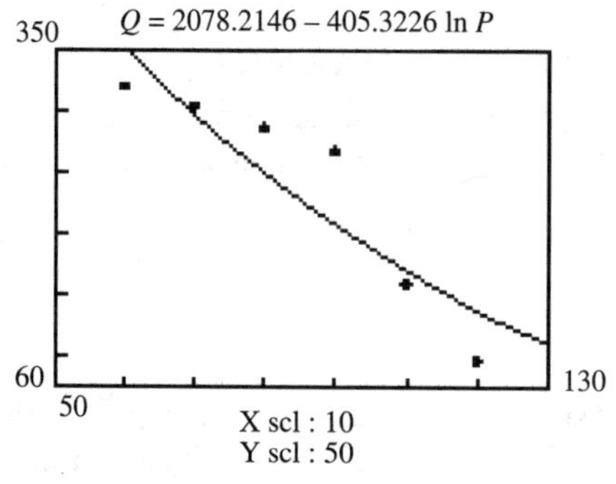

Figure 8

(b) By using a grapher, we find the logarithmic regression model to be

$$Q = 2078.2146 - 405.3226 \ln P.$$

(c) The graph of the logarithmic regression equation in part (b) is shown in Figure 8, along with the scattergram of the given data.

(d) To find the number of calculators needed to be purchased if the price is to be \$115, \$95, or \$75, we substitute $P = 115$, 95, and 75, respectively, into the regression equation to obtain the following approximations:

155 for the price of \$115

232 for the price of \$95

and 328 for the price of \$75 

PROBLEM SET 4.4

Mastering the Concepts

In problems 1–4, find the inverse of the given function, and then use symmetry to graph both functions on the same coordinate system. Also find the domain and range of each inverse function.

1. $f(x) = 5^x$ **2.** $g(x) = \left(\frac{1}{7}\right)^x$

3. $h(x) = \log_{1/10}x$ **4.** $f(x) = \log x$

5. Each curve in Figure 9 is the graph of a function of the form $y = \log_b x$. Find the base if its graph contains the given point.

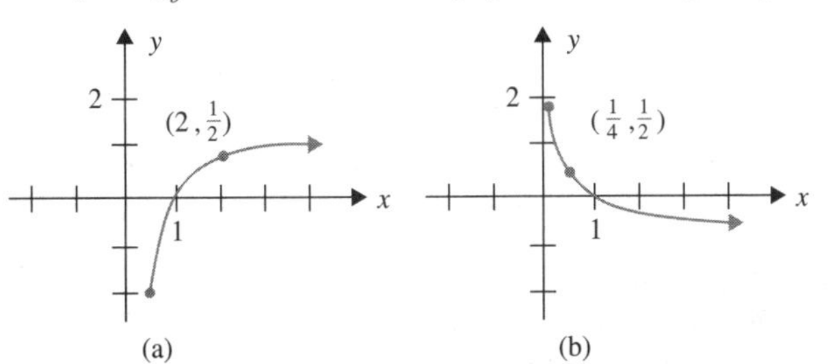

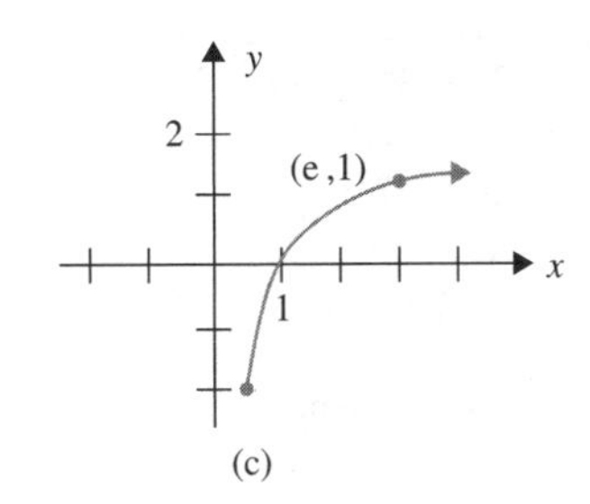

Figure 9

6. Simplify each expression.

(a) $\log 10^{3+2x}$

(b) $\ln\left(\frac{1}{e^{2x}}\right)$

In problems 7–16, find the domain of each function.

7. $f(x) = \ln(2 - x)$

8. $g(x) = \log(x + 3)$

9. $h(x) = \ln|x + 1|$

10. $F(x) = \ln\left(\frac{1}{\sqrt{x-3}}\right)$

11. $f(x) = \log_5(2x + 4)$

12. $g(x) = \log\left(\frac{\sqrt{x+1}}{x}\right)$

13. $g(x) = \log(x^2 - x)$

14. $h(x) = \ln(x^2 - 2x + 1)$

15. $f(x) = \log\left(\frac{1}{x^2 - 1}\right)$

16. $f(x) = \log_{\frac{1}{3}}\left(\frac{1}{x^2 - 9}\right)$

In problems 17 and 18, graph all three given logarithmic functions on the same coordinate system. Use the graphs to describe the relative steepness and limit behavior of the functions in terms of the different bases.

17. $f(x) = \log_2 x$; $g(x) = \log_{2.5} x$; $h(x) = \log_3 x$

18. $f(x) = \log_{1/3} x$; $g(x) = \log_{1/4} x$; $h(x) = \log_{1/5} x$

In problems 19–22, describe how the graph of f can be obtained from the graph of g. Find the domain of f, and write the equation of the vertical asymptote.

19. $g(x) = \log_5 x$; $f(x) = \log_5(x - 2)$

20. $g(x) = \log_4 x$; $f(x) = \log_4(-x)$

21. $g(x) = \ln x$; $f(x) = -\ln x$

22. $g(x) = \log x$; $f(x) = 2 + \log x$

In problems 23 and 24, find an equation of each curve that has been obtained by transforming the curve with the equation $y = \log_3 x$.

23.

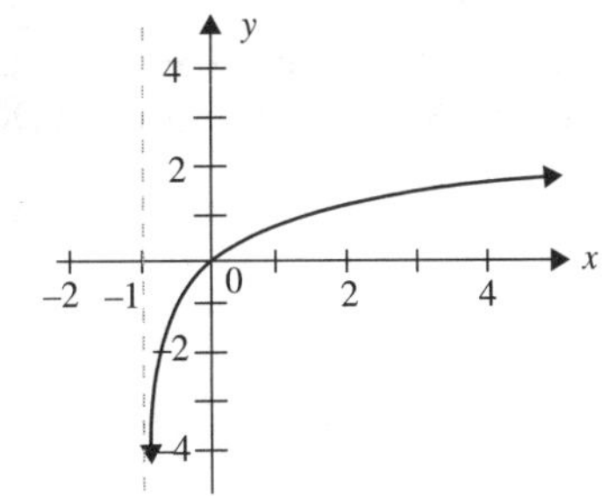

24.

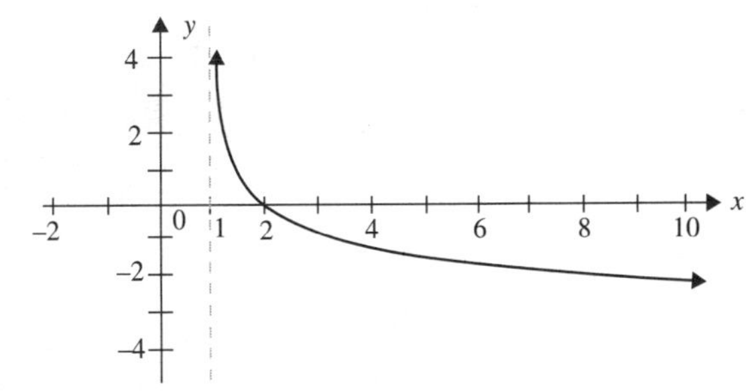

In problems 25–32, graph the given function by using transformations of the graph of $y = \log_b x$ with the same base. Determine the domain and vertical asymptote.

25. $f(x) = 1 + \ln x$

26. $g(x) = 3 \log x$

27. $g(x) = \frac{1}{3}\log_5 x - 2$

28. $h(x) = 1 - \log_7 x$

29. $f(x) = \log(x + 3)$

30. $g(x) = \log_5(x - 2)$

31. $f(x) = -2 \ln(x - e)$

32. $g(x) = \ln |1 - x|$

In problems 33–38, use the change-of-base formula along with a grapher to sketch the graph of each function. Also determine the domain and any vertical asymptotes.

33. $g(x) = \log_3 x$

34. $h(x) = \log_{1/3} x$

35. $g(x) = \log_4(5x + 1)$

36. $f(x) = \log_2(3 - 4x)$

37. $f(x) = \log_7(4x^2 - 1)$

38. $h(x) = \log_5(x^2 + 1)$

Applying the Concepts

39. **Advertising:** A company has determined that when its monthly advertising expenditure is x (in thousands of dollars), then the total monthly sales S (in thousands of dollars) is given by the model

$$S = \frac{2000 \ln(x + 5)}{\ln 10}$$

What are the expected monthly sales (to the nearest thousand dollars) if the monthly expenditure for advertising is \$10,000? \$50,000? \$100,000?

40. **Manufacturing:** A small auto parts supply company's weekly profit P (in dollars) is given by the function

$$P = 1828 + 914 \log x$$

where x is the number of parts produced each minute. Find the weekly profit (to the nearest dollar) if the number of parts produced each minute is 10, 50, 100, 500, and 1000.

Developing and Extending the Concepts

41. If $P_1 = -\log A$, $P_2 = -\log B$, and $0 < A < B$, then which is larger, P_1 or P_2? Explain.

42. Let $f(x) = 10 \ln\left(\frac{50}{50 - x}\right)$.

(a) For what value(s) of x is:

(i) $f(x) = 0$

(ii) $f(x) = 20$

Round off the answers to two decimal places.

(b) What is the domain of f?

In problems 43–48, use a grapher to sketch the graph of each function. Use the graph to find the domain and any vertical asymptotes. Confirm the domain algebraically.

43. $f(x) = x \ln(x - 2)$

44. $h(x) = 3[\log(3 - 2x)]^2$

45. $g(x) = \log(x^2 - 5x + 4)$

46. $f(x) = \ln(x^2 + 3x + 2)$

47. $f(x) = \log(2x + 3) + \log(2x - 3)$

48. $g(x) = \ln(x^2 - 9) - \ln(x + 3)$

49. Sketch the graph of $f(x) = 2 + \log_2 x$ and its inverse on the some coordinate system.

50. Show that for $(f \circ g)(x) = (g \circ f)(x) = x$ for $f(x) = \log_2(2x + 4)$ and $g(x) = 2^{x-1} - 2$. Also sketch the graphs of f and g on the same coordinate system.

OBJECTIVES

1. Solve Exponential Equations
2. Solve Exponential Inequalities
3. Solve Logarithmic Equations
4. Solve Applied Problems

4.5 Exponential and Logarithmic Equations and Inequalities

The properties of logarithms play an important role in solving certain equations and inequalities involving exponential and logarithmic functions such as

$$4^{2x-1} = 3, \; e^{-3t} < 0.5 \quad \text{and} \quad \log(x+2) - \log x = 1$$

are examples of these situations.

As we shall see, it is not always possible to solve these types of equations and inequalities algebraically. If this is the case, we rely on graphical solutions.

Solving Exponential Equations

Recall from Section 4.2 that we solved equations such as

$$2^x = 4^{2x-1}$$

by first writing each side as a power of the same base (2, in this example), equating the exponents and then solving the resulting equation. However, to solve an exponential equation such as

$$2^x = 7$$

where it is not easy to initially express each side in terms of the same base, we use the properties of logarithms, as the next example illustrates.

EXAMPLE 1 **Solving Equations Involving One Exponential Expression**

Use logarithms to solve each equation. Round off the answer to two decimal places.

(a) $4^{2x-1} = 3$ (b) $e^{-3t} = 0.5$

Algebraic Solution

(a) Taking the common logarithm of each side of the given equation, we have

$$\log 4^{2x-1} = \log 3$$

$$(2x-1)\log 4 = \log 3 \qquad \log N^r = r\log N$$

$$\left.\begin{aligned} 2x-1 &= \frac{\log 3}{\log 4} \\ 2x &= 1 + \frac{\log 3}{\log 4} \\ x &= \frac{1}{2}\left(1 + \frac{\log 3}{\log 4}\right) \end{aligned}\right\} \qquad \text{Solve for } x$$

$$x = 0.90 \text{ (approx.)} \qquad \text{Use a calculator}$$

(b) Taking the natural logarithm of each side of the equation $e^{-3t} = 0.5$, we have

$$\ln e^{-3t} = \ln 0.5$$

$$-3t\ln e = \ln 0.5 \qquad \ln N^r = r\ln N$$

$$t = \frac{-\ln 0.5}{3} \qquad \ln e = 1$$

$$t = 0.23 \text{ (approx.)} \qquad \text{Use a calculator}$$

Graphical Illustration

In each case, the solution represents the x intercept of a function.

(a) Since $4^{2x-1} = 3$ is equivelent to $4^{2x-1} - 3 = 0$, the solution is the same as the x intercept of $f(x) = 4^{2x-1} - 3$.

Figure 1a shows the x intercept of the graph of f (approximately 0.90), which is the solution of $4^{2x-1} = 3$.

(b) Figure 1b shows (approximately 0.23) the t intercept of the graph of $g(t) = e^{-3t} - 0.5$, which is the same as the solution of $e^{-3t} = 0.5$.

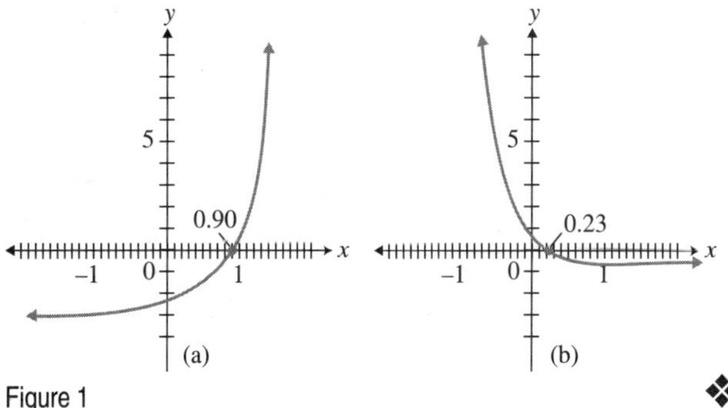

Figure 1

❖

EXAMPLE 2 **Solving an Equation Involving Two Exponential Expressions**

Solve the equation

$$5^{3x-2} = 3^x$$

Round off the answer to two decimal places.

Algebraic Solution

Taking the natural logarithm of each side, we have

$\ln(5^{3x-2}) = \ln 3^x$	Given
$(3x-2)\ln 5 = x\ln 3$	$\ln N^r = r\ln N$
$3x\ln 5 - 2\ln 5 = x\ln 3$	Distribute
$3x\ln 5 - x\ln 3 = 2\ln 5$	Rearrange to collect x terms on one side
$x(3\ln 5 - \ln 3) = 2\ln 5$	Factor out x
$x = \dfrac{2\ln 5}{3\ln 5 - \ln 3}$	Divide each side by $3\ln 5 - \ln 3$

the approximate solution, rounded to two decimal places, is given by

$$x = 0.86$$

Graphical Illustration

Figure 2 shows the graph of the function

$$y = 5^{3x-2} - 3^x.$$

The x intercept of the graph (Figure 2) (approximately 0.86) is the same as the solution of $5^{3x-2} = 3^x$.

Figure 2

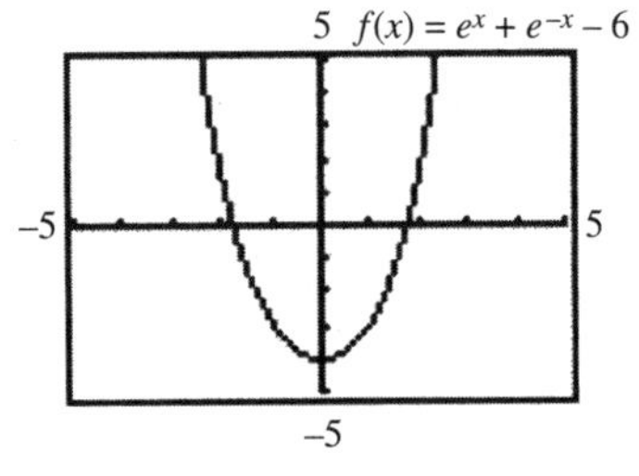

Figure 3

When encountering equations such as

$$9^x - 3^x - 12 = 0 \quad \text{or} \quad e^x + e^{-x} - 6 = 0,$$

we might not recognize that this type of equation cannot be solved by using algebraic methods. However, a graphical solution is possible.

For example, by graphing the function

$$f(x) = e^x + e^{-x} - 6$$

we observe that the x intercepts of the graph are approximately -1.76 and 1.76 (Figure 3). So $x = -1.76$ and $x = 1.76$ are the approximate solutions to the equation $e^x + e^{-x} - 6 = 0$.

Solving Exponential Inequalities

Earlier, we established that $f(x) = \log_b x$ is one-to-one. Therefore, if

$$b > 1 \quad u < v, \quad \text{then} \quad \log_b u < \log_b v.$$

This property is used in the next example.

EXAMPLE 3 **Solving an Inequality Involving an Exponential Expression**

Solve the inequality $2^{x-1} < 5$.

Express the solution in interval notation and show the solution on a number line. Round off the answer to two decimal places.

Algebraic Solution	Graphical Illustration
Taking the natural logarithm of each side, we have	Figure 4b shows the graph of the function

$$\log 2^{x-1} < \log 5$$

$$(x-1)\log 2 < \log 5 \qquad \log N^r = r\log N$$

$$x - 1 < \frac{\log 5}{\log 2} \qquad \text{Divide each side by } \log 2 \text{, which is a positive number.}$$

$$x < 1 + \frac{\log 5}{\log 2} \qquad \text{Add 1 to each side}$$

$$x < 3.32 \text{ (approx.)} \qquad \text{Use a calculator}$$

In interval notation, the solution set is $(-\infty, 3.32)$ and it is displayed in Figure 4a.

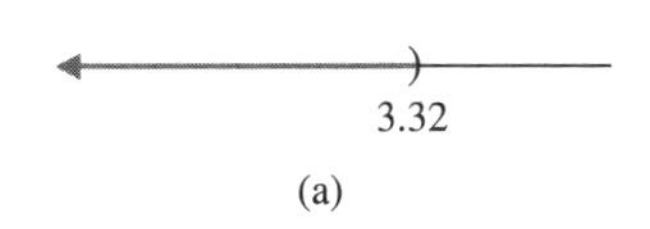

(a)

Figure 4

Figure 4b shows the graph of the function

$$y = 2^{x-1} - 5.$$

The x intercept of the graph is approximately 3.32. The solution of the inequality consists of all values of x, such that $y < 0$.

That is, $x < 3.32$ or $(-\infty, 3.32)$.

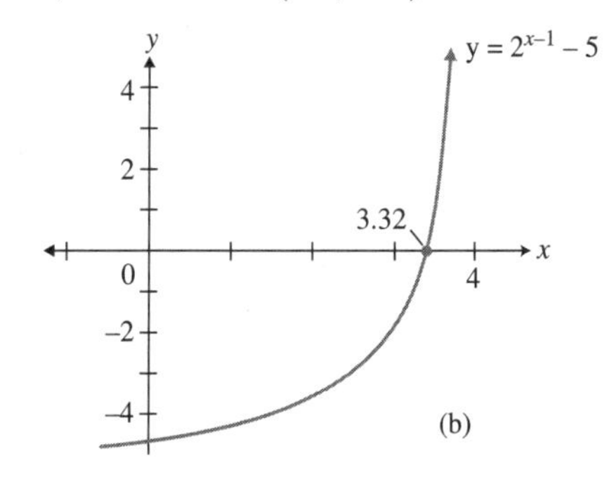

(b)

❖

Solving Logarithmic Equations

To solve an equation involving a multiple of the sum or difference of logarithms, we first combine the logarithmic expressions into a single logarithmic expression on one side of the equation and then use the following one-to-one property

$$\log_b M = \log_b N \quad \text{if and only if} \quad M = N$$

For instance, to solve an equation such as

$$\log x^2 - 2\log\sqrt{x} = 1$$

we combine the expressions on the left side as follows:

$$\log x^2 - \log\left(\sqrt{x}\right)^2 = 1 \qquad \text{Power property}$$

$$\log x^2 - \log x = 1 \qquad \text{Simplify}$$

$$\log\frac{x^2}{x} = 1 \qquad \text{Quotient property}$$

$$\log x = 1 \qquad \text{Simplify}$$

$$x = 10^1 = 10$$

EXAMPLE 4 Solving Logarithmic Equations

Solve the equation and check for extraneous roots.

(a) $\log_3(x+1) + \log_3(x+3) = 1$
(b) $\log(x^2 - 4) - \log(x - 2) = 1$

Algebraic Solution

(a) We proceed as follows:

$\log_3(x+1) + \log_3(x+3) = 1$	Given
$\log_3[(x+1)(x+3)] = 1$	Product property
$(x+1)(x+3) = 3^1$	Exponential form
$x^2 + 4x + 3 = 3$ $x^2 + 4x = 0$	Simplify
$x(x+4) = 0$	Factor
$x = 0$ or $x = -4$	Solve for x

Check:

For $x = -4$

$$\log_3(-4+1) + \log(-4+3) = \log_3(-3) + \log_3(-1).$$

Since the logarithms of negative numbers are undefined, -4 is an extraneous solution.

For $x = 0$ $\log_3 1 + \log_3 3 = 0 + 1 = 1$.

Thus, 0 is the only solution.

Graphical Illustration

(a) The given equation is equivalent to

$$\log_3(x+1) + \log_3(x+3) - 1 = 0.$$

So the solution of the equation is the same as the x intercept of the graph of

$$f(x) = \log_3(x+1) + \log_3(x+3) - 1.$$

Figure 5a shows the graph of the function f intercepts the x axis at the origin, which confirms that the solution is $x = 0$.

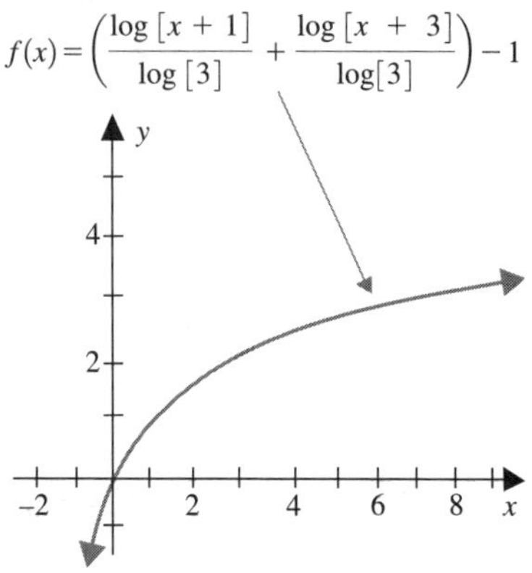

$$f(x) = \left(\frac{\log[x+1]}{\log[3]} + \frac{\log[x+3]}{\log[3]}\right) - 1$$

Figure 5a

Algebraic Solution

(b) We proceed as follows:

$\log(x^2 - 4) - \log(x-2) = 1$	Given.
$\log\left[\dfrac{x^2-4}{x-2}\right] = 1$	Use the quotient property
$\log\left[\dfrac{(x-2)(x+2)}{x-2}\right] = 1$	Factor
$\log(x+2) = 1$	Simplify
$x + 2 = 10$ $x = 8$	Convert to exponential form

Check:

For $x = 8$

$$\begin{aligned}\log(64 - 4) &- \log 6 \\ &= \log 60 - \log 6 \\ &= \log\frac{60}{6} \\ &= \log 10 \\ &= 1\end{aligned}$$

Thus, 8 is the solution.

Graphical Illustration

(b) Figure 5b shows the graph of

$$f(x) = \log(x^2 - 4) - \log(x-2) - 1.$$

The x intercept of the graph is 8, which confirms that 8 is the solution of the given equation

$$\log(x^2 - 4) - \log(x-2) - 1$$

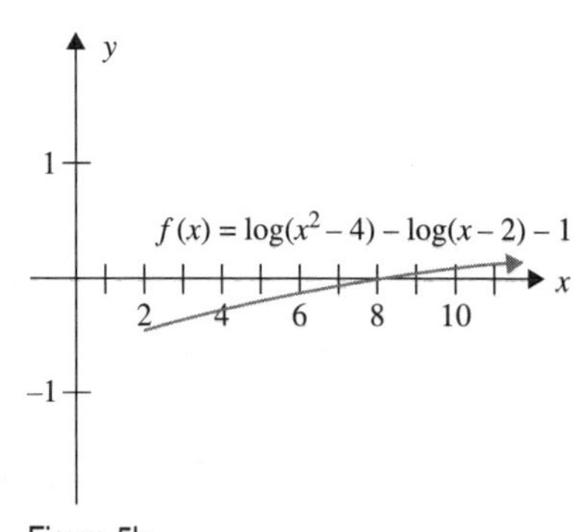

Figure 5b

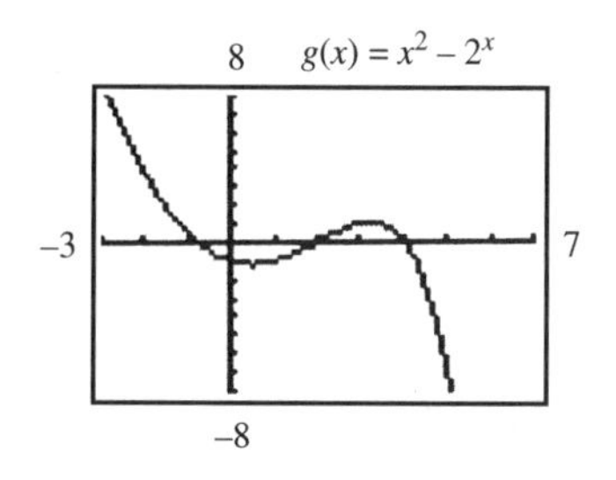

Figure 6

Not all exponential and logarithmic equations and inequalities can be solved using the algebraic methods studied so far. Such problems can sometimes be solved by graphical means. For example, we solve the inequality $x^2 > 2^x$ graphically as follows:

The inequality $x^2 > 2^x$ is equivalent to $x^2 - 2^x > 0$. Figure 6 shows the graph of $g(x) = x^2 - 2^x$, using the ZERO -finding feature we get the approximate locations of the three x intercepts, namely,

$$-0.77, 2, \quad \text{and} \quad 4.$$

By reading the graph, we see that $x^2 - 2^x > 0$, where the graph of g is above the x axis—that is, whenever x is in the interval $(-\infty, -0.77)$ or in the interval $(2, 4)$. So the solution includes all values of x such that

$$x < -0.77 \quad \text{or} \quad 2 < x < 4, \quad \text{approximately.}$$

Solving Applied Problems

We know that a radioactive substance decomposes according to the decay model

$$A = A_0 e^{-kt} \quad \text{where } k > 0$$

and t represents time. Logarithms can be used to find how long it takes for a substance to decay to a given level.

EXAMPLE 5 Using the Decay Model of an Element

Polonium-210 is a radioactive element that decays according to the model

$$A = A_0 e^{-0.005t}$$

where A_0 is the initial weight of the sample. How long will it take for a sample of this material to decay by 30% if t represents elapsed time in days?

Solution If the sample weighs A_0 grams initially, then after decaying by 30%, its weight A is given by

$$\begin{aligned} A &= A_0 - 0.30A_0 \\ &= 0.70A_0. \end{aligned}$$

Substituting into the decay model, we get

$$\begin{aligned} 0.70A_0 &= A_0 e^{-0.005t} && \text{Divide each side by } A_0. \\ 0.70 &= e^{-0.005t} && \text{Take the natural logarithm of each side} \\ \ln 0.70 &= \ln e^{-0.005t} && \ln N^r = r \ln N \\ \ln 0.70 &= -0.005t \ln e && \ln e = 1 \\ \ln 0.70 &= -0.005t \\ t &= \frac{\ln 0.70}{-0.005} && \text{Use a calculator} \\ t &= 71.33 \text{ (approx.)} \end{aligned}$$

Thus it takes a little over 71 days to decay by 30%. 

PROBLEM SET 4.5

Mastering the Concepts

In problems 1–14, use logarithms to solve each equation. Round off the answers to four decimal places.

1. $6^x = 12$
2. $3^{5x} = 2$
3. $e^{-5x} = 7$
4. $e^{4x} = 3$
5. $7^{3x-1} = 5$
6. $10^{x+1} = 4$
7. $e^{x+1} = 10$
8. $e^{2-3x} = 8$
9. $e^{2x-1} = 5^x$
10. $3^{x+1} = 17^{2x}$
11. $e^{2x-1} = 10^x$
12. $e^x = 10^{x+1}$
13. $(1.08)^{2x+3} = (1.7)^x$
14. $(0.97)^{3x-1} = (3.1)^{2x}$

In problems 15–24, solve the given inequality. Express the solution in interval notation and show it on a number line. Round off the answers to two decimal places.

15. $3^x < 21$
16. $4^x > 3$
17. $2^{4-x} > 5$
18. $2^{-x} < 5$
19. $e^{3x} < 7$
20. $e^{-2x} > 4$
21. $3^{x+2} < 2^{1-3x}$
22. $e^{2x+3} < 3^{x-2}$
23. $(1.06)^x > 2$
24. $14.7e^{-0.21x} < 7.35$

In problems 25–34, solve each equation and check for extraneous roots.

25. $\log 2x = \log 3 + \log(x - 1)$
26. $\log_5 y + \log_5(y - 4) = 1$
27. $\ln x + \ln(x - 2) = \ln(x + 4)$
28. $2 \ln(t + 1) - \ln(t + 4) = \ln(t - 1)$
29. $\log_2(w^2 - 9) - \log_2(w + 3) = 2$
30. $\log(v + 1) - \log v = 1$
31. $\log_4 x + \log_4(6x + 10) = 1$
32. $\log_3 t + \log_3(t - 6) = \log_3 7$
33. $\log(x^2 - 144) - \log(x + 12) = 1$
34. $\log x^2 = 2 \log x$

In problems 35–42, use a grapher to solve each equation. Round off the answers to three decimal places.

35. $2^x - 1 = 3x$
36. $3e^x - 7 = 5x$
37. $e^{-x^2} + x - 1 = 0$
38. $3^x - 4 = 4x$
39. $e^{x^2} - 3x - 5 = 0$
40. $3^{2x} = x^2$
41. $\ln x + x^2 = 2$
42. $\ln(x + 1) + x = 3$

In problems 43–50, use a grapher to solve each inequality. Round off the answers to two decimal places.

43. $3^x < x^2$
44. $e^{2x} > 3 - x$
45. $\ln x > \log 2x$
46. $\ln(x - 3) < \log x$
47. $e^{-x} \le 3 + \ln x$
48. $e^{3x} - \ln x \le 5$
49. $\log(x - 1) < 3x$
50. $3 \ln x - x^3 > 4$

Applying the Concepts

In problems 51–60, round off each answer to two decimal places.

51. Compound Interest: Suppose that \$10,000 is invested in a certificate of deposit at a 4.75% annual interest rate. How many years will it take for the money to double if the interest is compounded:
(a) Semiannually?
(b) Quarterly?
(c) Monthly?
(d) Continuously?

52. Compound Interest: Suppose that \$1500 is invested at a 4.65% annual interest rate. How many years will it take for the money to triple if the interest is compounded:
(a) Semiannually?
(b) Quarterly?
(c) Monthly?
(d) Continuously?

53. Investment Growth Model: Suppose that P dollars are invested in a money market fund that pays a 4.85% annual interest rate compounded quarterly, and that the accumulated amount S (in dollars) after t years is a multiple of P, that is, $S = kP$.
(a) Use common logarithms to express t as a function of k.
(b) Use the function in part (a) to find out how long it takes for the investment to double, to triple, and to quadruple.

54. Investment Growth Model: Suppose that P dollars are invested in a credit union paying a 5.03% annual interest rate compounded continuously, and that the accumulated amount S (in dollars) after t years is a multiple of P, that is, $S = kP$.
(a) Use natural logarithms to express t as a function of k.
(b) Use the function in part (a) to find out how long it takes for the investment to double, to triple, and to quadruple.

55. Productivity: In an experiment conducted by a company, it was determined that the number N of days of training needed for a factory worker to produce x automobile parts per day is given by the model

$$N = 100 - 25 \ln(60 - x).$$

How many parts per day is a factory worker able to produce if 25 training days are needed?

56. Advertising: A company predicts that sales will increase during a 30-day promotional television campaign according to the logistic model

$$N = \frac{400}{1 + 300e^{-0.5t}}$$

where N in is the number of daily sales t days after the campaign begins. How many days of advertising will result in daily sales of 300 units?

57. **Deer Population:** Suppose that the population P of a herd of deer newly introduced into a game preserve grows according to the model

$$P = 300 + 100 \ln(t + 1),$$

where t is the time in years.

(a) How many deer are present in the preserve initially?

(b) In how many years will the deer population be 500?

58. **Radioactive Decay:** Potassium-42, a radioactive element used by cardiologists as a tracer, has a half-life of about 12.5 hours.

(a) Find the decay model for this element.

(b) How long will it take for 80% of the original sample to decay?

59. **Heat-Treating:** A steel panel is tested for stress by heating it to 385°F and then immediately cooling it in a cooling chamber where the temperature is held at 32°F. The panel cools according to Newton's law of cooling:

$$T = 32 + 353e^{-kt}$$

where T is the temperature in degrees Fahrenheit after t minutes and k is a constant.

(a) Find k if the temperature of the panel is 198°F after 5 minutes.

(b) How long after the steel panel is placed in the cooling chamber will its temperature be 85°F?

60. **Drug Dosage Model:** The concentration C (in milligrams) of a certain drug after t hours in a patient's bloodstream is given by

$$C = 100e^{-0.442t}.$$

How long does it take for the drug concentration to reduce by 50%?

Developing and Extending the Concepts

61. Solve the equation

$$x^2 + 5x + 6 = 0.$$

Substitute e^t for x in the equation to get $e^{2t} + 5e^t + 6 = 0$. Explain why the latter equation has no solution.

62. Solve the equation

$$3y^2 - 28y + 9 = 0.$$

Substitute 3^x for y in the equation to get

$$3(3^{2x}) - 28(3^x) + 9 = 0.$$

63. Solve each equation.

(a) $\ln x^2 = (\ln x)^2$

(b) $\log x^3 = (\log x)^3$

64. Suppose that

$$y = 3 + 10(1 - e^{-0.2x}).$$

Solve for x in terms of y.

65. Recall from Section 4.2 that the hyperbolic sine function is denoted by $\sinh x$ and defined by

$$f(x) = \sinh x = \frac{e^x - e^{-x}}{2}.$$

Use a grapher to graph f. Then use the graph to solve the equation $\sinh x = 1$. Round off the answer to two decimal places.

66. Is it possible to solve the equation

$$\log_3(x^4 + 1) + \log_2(x^2 + 1) = 5$$

algebraically? If so, solve it. If not, use a grapher to approximate the solution for $-3 \leq x \leq 3$. Round off to two decimal places.

Interlude: Progress Check 4.5, Part 1

1. Solve for x: $7^{3x-2} = 15^{2x+1}$. Give the exact solution in terms of common logarithms.

2. A certain substance is decaying exponentially so that two hours from now, 85% of the substance will remain. Determine the half-life of the substance, in hours, accurate to two decimal places.

3. The height h above sea level (in meters) is related to air temperature t (in degrees Celsius), the atmospheric pressure p (in centimeters of mercury at height h), and the atmospheric pressure c at sea level according to

$$h = (30t + 8000)\ln(c/p).$$

Mount Everest is about 8850 meters high. What is the atmospheric pressure at the top of Mount Everest on a day when the temperature is $-25°C$ and the atmospheric pressure at sea level is 75 cm? Give your result accurate to two decimal places and label appropriately.

4. Use an algebraic technique to solve for x: $216^{5x} = \left(\frac{1}{36}\right)^{2x+8}$. Give the exact solution.

Interlude: Progress Check 4.5, Part 2

1. Solve the equation $3^{10x} - 3 \cdot (3^{5x}) - 10 = 0$. Give exact solutions in terms of natural logarithms.

2. A department store predicts that t weeks after the end of a sales promotion the volume of sales (in dollars) will be described by a function of the form $S(t) = 50{,}000 + Ae^{-kt}$ for some appropriate constants A and k. The sales volumes at the end of the first and third weeks, respectively, were \$83,515 and \$65,055. Determine the expected sales volume at the end of the fifth week.

3. A virus is spreading through a community of 200,000 in such a way that the number of people that have become infected t days after the virus was discovered is given by

$$N(t) = \frac{200,000}{1+39e^{-kt}}.$$

Suppose that 10,000 people have been infected four days after the virus was discovered.

(a) Determine the value of k. Give both the exact value of k as well as an approximation accurate to five decimal places.

(b) How long will it take for half of the community to be infected? Give your result as a number of days, accurate to two decimal places. Use algebraic techniques to obtain your solution and then illustrate your result graphically.

CHAPTER 4 READINESS: PART 1

Provide complete solutions in the space provided.

1. Suppose that $f(x)$ is a linear function such that $f(-4) = 12$ and $f^{-1}(6) = -5$. Calculate $f(2)$.

2. Let $f(x) = \left(\frac{1}{2}\right)^x$. Sketch a graph of $f^{-1}(x)$ on the provided grid. Label at least five points on the graph with exact coordinates and make your scale as accurate as possible.

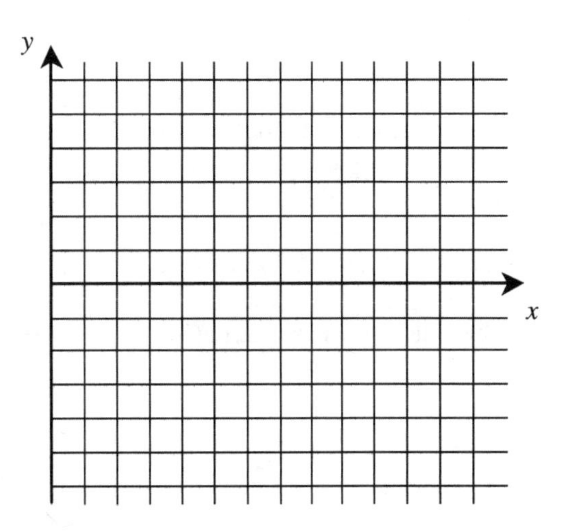

3. Fill in the missing entries in the table below.

x		$\frac{1}{9}$		243	6561
$f(x) = \log_3 x$	-4		0		

4. The population of Cesspool Villa is currently 32,600 people and is growing at a rate of about 400 people per year. The population of Armpit Junction is currently 4200 people and is growing at a rate of about 2.5% per year. Is there a point in the future where the population of Armpit Junction will match the population of Cesspool Villa? If so, then find that time. If not, then explain why.

5. Each curve in Figure 1 represents the balance in a bank account into which a single deposit was made at time zero. Assuming that interest is compounded continuously, find

   (a) The curve representing the largest initial deposit.

   (b) The curve representing the largest interest rate.

   (c) Two curves representing the same initial deposit.

   (d) Two curves representing the same interest rate.

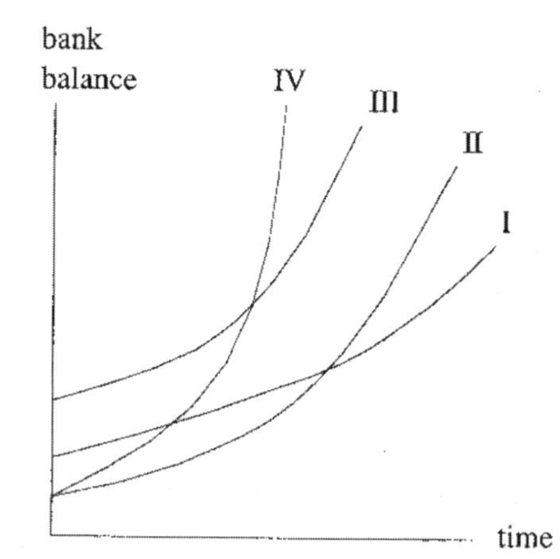

Figure 1

CHAPTER 4 READINESS: PART 2

Provide complete solutions in the space provided.

1. Give the exact solution to $15^{4x-3} = 4^{7x}$ in terms of natural logarithms.

2. Use an algebraic technique to solve the equation $\log_2 (x + 1) + \log_2 (4 - x) = \log_2 (6x)$.

3. Suppose that $\log_b 3 = 0.9597$, $\log_b 7 = 1.6999$, and $\log_b 10 = 2.0115$. Calculate $\log_b \sqrt[4]{9/70}$ without attempting to find b.

4. A car that was purchased new eight years ago for $52,000 now has an estimated market value of $16,000. If the value, V, of the car is an exponential function of the number of years, t, since the car was purchased new, then find a formula for V in terms of t and indicate by what percentage the value of the car decreases each year. Express your result as a percentage, accurate to two decimal places.

5. Under certain conditions, it is anticipated that a bacteria culture will increase in size at a continuous rate of 9.25% per hour. Given this, how long will it take for a culture of 130,000 bacteria to quadruple in size? Give your answer as a number of hours, correct to two decimal places.

CHAPTER 4 READINESS: PART 3

Provide complete solutions in the space provided.

1. Approximate all solutions to the equation $\log_8 x + \log_9 x = 2$, correct to three decimal places.

2. Find a number b such that the graph of $f(x) = \log_b x$ passes through the point (8, 16).

3. A class in underwater basket weaving is tested at the end of the semester and weekly thereafter on the same material. The average score on the exam taken after t weeks is given by the "forgetting function" $g(t) = 84 - 10 \cdot \ln(t + 1)$.

   (a) What was the average score on the original exam?

   (b) What was the average score after 6 weeks, accurate to one decimal place?

   (c) When did the average score drop below 50?

4. The *decibel* is a unit of measure of sound intensity based on the measured intensity of a sound and the intensity of the least audible sound a human can be expected to hear, and is given by

$$D = 10\log\left(\frac{I}{I_0}\right),$$

where

D = decibel level of the sound

I = the measured intensity of the sound (in watts/m^2)

I_0 = the intensity of the least audible sound a human can hear (in watts/m^2)

It is claimed that in 1994 the heavy metal band Manowar once achieved, during a performance, a decibel level of 129.5. By comparison, the decibel level of a typical vacuum cleaner is 80. Given this determine how many times more intense the sound was in the musical performance than the intensity of the sound from a typical vacuum cleaner, rounded to the nearest whole number.

5. Solve the inequality $\log_7 5 - \log_7 (x + 3) \geq \log_7 3 - \log_7 x$ and express the solution using interval notation.

Answers to Odd-Numbered Problems

PROBLEM SET 4.1

1.

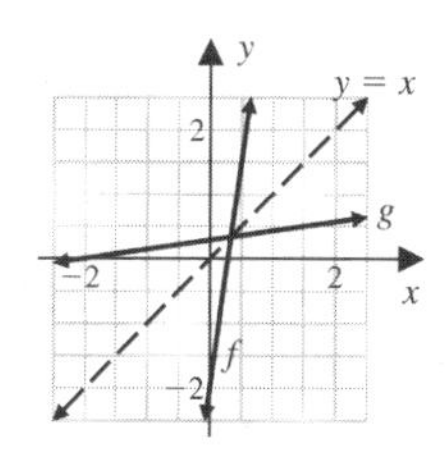

3.

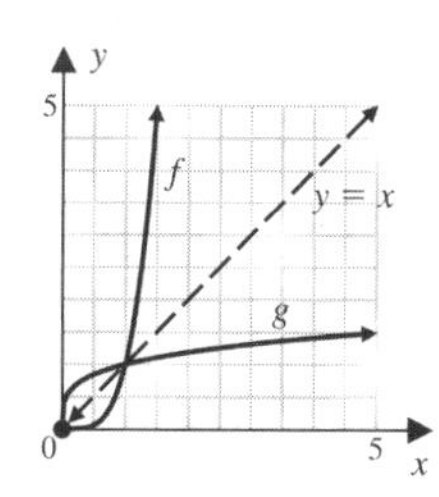

5.

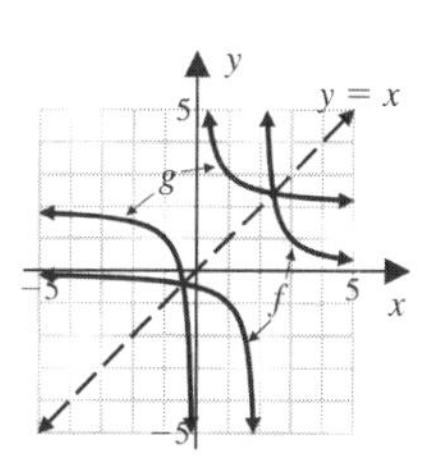

7. **(a)** Domain of f and Range of f^{-1}: $\mathbb{R}$;
Domain of f^{-1} and Range of f: $(0, \infty)$

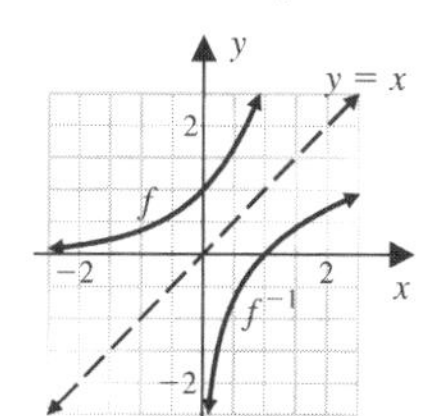

(b) Domain of f and Range of f^{-1}: $[-2, \infty)$;
Domain of f^{-1} and Range of f: $[0, \infty)$

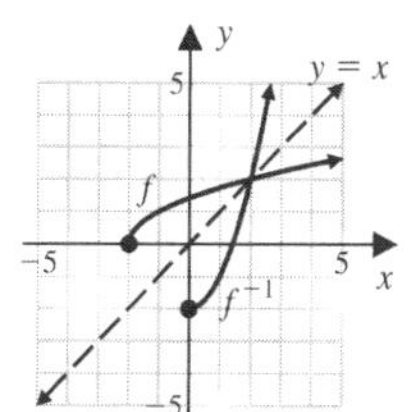

9. **(a)** Yes **(b)** No

11. One-to-one

13. Not one-to-one

15. Not one-to-one

17.

x	2	−3	3	5	7	100	−4	1
$f^{-1}(x)$	1	2	3	4	5	6	7	8

Domain of f and Range of f^{-1}: 1, 2, 3, 4, 5, 6, 7, 8;
Domain of f^{-1} and Range of f: −4, −3, 1, 2, 3, 5, 7, 100

19. $\dfrac{x-5}{7}$

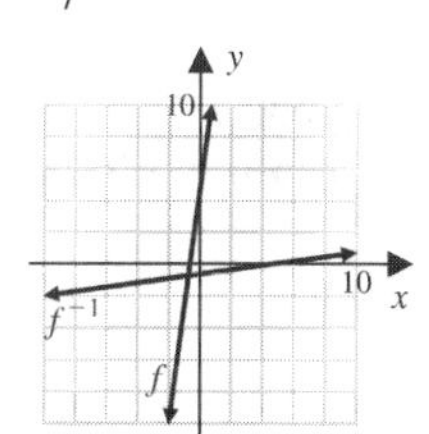

21. $\dfrac{3+x}{x}$

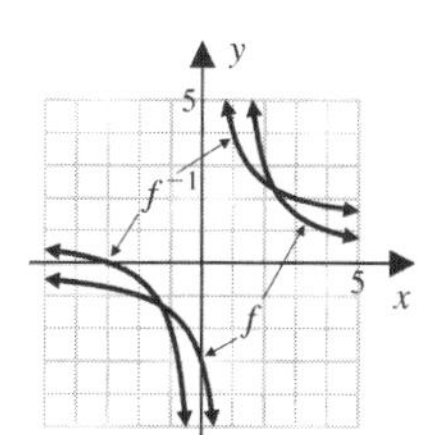

23. $x^2 + 3, x \geq 0$

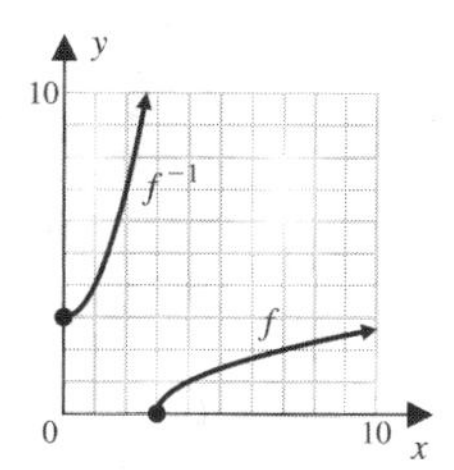

25. $\sqrt[3]{x+8}$

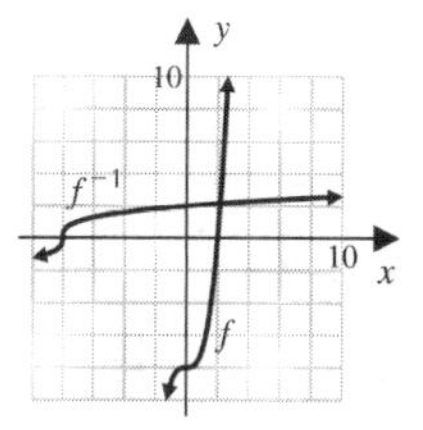

27. f^{-1} does not exist—f fails the horizontal-line test.

29. $(2-x)^3$

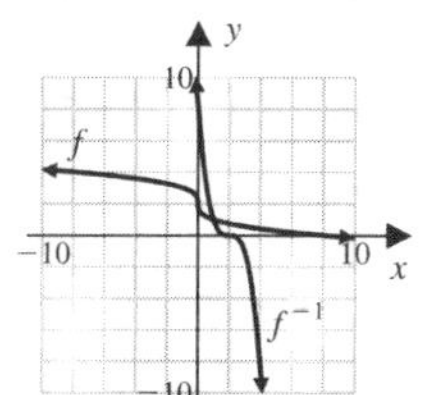

31. $\sqrt{-x}, x \leq 0$

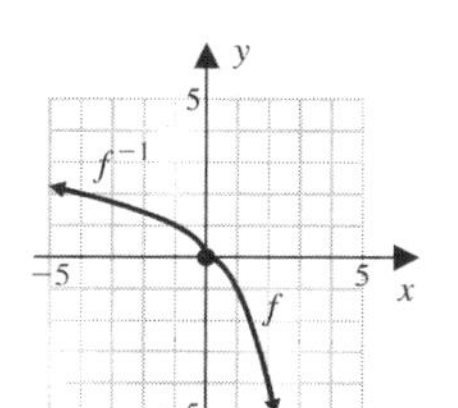

33. **(a)** f does not have an inverse: g has an inverse

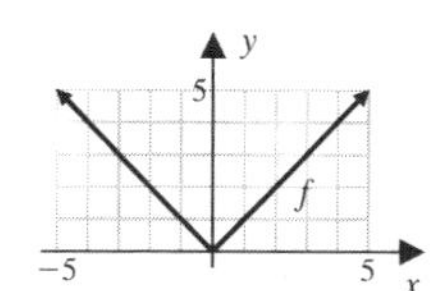

(b) $g^{-1}(x) = -x, x > 0$

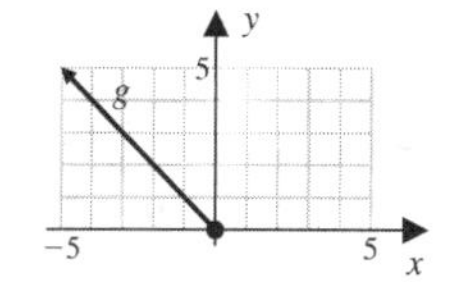

35. **(a)** f does not have an inverse; **(a)** $g^{-1}(x) = \sqrt{1 - x^2}$, g has an inverse $0 < x < 0$

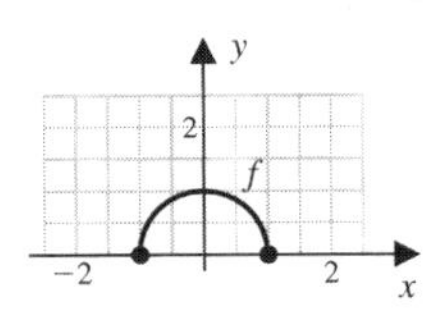

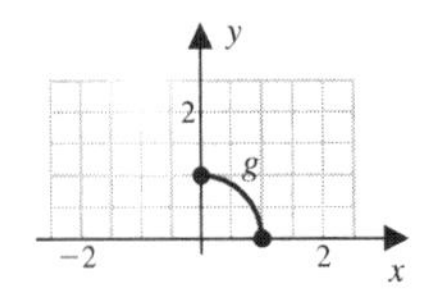

37. **(a)**

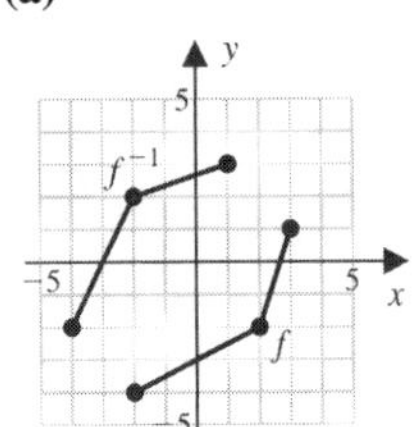

(b) $y = f^{-1}(x)$

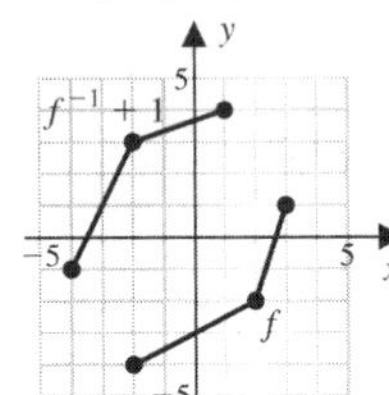

(c)

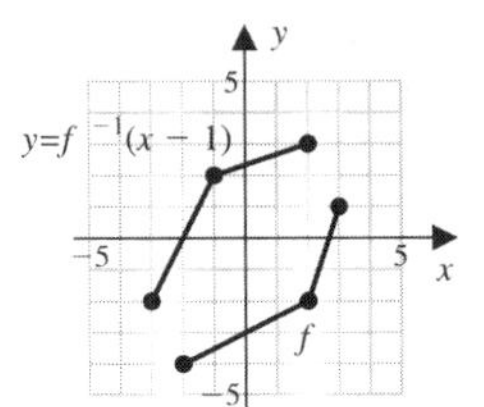

(d)

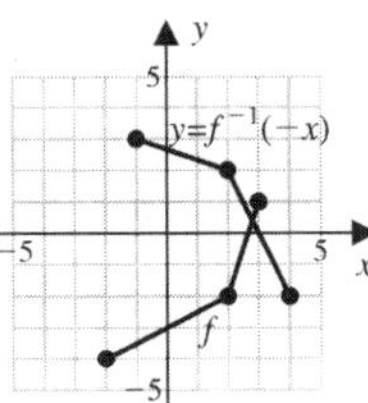

39. No; fails the horizontal-line test.

41. No; fails the horizontal-line test.

43. **(a)** $W = f(P) = 0.85P$

(b) $P = \dfrac{20}{17}W$

45. **(a)** $C = f^{-1}(F) = 5/9\,(F - 32)$

(b)

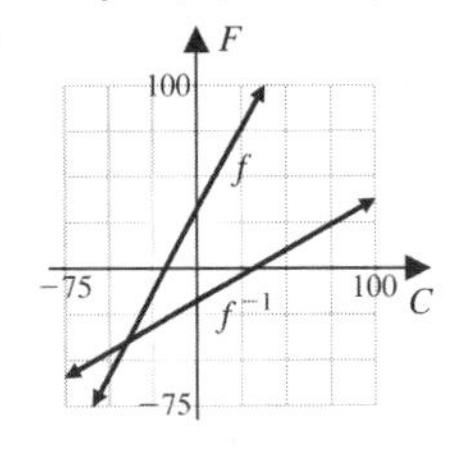

(c) $29\frac{4}{9}$°C; 45°C

47. **(a)** $y = \$25{,}000 + 0.40x,\ x > 0$

(b) $f^{-1}(x) = 2.5\,x - 62.500$

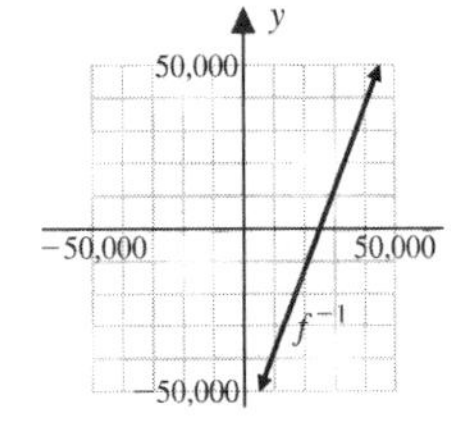

(c) f^{-} represents sales in terms of the salesperson's income

49. **(a)** $f^{-1}(x) = \begin{cases} 2x + 8 & \text{if } x < -4 \\ x + 4 & \text{if } x \geq -4 \end{cases}$

Domain and Range of f and f^{-1}: $\mathbb{R}$

(b) f^{-1} does not exist (fails the horizontal line test).

51. Reflect $y = f(x)$ about $y = x$, then reflect the result about $y = x$ to get $f(x)$.

53. **(a)**

$$f^{-1}(x) = \frac{1}{2}x - 2$$

$$g^{-1}(x) = \frac{x+1}{3}$$

$$(f \circ g)^{-1}(x) = \frac{x-2}{6}$$

(d) $(g \circ f)^{-1} = f^{-1} \circ g^{-1}$ this is the same result as for $f^{-1} \circ g^{-1}$ in part **(b)**.

55. **(a)** B **(b)** A **(c)** C **(d)** D

PROBLEM SET 4.2

1. **(a)** 8.8250 **(b)** 0.3752 **(c)** 6.3697 **(d)** 85.3490
3. **(a)** 29.9641 **(b)** 0.0432 **(c)** 4.1133 **(d)** -1.3594
5. **(a)** 2 **(b)** 0.30 **(c)** 2.25 **(d)** 0.74
7. For g shift the graph of $f(x) = e^x$ 2 units up. For h shift the graph 2 units down. **9.** **(a)** 3 **(b)** $\frac{1}{2}$

11. **(a)** $A = 3, b = 3$ **(b)** $A = 1, b = \frac{1}{2}$ **13.** **(a)** B **(b)** C **(c)** D **(d)** A **15.** 2

17. -1 and 6 **19.** -1

21. Shift the graph of $y = 5^x$ left 1 unit. Domain: $\mathbb{R}$; Range: $(0, \infty)$; increasing; horizontal asymptote: x axis

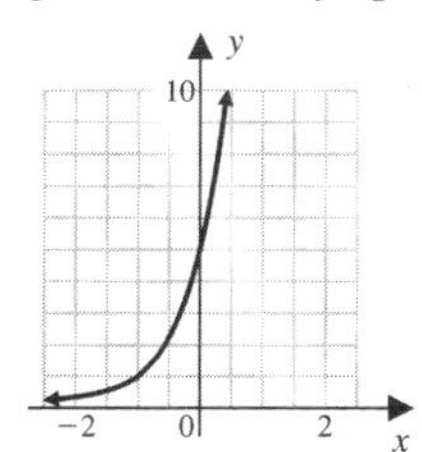

23. Shift the graph of $y = 2^x$ up 3 units. Domain: $\mathbb{R}$; Range: $(3, \infty)$; increasing; horizontal asymptote: $y = 3$

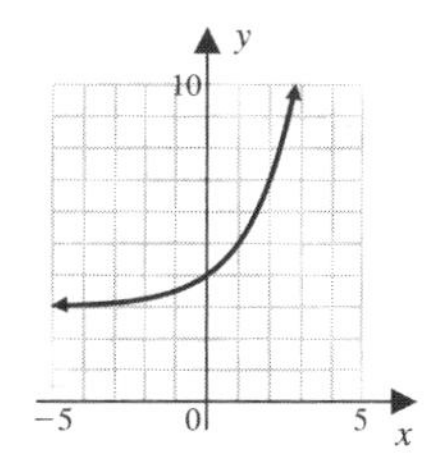

25. Stretch the graph of $y = e^x$ vertically by a factor of 2, reflect about the x axis. Domain: $\mathbb{R}$; Range: $(0, \infty)$; decreasing; horizontal asymptote: x axis

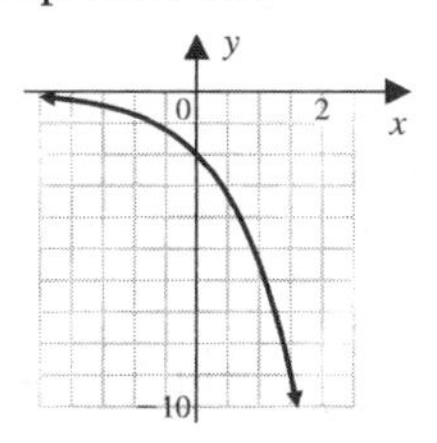

27. Reflect the graph of $y = e^x$ across the y axis and shift down 2 units. Domain: $\mathbb{R}$; Range: $(-2, \infty)$; decreasing; horizontal asymptote: $y = -2$

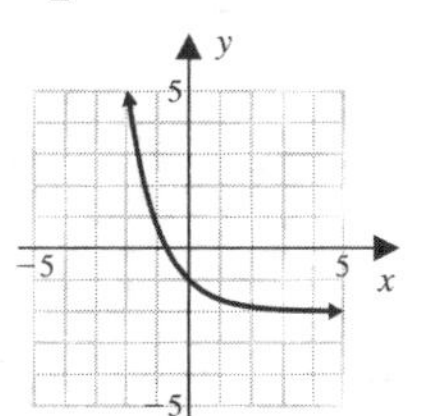

29. **(a)** \$6638.48 **(b)** \$6651.82 **(c)** \$6660.88 **(d)** \$6664.39
31. **(a)** $S = 5000e^{0.0575t}$ **(b)** \$6293 **(c)** Less
33. **(a)** $P = 6e^{0.013t}$ **(b)** 6.83 billion
35. **(a)** 10% per day **(b)** 41,218
37. **(a)** $V = \$19{,}545(0.80)^t$ **(b)** \$15,636; \$10,007; \$6,405
39. 1.63 million; 0.69 million; 0.29 million
41. 83.86%; 76.80% **43.** 32g; 25.6g; 13.11g
45. **(b)** $N = 224.474 \cdot 1.021^t$ **(d)** 340
47. **(b)** When rounded to two decimal places, the regression function values are the same as the table values for each value of t.
(d) \$0.50
49. **(a)** $f(x) = 1^x = 1$ is a constant function. **(b)** $(-3)^{\frac{1}{2}}$ is not a real number.

51. **(a)** $x = 4$ and 3 **(b)** 4

53. **(a)** -11 **(b)** 1, 2, and 3.

PROBLEM SET 4.3

1. **(a)** $\log_5 125 = 3$ **(b)** $\log_{32} 2 = \frac{1}{5}$ **(c)** $\log_e 17 = t$ or $\ln 17 = t$ **(d)** $\log_b(13z + 1) = x$

3. **(a)** $9^2 = 81$ **(b)** $10^{-4} = 0.0001$ **(c)** $c^w = 9$ **(d)** $e^{-1-3x} = \frac{1}{2}$

5. **(a)** -3 **(b)** 4 **(c)** $\frac{1}{3}$ **(d)** $\frac{2}{3}$ **7.** **(a)** 0 **(b)** -2 **(c)** 3 **(d)** $\frac{1}{3}$

9. **(a)** 36 **(b)** 16 **11.** **(a)** 3 **(b)** 8 **13.** **(a)** 10^{3-y} **(b)** $\frac{1}{2}(e^{y-8} - 1)$

15. **(a)** 1.5933; $10^{1.5933} = 39.2$ **(b)** 6.8680; $e^{6.8680} = 961$ **(c)** -0.1249; $10^{-0.1249} = \frac{3}{4}$

(d) -1.4697; $e^{-1.4679} = 0.23$ **(e)** 2.3026; $e^{2.3026} = 10$

17. **(a)** 10.6974 **(b)** 2.5924 **(c)** 0.0821 **(d)** 0.5112 **19.** **(a)** 2.0478 **(b)** 2.638

21. **(a)** $\log_3 x + \log_3(x + 1)$ **(b)** $\log_3 18 - \log_3(x + 2)$ **23.** **(a)** $4 \log_b(x + 3)$ **(b)** $\log x + \frac{1}{2}\log(2x + 1)$

25. **(a)** $\ln y + \frac{2}{3}\ln(3x + 1)$ **(b)** $\ln x + \ln(x + 7)$ **27.** **(a)** $\log_5 \frac{8}{7}$ **(b)** $\log_2\left(\frac{55}{2}\right)$

29. **(a)** $\ln\left(\frac{x + 3}{x - 3}\right)$ **(b)** $\ln\left(\frac{3x^2 + 7x + 4}{3x^2 - 5x - 12}\right)$ **31.** **(a)** 5 **(b)** $\frac{1}{9}$ **33.** **(a)** $\frac{e^{-7}}{x}$ **(b)** $x^2 - 4$

35. **(a)** 2.322 **(b)** 0.545 **(c)** -1.469 **(d)** 0.910

37. **(a)** $A = \frac{Pr(1 + r)^n}{(1 + r)^n - 1}$ **(b)** $A = 18{,}000\frac{\frac{0.1}{12}\left(1 + \frac{0.1}{12}\right)^n}{\left(1 + \frac{0.1}{12}\right)^n - 1}$; \$580.81; \$456.53; \$382.45

39. **(a)** $N = \frac{230}{1 + 6.931e^{-0.1702t}}$ **(b)** 29; 102; 158; 221 **41.** 7.7 **43.** **(a)** 2.5 **(b)** 7.8 **(c)** 6.4

45. **(a)** 90 db **(b)** 149.19 db **49.** **(a)** Try $a = b = 2$ **53.** Domain: $(2, \infty)$; x intercept: $e + 2$ **55.** 9

PROBLEM SET 4.4

1. $f^{-1}(x) = \log_5 x$;
Domain: $(0, \infty)$;
Range; $\mathbb{R}$

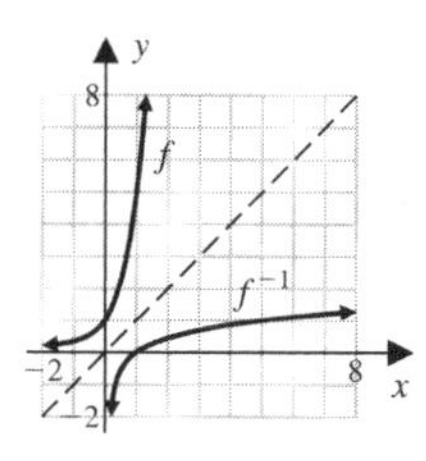

3. $h^{-1}(x) = \left(\dfrac{1}{10}\right)^x$;

Domain: $\mathbb{R}$;
Range; $(0, \infty)$

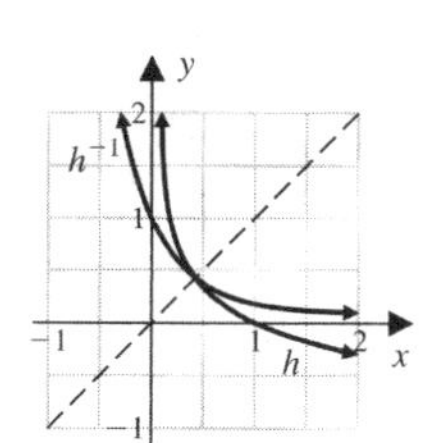

5. **(a)** $b = 4$ **(b)** $b = \dfrac{1}{16}$ **(c)** $b = e$

7. $(-\infty, 2)$ **9.** $(-\infty, -1)$ or $(-1, \infty)$ **11.** $(-2, \infty)$ **13.** $(-\infty, 0)$ or $(1, \infty)$ **15.** $(-\infty, -1)$ or $(1, \infty)$

17. $f(x)$ is steeper than $g(x)$; $g(x)$ is steeper than $h(x)$; $f(x) \to +\infty$, $g(x) \to +\infty$ and $h(x) \to +\infty$ as $x \to +\infty$

19. Shift 2 units to the right; Domain: $(2, \infty)$; Vertical asymptote: $x = 2$

21. Reflect the graph g across the x axis; Domain: $(0, \infty)$; Vertical asymptote: $x = 0$

23. $y = \log_3 (x + 1)$

25. Domain: $(0, \infty)$; vertical asymptote: y axis

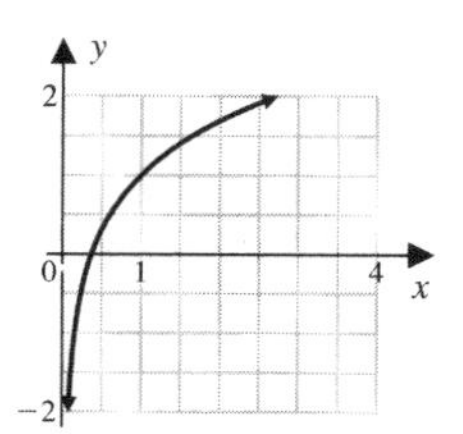

27. Domain: $(0, \infty)$;
vertical asymptote: y axis

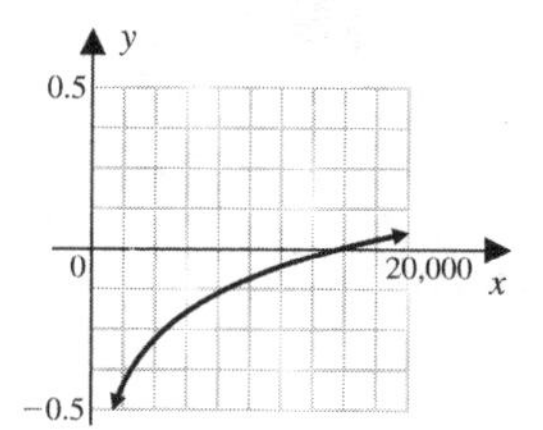

29. Domain: $(-3, \infty)$;
vertical asymptote: $x = -3$

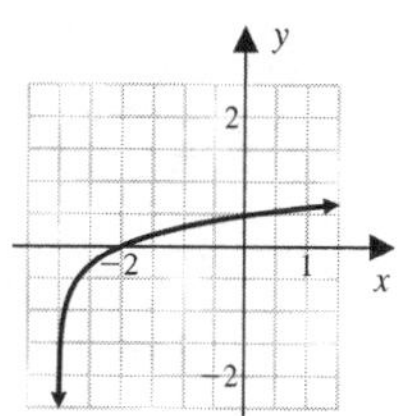

31. Domain: (e, ∞); vertical asymptote: $x = e$

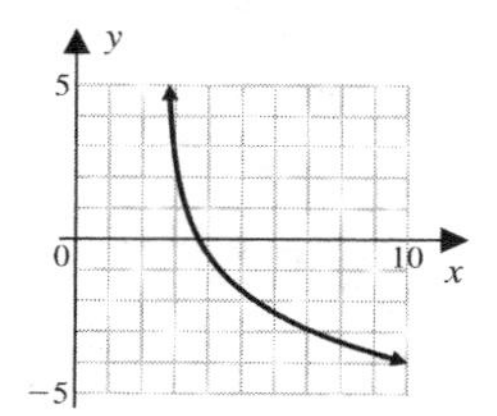

33. Domain: $(0, \infty)$; vertical asymptote: y axis

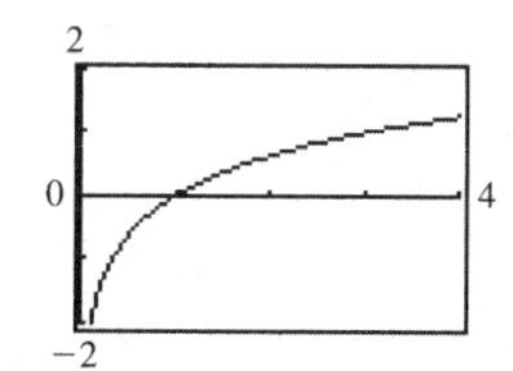

35. Domain:  $\left(-\dfrac{1}{5}, \infty\right)$

vertical asymptote: $x = -\dfrac{1}{5}$

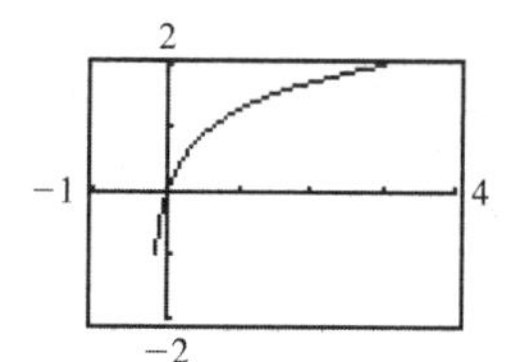

36. Domain: $\left(-\infty, -\frac{1}{2}\right)$ or $\left(\frac{1}{2}, \infty\right)$

vertical asymptotes: $x = -\frac{1}{2}, x = \frac{1}{2}$

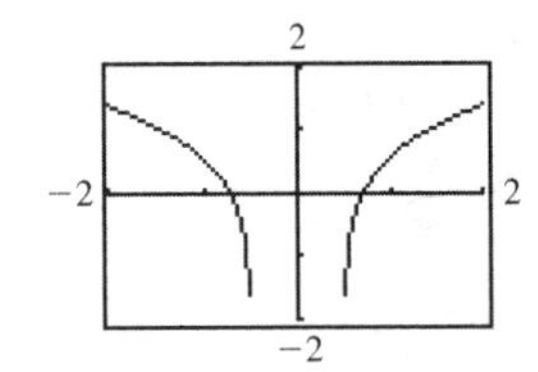

39. \$2,352,000; \$3,481,000; \$4,042,000

41. $P_1 > P_2$

43. Domain: $(2, \infty)$; vertical asymptote: $x = 2$

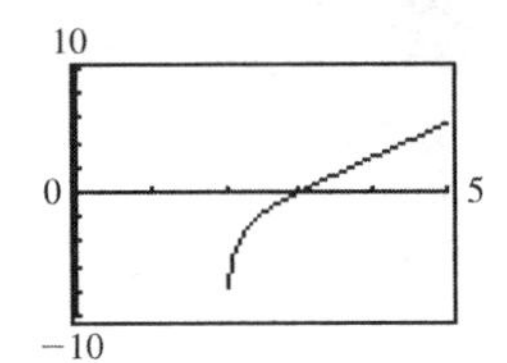

45. Domain: $(-\infty, 1)$ or $(4, \infty)$; vertical asymptotes: $x = 1, x = 4$

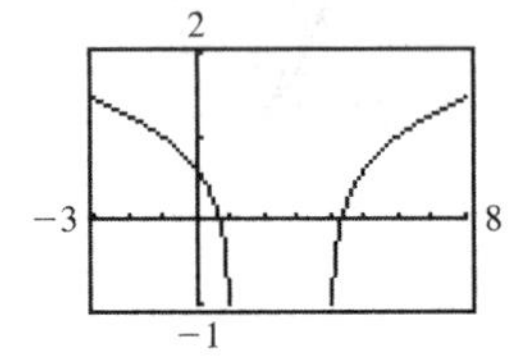

47. Domain: $\left(\frac{3}{2}, \infty\right)$, vertical asymptote: $x = \frac{3}{2}$

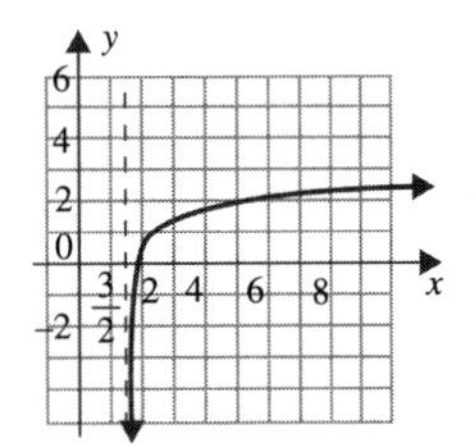

49.

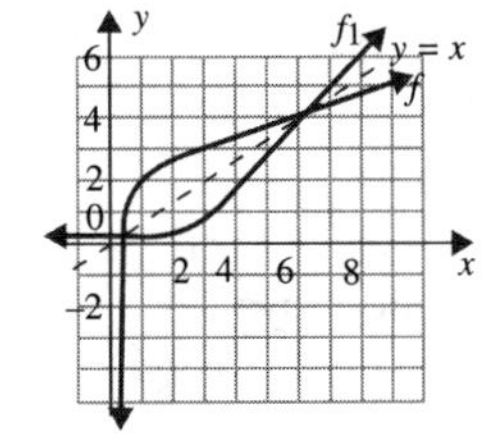

PROBLEM SET 4.5

1. 1.3869 **3.** -0.3892 **5.** 0.6090 **7.** 1.3026 **9.** 2.5604 **11.** -3.3049 **13.** 0.6129 **15.** $(-\infty, 2.77)$ **17.** $(-\infty, 1.68)$ **19.** $(-\infty, 0.65)$ **21.** $(-\infty, -0.47)$ **23.** $(11.90, \infty)$ **25.** 3 **27.** 4 **29.** 7 **31.** $\frac{1}{3}$ **33.** 22 **35.** 0; 3.538 **37.** 0 **39.** -0.907; 1.501 **41.** 1.314 **43.** $x < -0.69$ **45.** $x > 1.70$ **47.** $x \geqslant 0.12$ **49.** $x > 1$ **51.** **(a)** 14.77 yr. **(b)** 14.68 yr. **(c)** 14.62 yr. **(d)** 14.59 yr. **53.** **(a)** $t = \dfrac{\log k}{4 \log 1.012125}$ **(b)** 14.38 yr.; 22.79 yr.; 28.76 yr. **55.** 39.91 parts **57.** **(a)** 300 **(b)** 6.39 yr. **59.** **(a)** 0.15 **(b)** 12.64 min. **61.** no solution **63.** **(a)** $x > 0$ **(b)** $x > 0$ **65.** 0.88

Systems of Linear Equations

CHAPTER CONTENTS

Overview

In previous chapters, we discussed modeling a situation involving two related quantities where a function reasonably represented the situation. We now turn to other situations that are too complex to be modeled by a single function. In particular, we investigate situations where a collection of linear equations is used to model the relationships between some number of quantities.

OBJECTIVES

1. Solve Systems By Graphing, Substitution or Elimination
2. Solve Applied Problems

5.1 Systems of Linear Equations: A First Look

Consider the following problem.

> A student organization held a fundraiser dinner one evening. Adult dinners cost \$6 each, while children's dinners each cost \$3. The group raised a total of \$771 serving 189 dinners. How many adult dinners did they serve, and how many children's dinners did they serve?

We see here that there are two quantities that we need to identify. We will denote these quantities by a and c, where

$$a = \text{the number of adult dinners served}$$

$$c = \text{the number of children's dinners seved}$$

We also have two explicit relationships between a and c directly stated. First, we know that the total number of dinners served is 189, so

$$a + c = 189$$

In addition, we know that the total money raised from serving the 189 dinners is \$771. Since adult dinners cost \$6 and children's dinners cost \$3, we have

$$6a + 3c = 771$$

This gives us a pair of linear equations, which we now write together.

$$a + c = 189$$
$$6a + 3c = 771$$

We say here that we have a **system of linear equations**. Specifically, we have a system of two linear equations in two unknowns. Solving the problem posed requires finding appropriate values of a and c that simultaneously satisfy *both* equations. Any pair (a, c) that does this is called a **solution to the system**. Notice that the pair (63, 126) is *not* a solution to the system since when $a = 63$ and $c = 126$ we see that

$$63 + 126 = 189$$

but

$$6(63) + 3(126) = 756 \neq 771$$

That pair satisfies the first equation in the system but not the second.

We will set the problem at hand aside and begin looking at common methods for finding solutions to systems of linear equations in two unknowns. Later we will investigate methods that we can apply to systems of linear equations in any number of unknowns.

Graphical Solutions

The first example we consider involves a simple system and illustrates a geometric method for finding solutions to a system of linear equations in two unknowns.

EXAMPLE 1 **Solving a System Graphically**

Solve the system

$$3x - y = 11$$
$$x + 2y = -8$$

Solution The graph of each equation is a line. Points on a line represent pairs of values (x, y) that satisfy the equation for the line. So, when solving a system of equations graphically, we seek all points that the graphs of the equations have in common.

It is convenient to rewrite each of the equations in the system and use a graphing calculator to investigate their graphs. Working with the first equation, solving for y, we have

$$3x - y = 11$$
$$3x - 11 = y$$

Similarly, we rewrite the second equation.

$$x + 2y = -8$$
$$y = (-8 - x)/2$$

We simultaneously graph the equations

$$y_1 = 3x - 11$$
$$y_2 = (-8 - x)/2$$

on a graphing utility to obtain a graph as shown in Figure 1:

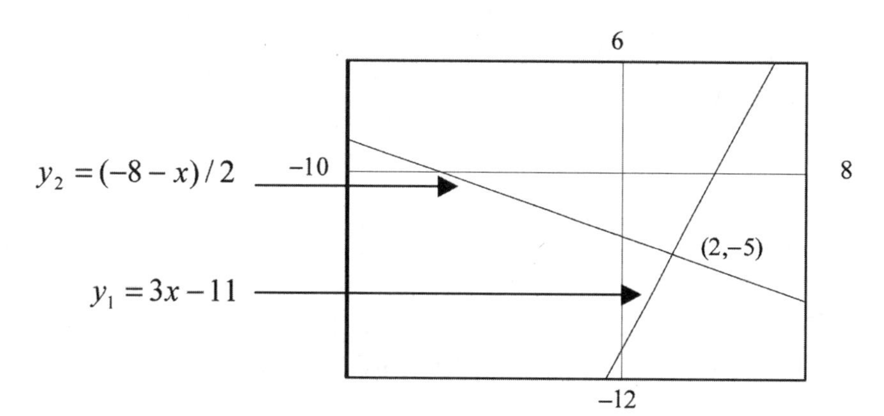

Figure 1

Using an intersection routine on the calculator, we find that the lines intersect at the point $(2, -5)$. Thus the pair $x = 2$ and $y = -5$ satisfies both of the equations in the system (check this) and hence forms a solution to the system.

Take note that in the previous example, the pair of numbers $x = 2$ and $y = -5$ forms *one* solution, not two solutions. ❖

Even when using a graphing calculator, the graphical method of finding solutions to a system of linear equations in two variables sometimes results in **approximate solutions** rather than in **exact solutions**. We can use algebraic methods to obtain exact solutions.

The Substitution Method

The basic approach to the method of substitution is easily summarized:

1. Solve for a variable in one of the equations in a system of equations.
2. Substitute for that variable in the other equation(s).

The following example illustrates this method.

EXAMPLE 2 Solving a System By Substitution

Solve the system

$$\begin{aligned} 3x - 11 &= y \\ x + 2y &= -8 \end{aligned}$$

Solution We may choose either variable to solve for and use either equation. In this case it is easy to solve for x in the second equation.

$$x = -8 - 2y$$

We substitute this expression for x in the first equation, $3x - y = 11$, as follows.

$$\begin{aligned} 3(-8 - 2y) - y &= 11 \\ -24 - 6y - y &= 11 \\ -24 - 7y &= 11 \\ -7y &= 35 \\ y &= -5 \end{aligned}$$

We now **back-substitute** to determine a value for x. Since $x = -8 - 2y$,

$$\begin{aligned} x &= -8 - 2(-5) \\ x &= -8 + 10 \\ x &= 2 \end{aligned}$$

We could have substituted $y = -5$ in either of the original equations. In any event, we find that the pair $x = 2$ and $y = -5$ is a solution to the given system, as we found in Example 1. ❖

The next algebraic method we look at is particularly important because later we will expand upon it to develop a relatively efficient approach to finding solutions to systems of linear equations in more than two variables.

The Method of Elimination

The method of elimination is a systematic approach to finding solutions to a system of linear equations by virtue of transforming the given system into a "simpler" but "equivalent" system. When we say that two systems of linear equations are **equivalent**, we mean that the systems have exactly the same solution(s). The objective in the method of elimination is to modify one (or more) of the equations in a given system to produce an equivalent system where one (or more) of the equations contains just one unknown, making it relatively easy to solve for that unknown.

There are only three legitimate operations that we can use to modify a system of linear equations to produce an equivalent system. However, we can perform these operations in any sequence for as many steps as we wish. These operations are:

Operations on Equations

- Interchange any two equations.
- Multiply (all terms of) an equation by any nonzero constant and replace that equation with the result.
- Multiply (all terms of) an equation by any constant and add (corresponding terms) to any other equation, replacing the second equation in this sum with the result.

Performing any sequence of these operations results in an equivalent system of equations.

We will use a type of shorthand notation for these operations. For a given system of linear equations, we let:

Notation for Operations on Equations

- $E_k \leftrightarrow E_j$ denote the operation "interchange equations k and j"
- $cE_k \rightarrow E_k$ denote the operation "multiply equation k by c and let the result replace equation k"
- $cE_k + E_j \rightarrow E_j$ denote the operation "multiply equation k by c, add to equation j and let the result replace equation j."

We illustrate the method of elimination in the next two examples.

EXAMPLE 3 Solving a System by Elimination

Solve the system

$$\begin{aligned} 3x - y &= 11 \\ x + 2y &= -8 \end{aligned}$$

Solution There are several reasonable ways to begin here. We start by rearranging the order is which the system is written; that is, we perform the operation $E_1 \leftrightarrow E_2$ to produce the equivalent system

$$\begin{aligned} x + 2y &= -8 \\ 3x - y &= 11 \end{aligned}$$

Since the coefficient of the x variable in the first equation is 1, we will "eliminate" the x variable from the second equation by performing the operation $-3E_1 + E_2 \to E_2$ on the most recent system.

$$\begin{array}{rl} -3x - 6y = 24 & (-3E_1) \\ 3x - y = 11 & \\ \hline -7y = 35 & \text{(replaces second equation)} \end{array}$$

This yields the equivalent system

$$\begin{aligned} x + 2y &= -8 \\ -7y &= 35 \end{aligned}$$

We now perform the operation $\frac{-1}{7} E_2 \to E_2$ on this system to obtain

$$\begin{aligned} x + 2y &= -8 \\ y &= -5 \end{aligned}$$

We immediately see that $y = -5$ and back-substitution into $x + 2y = -8$ gives

$$\begin{aligned} x + 2(-5) &= -8 \\ x &= 2 \end{aligned}$$

So, the pair $x = 2$ and $y = -5$ is the solution to the original system, as expected. ❖

EXAMPLE 4 Solving a System by Elimination

Solve the system

$$\begin{aligned} 3x + 4y &= 9 \\ 2x - 5y &= 15 \end{aligned}$$

Solution Although it is apparent that the arithmetic involved will be somewhat more difficult here compared to the last example, we proceed in a similar manner. We will eliminate the x variable from the second equation via the operation $\frac{-2}{3} E_1 + E_2 \to E_2$, so that

$$\begin{array}{rl} -2x - \frac{8}{3}y = -6 & (\frac{-2}{3}E_1) \\ 2x - 5y = 15 & \\ \hline \frac{-23}{3}y = 9 & \text{(replaces second equation)} \end{array}$$

This gives the equivalent system

$$\begin{aligned} 3x + 4y &= 9 \\ \tfrac{-23}{3}y &= 9 \end{aligned}$$

We modify the second equation in this system by the operation $\frac{-3}{23} E_2 \to E_2$, which yields the equivalent system

$$\begin{aligned} 3x + 4y &= 9 \\ y &= \tfrac{-27}{23} \end{aligned}$$

Since $y = \frac{-27}{23}$, we immediately back-substitute into $3x + 4y = 9$ to obtain

$$\begin{aligned} 3x + 4(\tfrac{-27}{23}) &= 9 \\ 3x - \tfrac{108}{23} &= 9 \\ 3x &= \tfrac{315}{23} \\ x &= \tfrac{105}{23} \end{aligned}$$

Thus the pair $x = \frac{105}{23}$ and $y = \frac{-27}{23}$ is the solution to the original system. ❖

Example 4 helps illustrate the difference between exact solutions and approximate solutions. Had we opted to look at the given system graphically, even using an intersection routine on a graphing calculator would have provided only a decimal approximation to the actual solution. In practice, a *good* approximation is often as useful as an exact answer, but it is useful at times to retain the option of finding exact solutions.

Thus far, we have seen systems of two linear equations in two unknowns where there was exactly one solution to the system. It is natural to wonder if there are other possible outcomes when we attempt to solve systems of equations. This is where we now turn our attention.

When Strange Things Happen

Consider the system

$$\begin{aligned} x + y &= 2 \\ 3x + 3y &= 9 \end{aligned}$$

We may elect to attempt to solve this system using the method of substitution, as it is easy to solve the first equation for x:

$$x = 2 - y$$

Substituting into the second equation finds

$$\begin{aligned} 3(2 - y) + 3y &= 9 \\ 6 - 3y + 3y &= 9 \\ 6 &= 9 \end{aligned}$$

The equation $6 = 9$ is certainly untrue (a **contradiction**). How should we interpret this? Recalling that solving a system of linear equations in two unknowns graphically boils down to finding all points where the graphs of the equations intersect, we observe that the nature of this situation is somewhat different than what we encountered earlier. If we rewrite the given linear equations in slope-intercept form, then the first equation becomes

$$y = -x + 2$$

while the second equation becomes

$$\begin{aligned} 3y &= -3x + 9 \\ y &= -x + 3 \end{aligned}$$

Graphically, this means that we have two parallel lines, as illustrated in Figure 2.

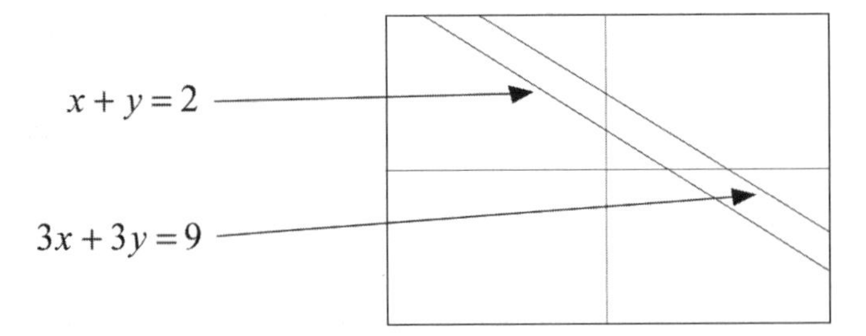

Figure 2

Since the lines corresponding to the equations in the system have no point of intersection, there is no pair (x, y) that simultaneously satisfies each equation—the system of equations does not have a solution. We call any system of equations having no solution an **inconsistent system**.

It is relatively easy to identify an inconsistent system of linear equations, no matter how many unknowns, because eventually a contradictory equation will appear when using algebraic methods to solve such a system, just as happened above. The next situation we consider, however, is much more subtle in nature.

Consider the system

$$\begin{aligned} 2x - y &= 4 \\ -4x + 2y &= -8 \end{aligned}$$

We will attempt to find a solution for this system using the method of elimination. In this case, the operation $2E_1 + E_2 \to E_2$ seems reasonable, as this will eliminate x from the second equation.

$$\begin{array}{rl} 4x - 2y = 8 & (2E_1) \\ \underline{-4x + 2y = -8} & \\ 0 = 0 & \text{(replaces second equation)} \end{array}$$

This gives the equivalent system

$$\begin{aligned} 2x - y &= 4 \\ 0 &= 0 \end{aligned}$$

The second equation in this system ($0 = 0$) is most certainly a true statement. However, it is not at all enlightening, nor does it appear to be helpful, as we are not able to solve for a single unknown. We again look at the original system graphically in an attempt to ascertain what is happening. Notice that when we rewrite each equation in slope-intercept form, we have

$$\begin{aligned} 2x - y &= 4 \\ y &= 2x - 4 \end{aligned}$$

and

$$\begin{aligned} -4x + 2y &= -8 \\ 2y &= 4x - 8 \\ y &= 2x - 4 \end{aligned}$$

The two equations in the original system have exactly the same slope-intercept form, so the graphs of each equation are the same line. As such, any point (x,y) on that line represents a solution to the system. Since there are an infinite number of points on a line, this means that the given system has an infinite number of solutions! See Figure 3.

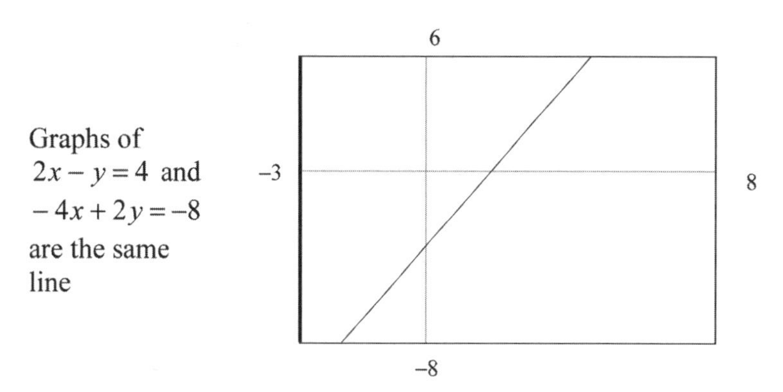

Figure 3

Thus far we have encountered three scenarios in terms of the number of solutions to a system of linear equations—no solutions, exactly one solution or an infinite number of solutions. In the case of a system of equations in two unknowns, it seems clear that these are the only possibilities (a pair of lines cannot have exactly two points of intersection, for example). As it turns out, *this is the case for systems of linear equations in any number of variables.*

We conclude this section with an application involving a system of linear equations where we will investigate a noteworthy occurrence within the solving process.

EXAMPLE 5 Application

Cybil has a combination of nickels and dimes in her pocket, 51 in all. The number of dimes is twice the number of nickels, and the total value of the change is \$4.25. How many of each kind of coin does she have in her pocket?

Solution We begin by declaring our unknowns. Let

d = the number of dimes Cybil has in her pocket

n = the number of nickels Cybil has in her pocket

We have several relationships between d and n. We know that Cybil has twice as many dimes as nickels, so

$$d = 2n$$

We also know that she has 51 coins in all, so

$$d + n = 51$$

Finally, the total value of the collection of coins is \$4.25, so

$$0.10d + 0.05n = 4.25$$

Rewriting $d = 2n$ as $d - 2n = 0$, we have modeled the problem with a system of linear equations:

$$\begin{aligned} d - 2n &= 0 \\ d + n &= 51 \\ 0.10d + 0.05n &= 4.25 \end{aligned}$$

We can take any approach to solving this system, but for the purpose of illustration we opt to use the method of elimination. We will eliminate d from the second and third equations by performing the operations $-1E_1 + E_2 \rightarrow E_2$ and $-0.10E_1 + E_3 \rightarrow E_3$. We see that

$$\begin{array}{rl} -d + 2n = 0 & (-1E_1) \\ \underline{d + n = 51} & \\ 3n = 51 & \text{(replaces second equation)} \end{array}$$

and

$$\begin{array}{rl} -0.10d + 0.2n = 0 & (-0.10E_1) \\ \underline{0.10d + 0.05n = 4.25} & \\ 0.25n = 4.25 & \text{(replaces third equation)} \end{array}$$

This gives us the equivalent system

$$\begin{aligned} d - 2n &= 0 \\ 3n &= 51 \\ 0.25n &= 4.25 \end{aligned}$$

Although this may seem superfluous, we now perform the operation $\frac{1}{3}E_2 \rightarrow E_2$ to obtain the system

$$\begin{aligned} d - 2n &= 0 \\ n &= 17 \\ 0.25n &= 4.25 \end{aligned}$$

And finally, we perform the operation $-0.25E_2 + E_3 \rightarrow E_3$

$$\begin{array}{rl} -0.25n = -4.25 & (-0.25E_2) \\ \underline{0.25n = 4.25} & \\ 0 = 0 & \text{(replaces third equation)} \end{array}$$

This results in the equivalent system

$$\begin{aligned} d - 2n &= 0 \\ n &= 17 \\ 0 &= 0 \end{aligned}$$

The third equation in this system ($0 = 0$) reminds us of the situation encountered earlier where a system had an infinite number of solutions. It may be tempting here to jump to that conclusion, but common sense suggests that that is not a reasonable outcome. After all, are there really an infinite number of ways to have a collection of a finite number of two different types of coins? Notice, though, that the second equation already gives us a value for n, and substituting that value of n into the first equation gives

$$\begin{aligned} d - 2(17) &= 0 \\ d - 34 &= 0 \\ d &= 34 \end{aligned}$$

Thus the pair $d = 34$ and $n = 17$ satisfies all three of the equations in the original system (check), and we conclude that Cybil has 34 dimes and 17 nickels in her pocket. ❖

PROBLEM SET 5.1

1. For each of the following systems, determine if the pair $x = 3$, $y = -2$ is a solution.

(a) $\begin{aligned} 2x - y &= 8 \\ x + y &= 1 \end{aligned}$

(b) $\begin{aligned} 5x + 27y &= 10 \\ -3x + 27y &= 7 \end{aligned}$

(c) $\begin{aligned} x - 4y &= 11 \\ 4x + 3y &= -12 \end{aligned}$

(d) $\begin{aligned} x - y &= 5 \\ 3x + 4y &= 1 \end{aligned}$

2. Suppose that $(-3, 1)$ and $(2, 4)$ are both solutions to a system of linear equations in two unknowns. What can you conclude about the number of solutions to that system?

In problems 3–8, solve the system **graphically**.

3. $\begin{aligned} 3x - 2y &= 3 \\ x - 3y &= -10 \end{aligned}$

4. $\begin{aligned} 2y - 10x &= -14 \\ 3y - 15x &= 3 \end{aligned}$

5. $\begin{aligned} 10s - 5t &= 25 \\ -8s + 4t &= -20 \end{aligned}$

6. $\begin{aligned} 7v - 4w &= -1 \\ 14v + 8w &= 10 \end{aligned}$

7. $\begin{aligned} 4x_1 - 3x_2 &= -4 \\ 2x_1 + x_2 &= 3 \end{aligned}$

8. $\begin{aligned} x - 3y &= 18 \\ 4x + 2y &= 5 \end{aligned}$

In problems 9–12, solve the system by **substitution**. Give exact answers.

9. $\begin{aligned} -5x + 2y &= 10 \\ 3x - y &= 4 \end{aligned}$

10. $\begin{aligned} 3k - 5l &= -27 \\ k + 2l &= 13 \end{aligned}$

11. $\begin{aligned} 4m - 3n &= 7 \\ 2m + 5n &= 14 \end{aligned}$

12. $\begin{aligned} -3x + 2y &= 6 \\ -x - y &= 10 \end{aligned}$

In problems 13–18, solve the system by **elimination**. Give exact answers.

13. $\begin{aligned} 3x - 2y &= 12 \\ 4x + 2y &= 2 \end{aligned}$

14. $\begin{aligned} 3x - 4y &= -10 \\ 5x + y &= 14 \end{aligned}$

15. $\begin{aligned} -\frac{1}{2}s + t &= 5 \\ 2s - 4t &= -20 \end{aligned}$

16. $\begin{aligned} \frac{2}{3}x - 4y &= 6 \\ 2x + \frac{1}{3}y &= 4 \end{aligned}$

17. $\begin{aligned} -7x_1 + 5x_2 &= 16 \\ 2x_1 + 3x_2 &= 10 \end{aligned}$

18. $\begin{aligned} -6x + 2y &= 10 \\ 9x - 3y &= 12 \end{aligned}$

19. The Completely Cookoo Chocolate Company makes two varieties of chocolate milk. Regular chocolate milk uses 1 gallon of syrup to 12 gallons of milk. Extra-thick chocolate milk uses 2 gallons of syrup to 19 gallons of milk. The company has 260 gallons of syrup and 2,720 gallons of milk. Determine the number of gallons of each type of chocolate milk the company should produce in order to use up all of the syrup and all of the milk.

20. Randolph has invested \$1,000 in two different certificates of deposit (CDs). One gives a return of 6% and the other gives a return of 7%. The total return on his investments is \$64. How much did he invest in each CD?

21. A year ago, Kris had twice as much money hidden in a shoe box as Shelley has hidden under a mattress. Kris stashed an additional \$100 in his shoe box this last year. However, Shelley stuck another \$200 under the mattress in that same period of time, and now she has \$50 more hidden away than Kris. How much does each of them have hidden away right now?

22. A batch of Sinfully Silky gourmet ice cream calls for 200 gallons of a milk-cream mixture. Milk costs the producer \$1.50 per gallon and cream costs \$2.75 per gallon. Profit studies suggest that the company should spend only \$450 for the milk-cream mixture for a batch of ice cream. How many gallons each of milk and cream should it mix together to make a mixture that costs \$450?

23. A recent rodeo event sold a total of 4,200 adult and children's tickets. The gate revenue was \$39,900. Adult tickets sell for \$10.50 apiece and children's tickets sell for \$7.50 apiece. Determine how many adult tickets sold and how many children's tickets sold.

Interlude: Progress Check 5.1, Part 1

1. For an experiment a chemist needs 500 liters of a 25% acid solution. On hand are solutions that are 18% and 30% acid, respectively. You need to determine how many liters of each of the available solutions should be mixed together to achieve the desired result.

(a) Set up a system of linear equations that can be used to solve the problem. Be sure to indicate what your variables represent.

(b) Solve the system of equations from (a) graphically. Show the graphs you used, label the point of intersection appropriately, and indicate how much of each solution is used.

2. Consider the system
$$\begin{aligned} -5x + 2y &= 3 \\ 20x + ky &= -27. \end{aligned}$$

(a) Determine all possible values of k so that the system has exactly one solution. Show how you arrived at your conclusion.

(b) Determine all possible values of k so that the system has an infinite number of solutions. Show how you arrived at your conclusion.

(c) Determine all possible values of k so that the system has no solutions. Show how you arrived at your conclusion.

Interlude: Progress Check 5.1, Part 2

1. A jar contains a combination of 60 nickels and dimes worth a total of \$4.30. Suppose that it is deemed desirable to determine how many of each kind of coin are in the jar.

(a) Set up a system of linear equations that can be used to solve the problem. Be sure to indicate what your variables represent.

(b) Solve the system of equations from (a) graphically. Show the graphs you used, label the point of intersection appropriately, and indicate how many of each coin are in the jar.

(c) Solve the system of equations from (a) algebraically. Show your method, and indicate how many of each coin are in the jar.

2. A bake shop makes two different sizes of blueberry muffins using prepackaged dough and blueberries. A large muffin requires 5 ounces of dough and 2 ounces of blueberries. A small muffin requires 2 ounces of dough and 1 ounce of blueberries. Each day, the shop takes delivery of 450 ounces of dough and 200 ounces of blueberries. The bake shop would like to know how many of each size muffin to bake each day to use up all of the dough and blueberries.

(a) Set up a system of linear equations that can be used to solve the problem. Be sure to indicate what your variables represent.

(b) Solve the system of equations from (a) graphically. Show the graphs you used, label the point of intersection appropriately, and indicate how many of each size of muffin to bake.

(c) Solve the system of equations from (a) algebraically. Show your method, and indicate how many of each size of muffin to bake.

OBJECTIVES

1. Represent Linear Systems With Matrices
2. Solve Linear Systems Using Row Reduction
3. Solve Applied Problems

5.2 Larger Systems: Matrix Representation and Gauss-Jordan Elimination

We can model many situations quite naturally by a system of linear equations involving more than one unknown. Consider the following problem.

> A testing lab in a pharmaceutical company has a group of young mice, adult female mice and adult male mice. There are 33 mice in all. On average, the mice consume 194 mg of food and 53.7 ml of water daily. Typically, young mice will consume 2 mg of food and 1.0 ml of water in a day. Similarly, adult female mice and adult male mice will consume 6 mg of food and 1.75 ml of water, and 8 mg of food and 1.85 ml of water, respectively. Given this, how many young mice, adult female mice and adult male mice are there?

We see that the problem posed has three unknown but sought quantities. Denote these quantities by x, y and z, where

$$x = \text{the number of young mice in the group}$$

$$y = \text{the number of adult female mice in the group}$$

$$z = \text{the number of adult male mice in the group}$$

We have several pieces of information about these quantities, which we summarize in a table.

	Young	Female	Male	Totals
Number	x	y	z	33
Food (mg)	2	6	8	194
Water (ml)	1.0	1.75	1.85	53.7

This information readily leads to equations relating x, y and z:

$$\begin{aligned} x + y + z &= 33 \quad \text{(total number of mice)} \\ 2x + 6y + 8z &= 194 \quad \text{(total mg of food consumed)} \\ 1.0x + 1.75y + 1.85z &= 53.7 \quad \text{(total ml of water consumed)} \end{aligned}$$

We now turn to the task of finding a solution to a system such as this, where there are more than two unknowns. In general, with systems of this nature, a graphical approach is not feasible, so we will look at algebraic methods.

The Method of Elimination, Revisited

Although it looks to be a labor-intensive process, we can certainly use the method of elimination to attempt to find a solution to the system of equations we just generated. To start, we will eliminate x from the second and third equations. Since the coefficient of x is 1 in the first equation, it is natural to use the first equation in the subsequent eliminations. We perform the operations $-2E_1 + E_2 \to E_2$ and $-1.0E_1 + E_3 \to E_3$.

$$\begin{array}{rl} -2x - 2y - 2z = -66 & (-2E_1) \\ 2x + 6y + 8z = 194 & \\ \hline 4y + 6z = 128 & \text{(replaces second equation)} \end{array}$$

$$\begin{array}{rl} -1.0x - 1.0y - 1.0z = -33 & (-2E_1) \\ 1.0x + 1.75y + 1.85z = 53.7 & \\ \hline 0.75y + 0.85z = 20.7 & \text{(replaces third equation)} \end{array}$$

This results in the equivalent system

$$\begin{aligned} x + y + z &= 33 \\ 4y + 6z &= 128 \\ 0.75y + 0.85z &= 20.7 \end{aligned}$$

Notice that the last two equations only involve the variables y and z. We focus on those two equations. We first perform the operation $\frac{1}{4} E_2 \to E_2$ to obtain

$$\begin{aligned} x + y + z &= 33 \\ y + 1.5z &= 32 \\ 0.75y + 0.85z &= 20.7 \end{aligned}$$

and then use the new second equation in that system to eliminate y from the third equation via the operation $-0.75E_2 + E_3 \to E_3$.

$$\begin{array}{rl} -0.75y - 1.125z = -24 & (-0.75E_2) \\ 0.75y + 0.85z = 20.7 & \\ \hline -0.275z = -3.3 & \text{(replaces third equation)} \end{array}$$

This gives the equivalent system

$$\begin{aligned} x + y + z &= 33 \\ y + 1.5z &= 32 \\ -0.275z &= -3.3 \end{aligned}$$

The third equation in this system is readily solved for z: $z = 12$. Back-substitution into the equation $y + 1.5z = 32$ yields

$$\begin{aligned} y + 1.5(12) &= 32 \\ y + 18 &= 32 \\ y &= 14 \end{aligned}$$

Finally, back-substituting the pair $y = 14$ and $z = 12$ into $x + y + z = 33$ gives

$$x + 14 + 12 = 33$$
$$x = 7$$

So, the triple $x = 7$, $y = 14$ and $z = 12$ is a solution to the original system (check), indicating that the group of lab mice consisted of 7 young mice, 14 adult female mice and 12 adult male mice.

While we can use the method of elimination on systems of linear equations in any number of variables, it can become rather tedious. Notice, though, that when using this method, the coefficients of the variables in the equations determined the choice of multipliers chosen for the subsequent operations. We now turn our attention to a method of solving a system of linear equations that is equivalent to the method of elimination but uses an object that will allow us to work only with coefficients and constant terms in the system and not with the variables.

Matrices

A **matrix** is a rectangular array of numbers arranged in rows and columns. We call each number in this array an **element** of the matrix. When we write a matrix, we typically enclose the array in brackets.

Matrices (plural) come in many sizes, determined by the number of rows and the number of columns. If a matrix has m rows and n columns, then we say that the **size of the matrix** is $m \times n$, read "m by n". The following are examples of matrices of various sizes.

$$\begin{bmatrix} 1 & 5 & -6 \\ 2 & 3 & 0 \end{bmatrix} \qquad \begin{bmatrix} 0.5 & -3 & 4 \\ 0 & 1 & 9.4 \\ 6 & 0.25 & 99 \end{bmatrix} \qquad \begin{bmatrix} 7 \\ 13 \\ -24 \end{bmatrix} \qquad \begin{bmatrix} -8 & 2 & 0 & 0 & 4 \\ 9 & 6 & 3 & -5 & -1 \end{bmatrix}$$

$$2 \times 3 \qquad\qquad 3 \times 3 \qquad\qquad 3 \times 1 \qquad\qquad 2 \times 5$$

When we enumerate rows and columns of a given matrix, we count rows from top to bottom and count columns from left to right.

Since a matrix is an array of numbers, we often see matrices used to record information, especially if rows and columns of a matrix can be understood to represent categories. As such, we can certainly use matrices to record pertinent information about a system of linear equations—coefficients of the variables as well as constants on the right-hand side of equations in the system.

Consider the system

$$2x_1 + 6x_2 = 9$$
$$-3x_1 + 5x_2 = -12$$

We adopt a convention here of using subscripted variables rather than individual letter variables to avoid possible difficulties in the number of available letters. We construct a 2×3 matrix, called the **augmented matrix for the system**, where each row represents information for a particular equation and each column represents either coefficients of a variable or the constants on the right-hand side of the equations. We write this matrix as follows.

$$\left[\begin{array}{cc|c} 2 & 6 & 9 \\ -3 & 5 & -12 \end{array}\right]$$

Notice the correspondence between the rows of this matrix and the equations in the system as well as the correspondence between the columns of the matrix and the coefficients and constant terms in the equations. The vertical line has no real purpose except to serve as a visual reminder of the location of the equal signs in the system and hence a separation between the coefficients of the variables and the constants on the right-hand side of the equations.

Before delving into the use of these augmented matrices representing systems, we will pause to put forth some terminology and notation. Recall that in the method of elimination we had three operations that we could use to produce equivalent systems of linear equations. We have a similar collection of **row operations** that we perform on matrices. We say two matrices are **row-equivalent** if one is obtained from the other by some sequence of row operations. These operations are as follows:

Row Operations

- Interchange any two rows.
- Multiply (all elements in) a row by any nonzero constant and replace that row with the result.
- Multiply (all elements in) a row by any constant and add (corresponding elements) to any other row, replacing the second row in this sum with the result.

Performing any sequence of these operations results in a row-equivalent matrix.

Notice the similarity between these operations and the operations employed in the method of elimination. We use a similar shorthand notation to indicate performing a particular row operation as well.

Notation for Row Operations

- $R_k \leftrightarrow R_j$ denotes the operation "interchange rows k and j."
- $cR_k \to R_k$ denotes the operation "multiply row k by c and let the result replace row k."
- $cR_k + R_j \to R_j$ denotes the operation "multiply row k by c and add to row j and let the result replace row j."

In the case where a matrix is the augmented matrix representing a system of linear equations, performing a row operation on the matrix is equivalent to performing the corresponding operation on a system of equations. So, *row-equivalent matrices represent equivalent systems of linear equations*. To demonstrate how to use augmented matrices to find solutions to systems of linear equations, we will show parallel operations in the method of elimination and the corresponding row operations.

EXAMPLE 1 **Using Augmented Matrices to Solve a System**

Solve the system

$$\begin{aligned} x_1 + 2x_2 - x_3 &= 3 \\ x_1 + 3x_2 - x_3 &= 4 \\ 2x_1 - 2x_2 + 2x_3 &= 8 \end{aligned}$$

Solution Consider the following sequences of equivalent systems and row-equivalent matrices. Pay close attention to the correspondence between elements in the matrices and the terms in the systems of equations.

Sequence of Equivalent Systems	Subsequent Operations on Equations	Sequence of Row-equivalent Matrices	Subsequent Row Operations
Original System $x_1 + 2x_2 - x_3 = 3$ $x_1 + 3x_2 - x_3 = 4$ $2x_1 - 2x_2 + 2x_3 = 8$	$-1E_1 + E_2 \to E_2$ $-2E_1 + E_3 \to E_3$ (eliminate x_1 from the last two equations)	*Original Augmented Matrix* $\left[\begin{array}{ccc\|c} 1 & 2 & -1 & 3 \\ 1 & 3 & -1 & 4 \\ 2 & -2 & 2 & 8 \end{array}\right]$	$-1R_1 + R_2 \to R_2$ $-2R_1 + R_3 \to R_3$ (get 0s in row 2, col 1 and row 3, col 1)
$x_1 + 2x_2 - x_3 = 3$ $x_2 = 1$ $-6x_2 + 4x_3 = 2$	$-2E_2 + E_1 \to E_1$ $6E_2 + E_3 \to E_3$ (eliminate x_2 from first and third equations)	$\left[\begin{array}{ccc\|c} 1 & 2 & -1 & 3 \\ 0 & 1 & 0 & 1 \\ 0 & -6 & 4 & 2 \end{array}\right]$	$-2R_2 + R_1 \to R_1$ $6R_2 + R_3 \to R_3$ (get 0s in row 1, col 2 and row 3, col 2)
$x_1 - x_3 = 1$ $x_2 = 1$ $4x_3 = 8$	$\frac{1}{4}E_2 \to E_2$ (modify third equation to get x_3 coefficient 1)	$\left[\begin{array}{ccc\|c} 1 & 0 & -1 & 1 \\ 0 & 1 & 0 & 1 \\ 0 & 0 & 4 & 8 \end{array}\right]$	$\frac{1}{4}E_2 \to E_2$ (get 1 in row 3, col 3)
$x_1 - x_3 = 1$ $x_2 = 1$ $x_3 = 2$	$E_3 + E_1 \to E_1$ (eliminate x_3 from first equation)	$\left[\begin{array}{ccc\|c} 1 & 0 & -1 & 1 \\ 0 & 1 & 0 & 1 \\ 0 & 0 & 1 & 2 \end{array}\right]$	$R_3 + R_1 \to R_1$ (get 0 in row 1, col 3)
$x_1 = 3$ $x_2 = 1$ $x_3 = 2$		$\left[\begin{array}{ccc\|c} 1 & 0 & 0 & 3 \\ 0 & 1 & 0 & 1 \\ 0 & 0 & 1 & 2 \end{array}\right]$	❖

We see, then, that the last system expressly gives the solution to the system. Notice how we can also read the solution from the last matrix as well. We performed more operations in this case than previously to illustrate that, at least in this case, the extra operations didn't result in significantly more work than a series of back-substitutions and to illustrate how, in the final augmented matrix, we can easily read the solution to the system.

After arriving at the end of the sequence of row operations and resulting matrices in the previous example, we would say that we have **reduced** the augmented matrix that represented the system of equations being investigated to a simpler matrix which evidently displayed the solution to the system. This reduction process is called **Gauss-Jordan elimination**. When we engage in this process in an organized manner, it is somewhat more efficient than the method of elimination, at least relative to systems of equations in more than two variables. It may seem that the process will still be very tedious, but many graphing calculators and computer programs work very well for such a task, as the process is really just arithmetic on an array of numbers.

Bear in mind that when using this method, every matrix generated represents a system of linear equations, and any two matrices that are row-equivalent represent equivalent systems.

Before attempting to summarize the method of Gauss-Jordan elimination, we will introduce the notion of **reduced echelon form** of a matrix. We say a matrix is in reduced echelon form if all of the following are true:

Reduced Echelon Form

1. All rows consisting entirely of zeros are grouped at the bottom of the matrix.
2. The leftmost nonzero number in each row is 1 (this element is called the **leading 1 of the row**).
3. The leading 1 of a row is to the right of the leading 1 of the previous row.
4. All other elements in the same column as a leading 1 are zeros.

The following matrices are in reduced echelon form. Consider them as augmented matrices representing a system of linear equations and note that the corresponding system is also listed.

Reduced Echelon Form Matrix	Corresponding System of Linear Equations
$\left[\begin{array}{ccc\|c} 1 & 0 & 0 & 6 \\ 0 & 1 & 0 & -4 \\ 0 & 0 & 1 & 0 \end{array}\right]$	$x_1 = 6$ $x_2 = -4$ $x_3 = 0$
$\left[\begin{array}{cc\|c} 1 & 0 & \frac{3}{2} \\ 0 & 1 & -10 \\ 0 & 0 & 0 \end{array}\right]$	$x_1 = \frac{3}{2}$ $x_2 = -10$ $0 = 0$

Matrix	System
$\left[\begin{array}{cccc\|c} 1 & 2 & 0 & 5 & 0 \\ 0 & 0 & 1 & -4 & 0 \\ 0 & 0 & 0 & 0 & 1 \\ 0 & 0 & 0 & 0 & 0 \end{array}\right]$	$x_1 + 2x_2 + 5x_4 = 0$ $x_3 - 4x_4 = 0$ $0 = 1$ $0 = 0$
$\left[\begin{array}{ccc\|c} 1 & -5 & 0 & 8 \\ 0 & 0 & 1 & -13 \\ 0 & 0 & 0 & 0 \end{array}\right]$	$x_1 - 5x_2 = 8$ $x_3 = -13$ $0 = 0$

It may seem superfluous to list equations like $0 = 0$, but it is important when interpreting what exactly is happening with solutions to linear systems to clearly list all information rather than to run the risk of jumping to a hasty conclusion or ignoring a piece that is vital.

We can summarize the method of Gauss-Jordan elimination in three statements:

Gauss-Jordan Elimination

1. Represent a system of linear equations with an augmented matrix.
2. Reduce the original augmented matrix to reduced echelon form.
3. Interpret the reduced echelon form matrix.

Naturally, this summary does not indicate the nature of detail in the process. For instance, leading 1s are an important feature of reduced echelon form. During the reduction process, it is beneficial to obtain leading 1s "quickly" and to use these in the process of reducing other elements in the matrix to zeros. Also, we need to deal with the issue of interpreting the information presented in the resulting reduced echelon form matrix. We explore some of this in the next examples.

EXAMPLE 2 Gauss-Jordan Elimination

Solve the following system using matrices, if possible.

$$x_2 - 3x_3 = 2$$
$$2x_1 + 4x_2 + 6x_3 = -4$$
$$3x_1 + 5x_2 + 2x_3 = 2$$

Solution We begin by writing the augmented matrix associated with the system and proceed to reduce that matrix to reduced echelon form. Notice that the coefficient of the x_1 variable in the first equation is 0.

Sequence of Matrices	Subsequent Row Operations	Comments
$\left[\begin{array}{ccc\|c} 0 & 1 & -3 & 2 \\ 2 & 4 & 6 & -4 \\ 3 & 5 & 2 & 2 \end{array}\right]$	$R_1 \leftrightarrow R_2$	We will be able to get a leading 1 in row 1
$\left[\begin{array}{ccc\|c} 2 & 4 & 6 & -4 \\ 0 & 1 & -3 & 2 \\ 3 & 5 & 2 & 2 \end{array}\right]$	$\frac{1}{2}E_2 \to E_2$	Get leading 1 in row 1
$\left[\begin{array}{ccc\|c} 1 & 2 & 3 & -2 \\ 0 & 1 & -3 & 2 \\ 3 & 5 & 2 & 2 \end{array}\right]$	$-3R_1 + R_3 \to R_3$	Use the leading 1 in row 1 to reduce the rest of column 1
$\left[\begin{array}{ccc\|c} 1 & 2 & 3 & -2 \\ 0 & 1 & -3 & 2 \\ 0 & -1 & -7 & 8 \end{array}\right]$	$-2R_2 + R_1 \to R_1$ $R_2 + R_3 \to R_3$	Row 2 already has a leading 1 Use the leading 1 in row 2 to reduce the rest of column 2
$\left[\begin{array}{ccc\|c} 1 & 0 & 9 & -6 \\ 0 & 1 & -3 & 2 \\ 0 & 0 & -10 & 10 \end{array}\right]$	$\frac{1}{10}R_3 \to R_3$	Get leading 1 in row 3
$\left[\begin{array}{ccc\|c} 1 & 0 & 9 & -6 \\ 0 & 1 & -3 & 2 \\ 0 & 0 & 1 & -1 \end{array}\right]$	$-9R_3 + R_1 \to R_1$ $3R_3 + R_2 \to R_2$	Use the leading 1 in row 3 to reduce the rest of column 3
$\left[\begin{array}{ccc\|c} 1 & 0 & 0 & 3 \\ 0 & 1 & 0 & -1 \\ 0 & 0 & 1 & -1 \end{array}\right]$		Reduction Complete ❖

The final matrix in the sequence indicates that the system of equations has a single solution, the triple $x_1 = 3$, $x_2 = -1$, and $x_3 = -1$ (check).

EXAMPLE 3 Gauss-Jordan Elimination

Solve the following system using matrices, if possible.

$$\begin{aligned} x_1 + x_2 - 2x_3 &= 3 \\ -2x_1 - x_2 + 5x_3 &= -2 \\ -3x_1 + 2x_2 + 11x_3 &= 11 \end{aligned}$$

Solution Again, we begin by writing the augmented matrix associated with the system and proceed to reduce that matrix to reduced echelon form.

Sequence of Matrices	Subsequent Row Operations	Comments
$\left[\begin{array}{ccc\|c} 1 & 1 & -2 & 3 \\ -2 & -1 & 5 & -2 \\ -3 & 2 & 11 & 11 \end{array}\right]$	$2R_1 + R_2 \rightarrow R_2$ $3R_1 + R_3 \rightarrow R_3$	Row 1 already has a leading 1 Use the leading 1 in row 1 to reduce the rest of column 1
$\left[\begin{array}{ccc\|c} 1 & 1 & -2 & 3 \\ 0 & 1 & 1 & 4 \\ 0 & 5 & 5 & 20 \end{array}\right]$	$-1R_2 + R_1 \rightarrow R_1$ $-5R_2 + R_3 \rightarrow R_3$	Row 2 already has a leading 1 Use the leading 1 in row 2 to reduce the rest of column 2
$\left[\begin{array}{ccc\|c} 1 & 0 & -3 & -1 \\ 0 & 1 & 1 & 4 \\ 0 & 0 & 0 & 0 \end{array}\right]$		Reduction is complete

At this point, we see that the final matrix represents the system

$$\begin{aligned} x_1 - 3x_3 &= -1 \\ x_2 + x_3 &= 4 \\ 0 &= 0 \end{aligned}$$

There are no contradictions present, but we cannot solve explicitly (find a real number result) for any of the variables. Consequently, the system of equations has an infinite number of solutions. This text is not going to examine techniques for describing these solutions. ❖

EXAMPLE 4 Gauss-Jordan Elimination

Solve the following system using matrices, if possible.

$$\begin{aligned} 2x_1 + 7x_2 + 12x_3 &= 1 \\ -x_1 + 4x_2 + 9x_3 &= 8 \\ x_1 + 2x_2 + 3x_3 &= -1 \end{aligned}$$

Solution Again, we begin by writing the augmented matrix associated with the system and proceed to reduce that matrix to reduced echelon form.

Sequence of Matrices	Subsequent Row Operations	Comments
$\left[\begin{array}{ccc\|c} 2 & 7 & 12 & 1 \\ -1 & 4 & 9 & 8 \\ 1 & 2 & 3 & -1 \end{array}\right]$	$R_1 \leftrightarrow R_3$	Get a leading 1 in row 1
$\left[\begin{array}{ccc\|c} 1 & 2 & 3 & -1 \\ -1 & 4 & 9 & 8 \\ 2 & 7 & 12 & 1 \end{array}\right]$	$R_1 + R_2 \rightarrow R_2$ $-2R_1 + R_3 \rightarrow R_3$	Use the leading 1 in row 1 to reduce the rest of column 1

$\left[\begin{array}{ccc\|c} 1 & 2 & 3 & -1 \\ 0 & 6 & 12 & 7 \\ 0 & 3 & 6 & 3 \end{array}\right]$	$\frac{1}{6}R_2 \rightarrow R_2$	Get a leading 1 in row 2
$\left[\begin{array}{ccc\|c} 1 & 2 & 3 & -1 \\ 0 & 1 & 2 & \frac{7}{6} \\ 0 & 3 & 6 & 3 \end{array}\right]$	$-2R_2 + R_1 \rightarrow R_1$ $-3R_2 + R_3 \rightarrow R_3$	Use the leading 1 in row 2 to reduce the rest of column 2
$\left[\begin{array}{ccc\|c} 1 & 0 & -1 & -\frac{10}{3} \\ 0 & 1 & 2 & \frac{7}{6} \\ 0 & 0 & 0 & -\frac{1}{2} \end{array}\right]$		There is no need to continue

The system represented by the last matrix shown is

$$\begin{aligned} x_1 - x_3 &= -\tfrac{10}{3} \\ x_2 + 2x_3 &= \tfrac{7}{6} \\ 0 &= -\tfrac{1}{2} \end{aligned}$$

The last equation in this system is a contradiction. Consequently, the original system of equations is inconsistent and has no solution. ❖

In summary, the nonzero rows of the reduced echelon form of the augmented matrix for a system of linear equations give all of the information needed to ascertain the nature of solutions to the system. There are only three possibilities.

1. **The system has no solution.** One row of the reduced echelon form matrix will indicate a contradiction—all zeros in the coefficient portion of the matrix and a nonzero constant in the last column position of that row.

   In the event that the system has solutions (no contradictions), there are two possibilities.
2. **The solution is unique.** The number of nonzero rows in the reduced echelon form matrix is exactly the same as the number of variables in the system. (This *does not* mean that the system must have the same number of equations as variables.)
3. **The system has an infinite number of solutions.** The number of nonzero rows in the reduced echelon form matrix is less than the number of variables in the system. (This *does not* mean that the system must have fewer equations than variables.)

PROBLEM SET 5.2

1. For each of the following augmented matrices, indicate whether or not it is in reduced echelon form. If it is not in reduced echelon form, then state why it is not.

(a) $\left[\begin{array}{ccc|c} 1 & 0 & 0 & 2 \\ 0 & 1 & 0 & 3 \\ 0 & 0 & 1 & 0 \end{array}\right]$

(b) $\left[\begin{array}{ccc|c} 1 & 2 & 0 & 0 \\ 0 & 0 & 1 & 0 \\ 0 & 0 & 0 & 1 \end{array}\right]$

(c) $\left[\begin{array}{cccc|c} 1 & 0 & 0 & 3 & -2 \\ 0 & 1 & 0 & 1 & 1 \\ 0 & 0 & 0 & 1 & -4 \\ 0 & 0 & 1 & 0 & 2 \end{array}\right]$

(d) $\left[\begin{array}{cc|c} 1 & 0 & -2 \\ 0 & 1 & 3 \\ 0 & 0 & 0 \end{array}\right]$

Let $M = \begin{bmatrix} 1 & 0 & 1 & 1 \\ 2 & 4 & 6 & 7 \\ 0 & 2 & -1 & 0 \end{bmatrix}$. In problems **2–5**, specify row operations that transform M into the given matrix.

2. $\begin{bmatrix} -2 & 0 & -2 & -2 \\ 2 & 4 & 6 & 7 \\ 0 & 2 & -1 & 0 \end{bmatrix}$

3. $\begin{bmatrix} 1 & 0 & 1 & 1 \\ 3 & 4 & 7 & 8 \\ 0 & 2 & -1 & 0 \end{bmatrix}$

4. $\begin{bmatrix} 1 & 0 & 1 & 1 \\ 0 & 2 & -1 & 0 \\ 2 & 4 & 6 & 7 \end{bmatrix}$

5. $\begin{bmatrix} 1 & 0 & 1 & 1 \\ 0 & 1 & 1 & 1.25 \\ 0 & 0 & -3 & -2.5 \end{bmatrix}$

6. For each system of linear equations, write the corresponding augmented matrix.

(a)
$$\begin{aligned} x + 2y - z &= 5 \\ 3x - y + z &= 6 \\ x \qquad + z &= 3 \end{aligned}$$

(b)
$$\begin{aligned} -x_1 + 3x_2 - x_3 + x_4 &= -2 \\ 2x_2 + 3x_3 - x_4 &= 7 \\ 3x_1 - \qquad x_3 + 5x_4 &= 1 \end{aligned}$$

(c)
$$\begin{aligned} 2x_1 - x_2 &= 4 \\ x_1 + x_2 &= 6 \end{aligned}$$

(d)
$$\begin{aligned} 4a + 3b - c &= -10 \\ 2a - b \qquad &= 5 \end{aligned}$$

7. For each of the following systems of linear equations, indicate whether or not the triple $x_1 = 3, x_2 = -1, x_3 = 4$ is a solution.

(a)
$$\begin{aligned} x_1 + x_2 + x_3 &= 6 \\ 2x_1 - x_2 - 2x_3 &= -1 \\ -3x_1 + 3x_2 - x_3 &= 5 \end{aligned}$$

(b)
$$\begin{aligned} 2x_1 - 3x_2 + 5x_3 &= 29 \\ x_1 - x_2 - x_3 &= 0 \\ 4x_1 + x_2 + 3x_3 &= 23 \end{aligned}$$

(c)
$$\begin{aligned} x_1 - 5x_2 + x_3 &= 9 \\ 2x_1 \qquad + 3x_3 &= 18 \end{aligned}$$

(d)
$$\begin{aligned} -3x_1 + 2x_2 - x_3 &= -15 \\ x_2 + 4x_3 &= 15 \end{aligned}$$

In problems 8–14, solve the system using Gauss-Jordan elimination.

8.
$$\begin{aligned} x - 2y &= 9 \\ 2x + y &= -2 \end{aligned}$$

9.
$$\begin{aligned} 3x_1 - 2x_2 + x_3 &= 5 \\ x_1 + x_2 - x_3 &= 3 \\ 2x_1 - x_2 \qquad &= 2 \end{aligned}$$

10.
$$\begin{aligned} -x_1 + 2x_2 + 3x_3 &= 4 \\ x_1 - 3x_2 + 2x_3 &= -1 \\ 3x_1 - 9x_2 + 6x_3 &= 1 \end{aligned}$$

11.
$$\begin{aligned} -6x_1 + 18x_2 - 12x_3 &= -3 \\ 3x_1 - 9x_2 + 6x_3 &= 1.5 \\ 4x_1 - 12x_2 + 8x_3 &= 2 \end{aligned}$$

12.
$$\begin{aligned} 2x_1 - x_2 + 4x_3 + x_4 &= 13 \\ -x_1 + x_2 - 2x_3 + 4x_4 &= -15 \\ 3x_1 + 3x_2 + x_3 - x_4 &= 14 \\ -2x_1 + x_2 - 7x_3 + 2x_4 &= -28 \end{aligned}$$

13.
$$\begin{aligned} 3x_1 - 2x_2 &= -5 \\ 6x_1 + x_2 &= 0 \\ -2x_1 - \tfrac{1}{3}x_2 &= 0 \end{aligned}$$

14.
$$\begin{aligned} 3x_1 + 5x_2 - x_3 + 2x_4 &= 2 \\ 2x_1 \qquad - x_3 \qquad &= 0 \\ 2x_2 \qquad\qquad - x_4 &= -3 \\ x_1 - 3x_2 - x_3 \qquad &= 1 \end{aligned}$$

15. Verify that the following system has an infinite number of solutions.

$$\begin{aligned} x_1 + 2x_2 - x_3 &= 5 \\ 2x_1 + 3x_2 - 3x_3 &= 8 \\ x_2 + x_3 &= 2 \end{aligned}$$

Find three specific solutions to this system. (**Hint**: Specify a value for x_3, then solve for x_1 and x_2.)

Interlude: Progress Check 5.2, Part 1

1. Given $\begin{bmatrix} -3 & -1 & 2 & 4 \\ 5 & 0 & -2 & 7 \\ 6 & -2 & 1 & 9 \end{bmatrix}$, I performed this row operation: $-3R_2 + R_3 \rightarrow R_3$.

(a) Show the resulting matrix.

(b) What row operation will transform the matrix from (a) back into the original matrix?

2. Write a system of three linear equations in three variables that has $(-3, 1, 4)$ as one of an infinite number of solutions. Your system should have all nonzero coefficients. Show clearly that your system actually satisfies the given requirement.

3. Consider the matrix $A = \left[\begin{array}{cccc|c} 1 & 1 & 1 & 1 & 20 \\ 0 & 1 & 1 & 1 & 0 \\ 0 & 0 & 1 & 1 & 13 \\ 0 & 2 & 0 & -2 & -10 \end{array}\right]$

(a) Write the system of equations represented by this matrix.

(b) Use row operations to completely row-reduce A and show the resulting matrix after each row operation.

(c) Give the solution to the system represented by A.

Interlude: Progress Check 5.2, Part 2

1. Consider the following matrices. For each, write the system of equations it represents, and determine how many solutions there are to that system.

$$A = \left[\begin{array}{ccc|c} 1 & 0 & 6 & 4 \\ 0 & 1 & -3 & 8 \\ 0 & 0 & 1 & 0 \end{array}\right] \qquad B = \left[\begin{array}{ccc|c} 1 & 5 & -2 & 0 \\ 0 & 0 & 1 & -1 \\ 0 & 0 & 5 & -5 \end{array}\right]$$

2. Consider the matrices $A = \begin{bmatrix} 3 & -4 & -5 & 8 \\ 1 & -2 & 1 & 0 \\ -4 & 5 & 9 & -9 \end{bmatrix}$ and $B = \begin{bmatrix} 3 & -24 & 12 & -27 \\ 0 & 1 & -2 & -42 \\ 0 & 0 & 2 & 58 \end{bmatrix}$. Are A and B row-equivalent?

Explain why or why not.

3. Verify that the system

$$\begin{aligned} 5x_1 - x_2 - 2x_3 + 5x_4 &= -9 \\ -4x_1 + x_2 + 2x_3 - 4x_4 &= 11 \\ 2x_1 - x_2 - x_3 &= -10 \end{aligned}$$

has an infinite number of solutions, and find three distinct solutions to the system.

OBJECTIVE

1. Solve Applied Problems

5.3 More Modeling and Applications

Thus far, we have presented a few problems where a system of linear equations modeled the situation at hand. We will now concentrate on more types of problems that systems of linear equations model naturally.

Familiarizing yourself with the problem situation and setting up the subsequent model is, typically, the most challenging part of a problem. In the types of problems we will be considering here, it is often helpful to summarize and organize the information, perhaps using a table, before starting the modeling process.

EXAMPLE 1 A Dietician Problem

A dietician must make a meal consisting of 700 calories, 450 grams of protein and 220 milligrams of vitamin A from three types of food: ground beef, baked potato and mixed vegetables. One serving of ground beef contains 200 calories, 150 grams of protein and 10 milligrams of vitamin A. One serving of baked potato contains 150 calories, 100 grams of protein and 30 milligrams of vitamin A. One serving of mixed vegetables contains 100 calories, 50 grams of protein and 75 milligrams of vitamin A. How many servings of each type of food should the meal contain?

Solution Since we seek the number of servings of the three types of food needed to satisfy various dietary requirements, we begin by summarizing information about servings of ground beef, baked potato and mixed vegetables in a table. We also list the dietary requirements.

	Ground Beef	Baked Potato	Mixed Vegetables	Requirement
calories	200	150	100	700
protein (g)	150	100	50	450
vitamin A (mg)	10	30	75	220

We now declare our unknowns. Let

x_1 = the number of servings of ground beef needed

x_2 = the number of servings of baked potato needed

x_3 = the number of servings of mixed vegetables needed

We can see from the table constructed above that x_1, x_2 and x_3 must satisfy three equations, one for each dietary requirement.

$$200x_1 + 150x_2 + 100x_3 = 700 \quad \text{(calorie requirement)}$$
$$150x_1 + 100x_2 + 50x_3 = 450 \quad \text{(protein requirement)}$$
$$10x_1 + 30x_2 + 75x_3 = 220 \quad \text{(vitamin A requirement)}$$

Thus the model for the situation is a system of three linear equations in three unknowns. We will solve the system using the associated augmented matrix and Gauss-Jordan elimination. The augmented matrix for the system is

$$\left[\begin{array}{ccc|c} 200 & 150 & 100 & 700 \\ 150 & 100 & 50 & 450 \\ 10 & 30 & 75 & 220 \end{array}\right]$$

We can show that the reduced echelon form of this matrix is

$$\left[\begin{array}{ccc|c} 1 & 0 & 0 & 1 \\ 0 & 1 & 0 & 2 \\ 0 & 0 & 1 & 2 \end{array}\right]$$

This yields the corresponding system

$$x_1 = 1$$
$$x_2 = 2$$
$$x_3 = 2$$

We conclude, then, that the meal should consist of one serving of ground beef, two servings of baked potato and two servings of mixed vegetables (check this in the original system). ❖

EXAMPLE 2 Portfolio Analysis

The following table describes the expected annual appreciation and potential loss of three types of investments.

	Growth Stocks	Dividend Stocks	Tax-free Bonds
Expected Appreciation	12%	8%	5%
Potential Loss	10%	8%	3%

A total of \$250,000 is to be invested among the three types of investments listed above. The desired overall annual appreciation for the portfolio is 7%, with a potential loss of 5%. How should the \$250,000 be allocated among the three types of investments to meet these goals?

Solution Given a \$250,000 portfolio, an annual appreciation of 7% would be growth of 250,000(0.07) = \$17,500. A loss of 5% would be a loss of 250,000(0.05) = \$12,500. Since we need to decide how much money to allocate to each of the three types of investments, we have three unknowns, which we declare now. Let

x_1 = the number of dollars allocated to growth stocks

x_2 = the number of dollars allocated to dividend stocks

x_3 = the number of dollars allocated to tax-free bonds

We summarize the situation in a table.

	Growth Stocks	Dividend Stocks	Tax-free Bonds	Total
Amount Allocated	x_1	x_2	x_3	$250,000
Expected Appreciation	12%	8%	5%	$17,500
Potential Loss	10%	8%	3%	$12,500

This naturally leads to a system of three equations:

$$\begin{aligned} x_1 + x_2 + x_3 &= 250000 \quad \text{(total allocation)} \\ 0.12x_1 + 0.08x_2 + 0.05x_3 &= 17500 \quad \text{(appreciation goal)} \\ 0.1x_1 + 0.08x_2 + 0.03x_3 &= 12500 \quad \text{(loss maximum)} \end{aligned}$$

Again, we reduce the augmented matrix for the system. The matrix

$$\left[\begin{array}{ccc|c} 1 & 1 & 1 & 250,000 \\ 0.12 & 0.08 & 0.05 & 17,500 \\ 0.1 & 0.08 & 0.03 & 12,500 \end{array}\right]$$

reduces to

$$\left[\begin{array}{ccc|c} 1 & 0 & 0 & 71,428.57 \\ 0 & 1 & 0 & 0 \\ 0 & 0 & 1 & 178,571.43 \end{array}\right]$$

The entries in the last column have been rounded to two decimal places. This gives the system

$$\begin{aligned} x_1 &= 71{,}428.57 \\ x_2 &= 0 \\ x_3 &= 178{,}571.43 \end{aligned}$$

We conclude that we should allocate $71,428.57 to growth stocks, $178,571.43 to tax-free bonds and no money to dividend stocks in order to meet the established objectives (check this in the original system). ❖

EXAMPLE 3 **Agriculture**

According to data from a 1984 Texas agricultural report, the following table shows the amount of nitrogen (in lb/acre), phosphate (in lb/acre) and labor (in hr/acre) needed to grow honeydews, yellow onions, and lettuce.

	Honeydews	Onions	Lettuce
Nitrogen (lb/acre)	120	150	180
Phosphate (lb/acre)	180	80	80
Labor (hr/acre)	4.97	4.45	4.65

If a farmer has 220 acres of land, 29,100 lbs. of nitrogen, 32,600 lbs. of phosphate and money for 480 hours of labor, how many acres should he allot for each crop in order to use all of the available resources?

Solution Once again we must draw a conclusion about how to divide a quantity among three possible options. Let

x_1 = the number of acres alloted to honeydews

x_2 = the number of acres alloted to yellow onions

x_3 = the number of acres alloted to lettuce

We summarize the situation in a table.

	Honeydews	Onions	Lettuce	Total
Acres Allotted	x_1	x_2	x_3	220
Nitrogen (lb/acre)	120	150	180	29,100
Phosphate (lb/acre)	180	80	80	32,600
Labor (hr/acre)	4.97	4.45	4.65	480

This naturally leads to a system of linear equations:

$$\begin{aligned} x_1 + x_2 + x_3 &= 220 \\ 120x_1 + 150x_2 + 180x_3 &= 29{,}100 \\ 180x_1 + 80x_2 + 80x_3 &= 32{,}600 \\ 4.97x_1 + 4.45x_2 + 4.65x_3 &= 480 \end{aligned}$$

The augmented matrix for the system is

$$\left[\begin{array}{ccc|c} 1 & 1 & 1 & 220 \\ 120 & 150 & 180 & 29{,}100 \\ 180 & 80 & 80 & 32{,}600 \\ 4.97 & 4.45 & 4.65 & 480 \end{array}\right]$$

which reduces to

$$\left[\begin{array}{ccc|c} 1 & 0 & 0 & 0 \\ 0 & 1 & 0 & 0 \\ 0 & 0 & 1 & 0 \\ 0 & 0 & 0 & 1 \end{array}\right]$$

Notice that the last row of the reduced echelon form matrix corresponds to the equation $0 = 1$, which is a contradiction. Consequently, we conclude that it is **not** possible for the farmer to use all of the available resources, no matter how many acres he allots to each crop. ❖

EXAMPLE 4 Agriculture

Repeat Example 3, except this time assume that the farmer has resources available for 1,061 hours of labor.

Solution The same analysis as done above will lead to a system of four linear equations in four unknowns. The only difference in the systems will be in the last equation. Here we have

$$4.97x_1 + 4.45x_2 + 4.65x_3 = 1061$$

The associated augmented matrix becomes

$$\left[\begin{array}{ccc|c} 1 & 1 & 1 & 220 \\ 120 & 150 & 180 & 29100 \\ 180 & 80 & 80 & 32600 \\ 4.97 & 4.45 & 4.65 & 1061 \end{array}\right]$$

which reduces to

$$\left[\begin{array}{ccc|c} 1 & 0 & 0 & 150 \\ 0 & 1 & 0 & 50 \\ 0 & 0 & 1 & 20 \\ 0 & 0 & 0 & 0 \end{array}\right]$$

This gives the system

$$x_1 = 150$$
$$x_2 = 50$$
$$x_3 = 20$$
$$0 = 0$$

The last equation in this system does not give any useful information, but we do have explicit values for the three unknowns. Consequently, the farmer should plant 150 acres of honeydews, 50 acres of yellow onions and 20 acres of lettuce to use all of the available resources (check this in the original system). ❖

EXAMPLE 5 Traffic Control

During rush hour, drivers encounter substantial traffic congestion at the traffic intersections shown in Figure 1. The arrows indicate that the streets are one-way streets.

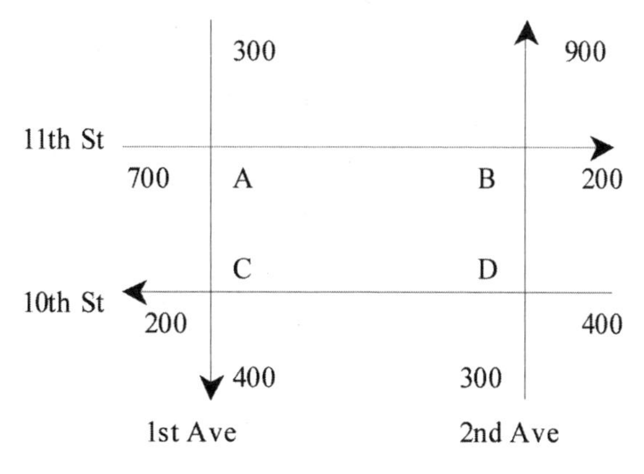

Figure 1

City engineers wish to speed the flow of traffic through these intersections by improving traffic signals. What the diagram shows is that 700 cars per hour come down 11th Street to intersection A and 300 cars per hour come down 1st Avenue to intersection A. Let x_1 denote the number of cars that leave intersection A along 11th Street and let x_2 denote the number of cars that leave intersection A along 1st Avenue. Since a total of 1,000 cars per hour enter intersection A, 1000 cars per hour must leave, so

$$x_1 + x_2 = 1000$$

Similarly, let x_3 denote the number of cars that leave intersection D along 2nd Avenue and let x_4 denote the number of cars that leave intersection D along 10th Street. We see that

$$x_3 + x_4 = 700$$

You should also verify that we must have

$$x_1 + x_3 = 1100 \text{ (intersection } B)$$

$$x_2 + x_4 = 600 \text{ (intersection } C)$$

Putting all of these linear equations together gives a system with augmented matrix

$$\left[\begin{array}{cccc|c} 1 & 1 & 0 & 0 & 1000 \\ 0 & 0 & 1 & 1 & 700 \\ 1 & 0 & 1 & 0 & 1100 \\ 0 & 1 & 0 & 1 & 600 \end{array}\right]$$

The reduced echelon form of this matrix is

$$\left[\begin{array}{cccc|c} 1 & 0 & 0 & -1 & 400 \\ 0 & 1 & 0 & 1 & 600 \\ 0 & 0 & 1 & 1 & 700 \\ 0 & 0 & 0 & 0 & 0 \end{array}\right]$$

This yields the system

$$x_1 - x_4 = 400$$

$$x_2 + x_4 = 600$$

$$x_3 + x_4 = 700$$

$$0 = 0$$

This system has an infinite number of solutions. The question, then, is what information these solutions provide. Although we can investigate many things here, we focus only on intersection A and what solutions to the system of equations can indicate there.

For instance, x_1 gives the number of cars leaving intersection *A* along 11th Street. From above we see that $x_1 - x_4 = 400$, so

$$x_1 = 400 + x_4$$

Since $x_4 \geq 0$, the smallest possible value for x_1 is 400, meaning that the least number of cars that can leave intersection *A* along 11th Street is 400 per hour. The largest number of cars that can leave intersection *A* along 11th Street is 1,100 per hour since $x_4 \leq 700$. Also, we see that $x_2 + x_4 = 600$, so

$$x_2 = 600 - x_4$$

Since $x_4 \geq 0$, the largest number of cars that can leave intersection *A* along 1st Avenue is 600 per hour (and the least, of course, is zero).

Similar analysis of the other intersections gives the city engineers some bounds on the number of cars that they can expect to approach or leave other intersections as well. Based on these bounds, they can make decisions on how to adjust traffic signals in order to speed the flow of traffic through the intersections. ❖

PROBLEM SET 5.3

1. Jack spends 22 hours per week doing three types of exercises: running, biking and swimming. He spends twice as much time biking as he does running, and he spends three more hours biking than swimming. How much time per week does Jack spend on each activity?
2. Sue was doing homework for two classes, math and chemistry. She can do 10 math problems per hour and 15 chemistry problems per hour. One night she worked for 4 hours and completed 49 problems. How many math problems and how many chemistry problems did she do?
3. A company that rents small moving trucks wants to purchase 25 trucks with a combined capacity of 28,000 cubic feet. Three different types of trucks are available: a 10-foot truck with a capacity of 350 cubic feet, a 14-foot truck with a capacity of 700 cubic feet and a 24-foot truck with a capacity of 1,400 cubic feet. How many of each type of truck should the company purchase? (Hint: There may be more than one solution. Find restrictions on the variables by using the fact that the number of each type of truck purchased must be a nonnegative, whole number.)
4. Every left-handed widget uses 5 ounces of plastic and 10 ounces of steel, while every right-handed widget uses 4 ounces of plastic and 11 ounces of steel. What is the total number of widgets that one can make from 300 ounces of plastic and 750 ounces of steel?
5. A manufacturer of women's blouses makes three types of blouses: sleeveless, short-sleeve and long-sleeve. The accompanying table shows the time required by each department to produce a dozen blouses of each type.

Department	Sleeveless	Short-sleeve	Long-sleeve
Cutting	9 min.	12 min.	15 min.
Sewing	22 min.	24 min.	28 min.
Packaging	6 min.	8 min.	8 min.

The cutting, sewing and packaging departments have available a maximum of 80, 160 and 48 hours, respectively, per day. How many dozens of each type of blouse can these departments produce each day if the plant operates at full capacity?

6. A man has $260,000 invested at 4.7 percent, 5.5 percent, 6 percent and 7 percent. His annual income from the investments is $16,058.40. His investments at 4.7 percent and 5.5 percent total $64,800 while his investment at 6 percent is $5,800 more than twice his investment at 4.7 percent. How much does he have invested at each rate?
7. A company produces three combinations of mixed vegetables that sell in 1 kg packages. Italian style combines 0.3 kg of zucchini, 0.3 kg of broccoli and 0.4 kg of carrots. French style combines 0.6 kg of broccoli and 0.4 kg of carrots. Oriental style combines 0.2 kg of zucchini, 0.5 kg of broccoli and 0.3 kg of carrots. The company has a stock of 16,200 kg of zucchini, 41,400 kg of broccoli and 29,400 kg of carrots. How many packages of each style should it prepare to use up existing supplies?
8. A dietician is preparing a meal consisting of foods A, B and C. Each ounce of food A contains 2 units of protein, 3 units of fat and 4 units of carbohydrate. Each ounce of food B contains 3 units of protein, 2 units of fat and 1 unit of carbohydrate. Each ounce of food C contains 3 units of protein, 3 units of fat and 2 units of carbohydrate. If the meal must provide exactly 25 units of protein, 24 units of fat and 21 units of carbohydrate, how many ounces of each type of food should the dietician use?
9. The Hiker and Biker Outfitting Shop makes packages of snacks to order. One customer likes peanuts, raisins and chocolate chips but wants only two ingredients per snack. In peanut-raisin mixes she likes twice as many peanuts as raisins; in peanut-chocolate chip mixes she likes twice as many chocolate chips as peanuts; and in raisin-chocolate chip mixes she likes equal amounts of the two foods. Her total purchase contains 4 ounces of peanuts, 6 ounces of raisins and 12 ounces of chocolate chips. How many ounces of each mix did she buy?
10. The head dietician at Gigantic State University has 330 pounds of fresh citrus fruit and 230 pounds of mixed fruit, which she must use today. She has recipes for salads that use both. Citrus salad requires 3 pounds of fresh citrus and 1 pound of mixed fruit to produce 4 pounds of salad, and fruit salad requires 1.5 pounds of citrus fruit and 2.5 pounds of mixed fruit to produce 4 pounds of salad. How many pounds of each type of salad must she make to use all of the fruit?
11. In 20 oz of one alloy there are 6 oz of copper, 4 oz of zinc and 10 oz of lead. In 20 oz of a second alloy there are 12 oz of copper, 5 oz of zinc and 3 oz of lead, while in 20 oz of a third alloy there are 8 oz of copper, 6 oz of zinc and 6 oz of lead. How many ounces of each alloy should be combined to make a new alloy containing 34 oz of copper, 17 oz of zinc and 19 oz of lead?

12. At a fast food restaurant, one group of customers bought 8 deluxe hamburgers, 6 orders of large fries and 6 large colas for $26.10. A second group ordered 10 deluxe hamburgers, 6 large fries and 8 large colas and paid $31.60. Is there sufficient information to determine the price of each food item? If not, construct a table showing the various possibilities. Assume that the hamburgers cost between $1.75 and $2.25, the fries between $0.75 and $1.00 and the colas between $0.60 and $0.90.

13. A Florida juice company completes the preparation of its products by sterilizing, filling and labeling bottles. Each case of orange juice requires 9 minutes for sterilizing, 6 minutes for filling and 1 minute for labeling. Each case of grapefruit juice requires 10 minutes for sterilizing, 4 minutes for filling and 2 minutes for labeling. Each case of tomato juice requires 12 minutes for sterilizing, 4 minutes for filling and 1 minute for labeling. If the company runs the sterilizing machine for 398 minutes, the filling machine for 164 minutes and the labeling machine for 58 minutes, how many cases of each type of juice will it prepare?

Interlude: Progress Check 5.3, Part 1

1. A farmer in Minnesota has a choice of three varieties of rye (call them varieties R_1, R_2, and R_3) to grow and wishes to devote part of his available acreage to each variety. The farmer needs to harvest 9,000 bushels of grain at the end of the growing season to ensure a reasonable profit. He has 45 acres available for rye. Variety R_1 requires 3 person-hours of labor per bushel during the growing season and is expected to yield 400 bushels per acre. Variety R_2 requires 2 person-hours of labor per bushel during the growing season but is only expected to yield 200 bushels per acre. Variety R_3 requires only 1 person-hour of labor per bushel during the growing season but is also the lowest-yielding variety, only expected to yield 100 bushels per acre. The farmer has enough of a work force to provide 20,000 person-hours of labor for the growing season. Determine how many bushels of each variety the farmer should plan to sow in order to use up all available labor-hours. Also determine how many acres will be devoted to each variety of rye. Set up an appropriate model for the problem, declaring what all variables represent.

2. In preparation for Halloween, Fantastically Freakish Figurines is scheduling production of its three specialty figurines: Astrocreep, Bubonican, and Cthuluazon. Each figurine requires some processing time using three different machines. Call them Machine I, II, and III, respectively. Manufacture of Astrocreep requires 2 minutes on Machine I, 1 minute on Machine II, and 2 minutes on Machine III. Manufacture of Bubonican requires 1 minute on Machine I, 3 minutes on Machine II, and 1 minute on Machine III. Manufacture of Cthuluazon requires 1 minute on Machine I, 2 minutes on Machine II, and 2 minutes on Machine III. Machine I is available for 3 hours a day, Machine II is available for 5 hours a day, and Machine III is available for 4 hours a day. Determine how many of each type of figurine can be produced each day by using all of the available machine time. Set up an appropriate model for the problem, declaring what all variables represent.

Interlude: Progress Check 5.3, Part 2

1. A large farming operation is in need of 35,000 pounds of potash, 68,000 pounds of nitrogen, and 25,000 pounds of phosphoric acid. The local distributor carries three different brands of fertilizer, Growex, Wyesure, and Zeeplot. Each contains some amount of potash, nitrogen, and phosphoric acid, as given in the following table. The operation needs to determine how many tons of each brand of fertilizer to order to meet its needs.

(Pounds / Ton)	Potash	Nitrogen	Phosphoric Acid
Growex	400	500	300
Wyesure	600	1000	400
Zeeplot	700	1600	500

(a) Write a system of equations modeling the farming operation's problem. Declare what the variables represent.

(b) Solve this problem.

2. Percy has invested \$40,000 in three stocks. The first year, stock A paid 6% dividends and increased 3% in value; stock B paid 7% dividends and increased 4% in value; stock C paid 8% dividends and increased 2% in value. If the total dividends were \$2730 and the total increase in value was \$1080, how much was invested in each stock? Set up an appropriate model declaring what all variables represent, and solve the problem. Clearly state the conclusion.

Interlude: Progress Check 5.3, Part 3

1. A chemist mixes three different solutions with concentrations of 20%, 30%, and 45% glucose to obtain 10 liters of a 38% glucose solution. If the amount of 30% solution used is one liter more than twice the amount of 20% solution used, then find the amount of each solution used. When solving the problem, write an appropriate model, declaring what all variables represent.

2. As part of a promotion, a local bank invites its customers to view a large sack full of \$5, \$10, and \$20 gold pieces, promising to give the sack to the first person able to state the number of coins for each denomination. Customers are told there are exactly 250 coins and with a total face value of \$1875. If there are also seven times as many \$5 gold pieces as \$20 gold pieces, then how many of each denomination are there? When solving the problem, write an appropriate model, declaring what all variables represent.

CHAPTER 5 READINESS, PART 1

1. Wholesale Accompaniments receives an order for a total of 900 tie tacks and lapel pins. The tie tacks cost \$5.85 each, and the lapel pins cost \$8.16 each. A check for \$6253.68 is enclosed, but the order fails to specify the number of each item desired. Write an appropriate system of linear equations and graphically solve this system to determine the number of each item desired. Declare what your variables represent. Make a sketch of your graphs. State your conclusion in a complete sentence.

2. Two farmers, John and Patty, have a wager going. Patty has challenged John to find a way to purchase exactly 100 chickens for \$100. Roosters cost \$5 each, hens cost \$3 each, and baby chicks cost \$0.05 each. Patty has agreed to pay for the cost of the chickens if John can figure out how to accomplish the purchase goal, provided he purchase at least one rooster, at least one hen, and at least one baby chick. If he cannot figure out how to do this (and hence lose the wager), then he must clean Patty's chicken coop. Can John actually win this wager? If so, determine how. If not, explain why not. Your analysis must include an appropriate mathematical model of the situation and justification for your conclusion.

3. Quinn's Quickmart is open from 8:00 in the morning until midnight, every day of the year. The store has three shifts: 8 A.M to 4 P.M, noon to 8 P.M, and 4 P.M. to midnight. These shifts must be apportioned among 60 workers. Management wishes to have 45 workers on hand during the overlap between the first two shifts and 40 workers on hand during the overlap between the last two shifts. Determine how the 60 workers should be allotted to the three shifts. Set up an appropriate model and declare what all variables represent.

CHAPTER 5 READINESS, PART 2

1. Centerville Community College is planning to offer courses in College Algebra, Applied Calculus, and Web Design. Each section of College Algebra holds 40 students and generates $40,000 in tuition revenue. Each section of Applied Calculus holds 40 students and generates $60,000 in tuition revenue. Each section of Web Design holds 10 students and generates $20,000 in tuition revenue. The school will offer a total of six sections of the courses, in order to accommodate 210 students and will need to total tuition revenue of $260,000 to run the courses. How many sections of each of the courses need to be offered? Set up an appropriate model, declare what all variables represent, and solve the problem. Clearly state the conclusion.

2. Dewey, Skrughem, and Howe is a real estate development firm that plans to build a new apartment complex close to campus consisting of one-bedroom apartments and two- and three-bedroom townhouses. A total of 192 units is planned. The total number of two- and three-bedroom townhouses will be the same as the number of one-bedroom apartments. The number of one-bedroom apartments will be three times the number of three-bedroom townhouses. Determine how many of each type of housing unit will be in the complex. Set up an appropriate model, declare what all variables represent, and solve the problem. Clearly state the conclusion.

Answers to Odd-Numbered Problems

Problem Set 5.1

1. **(a)** Yes **(b)** No **(c)** No (d) Yes

3. $x = -1, y = 3$

5. Infinite number of solutions

7. $x_1 = .5, x_2 = 2$

9. $x = 18, y = 50$

11. $m = \frac{77}{26}, n = \frac{21}{13}$

13. $x = 2, y = -3$

15. Infinite number of solutions

17. $x_1 = \frac{2}{31}, x_2 = \frac{102}{31}$

19. 1300 gallons of regular chocolate milk, 1680 gallons of extra-thick chocolate milk

21. Kris has $200, Shelley has $250

23. 2800 adult and 1400 children's tickets

Problem Set 5.2

1. **(a)** Yes **(b)** Yes **(c)** No, doesn't satisfy condition 2 of reduced echelon form, for example. **(d)** Yes

3. $R_1 + R_2 \to R_2$

5. $-2R_1 + R_2 \to R_2, \frac{1}{4}R_2 \to R_2, -2R_2 + R_3 \to R_3$

7. **(a)** No **(b)** Yes **(c)** No **(d)** Yes

9. $x_1 = 3, x_2 = 4, x_3 = 4$

11. Infinite number of solutions

13. $x_1 = -\frac{1}{3}, x_2 = 2$

15. $x_1 = 1, x_2 = 2, x_3 = 0; x_1 = 4, x_2 = 1, x_3 = 1; x_1 = 7, x_2 = 0, x_3 = 2$

Problem Set 5.3

1. 5 hrs. running, 10 hrs. biking, 7 hrs. swimming

3. 0 10-ft. trucks, 10 14-ft. trucks, and 15 24-ft. trucks; 2 10-ft. trucks, 7 14-ft. trucks, and 16 24-ft. trucks; 4 10-ft., 4 14-ft., and 17 24-ft.; 6 10-ft., 1 14-ft., and 18 24-ft.

5. 80 doz. sleeveless, 140 doz. short-sleeve, 160 doz. long-sleeve

7. 18,000 packages Italian style, 15,000 packages French style, 54,000 packages Oriental style

9. 1.2 oz. peanut-raisin, 9.6 oz. peanut-chocolate chip, 11.2 oz. raisin-chocolate chip

11. 20 oz. of the first alloy, 40 oz. of the second alloy, 10 oz. of the third alloy

13. 6 cases orange juice, 20 cases grapefruit juice, 12 cases tomato juice

Graphing Calculator Exercises

Your owner's manual is useful for finding the locations of the common commands and routines that you are expected to use. Another resource for Texas Instruments graphing calculators can be found at

http://education.ti.com/calculators/pd/us/Online-Learning/Tutorials

PART I: SETTING YOUR VIEWING WINDOW APPROPRIATELY (NO "ZOOMING" ALLOWED)

1. The graph of the equation

$$y = 22x^5 + 17.6x^4 - 68x^3 - 17.5x^2 + 27x - 34$$

has two "peaks" and two "valleys". Find appropriate viewing window settings to clearly view these features on your calculator. Sketch the picture you see and list your **Xmin, Xmax, Ymin,** and **Ymax** values.

Xmin:

Xmax:

Ymin:

Ymax:

2. The graph of $y = 0.2x^4 + 0.4x^3 - 79x^2 - 360x + 3600$ has one "peak" and two "valleys". Find appropriate viewing window settings to clearly view these features on your calculator. Sketch the picture you see and list your **Xmin, Xmax, Ymin,** and **Ymax** values.

Xmin:

Xmax:

Ymin:

Ymax:

3. Graph the equation $y = -0.5x^4 + x^2 - 0.3$. Find appropriate viewing window settings on your calculator to obtain a picture like that shown in Figure 1. List your **Xmin, Xmax, Ymin,** and **Ymax** values.

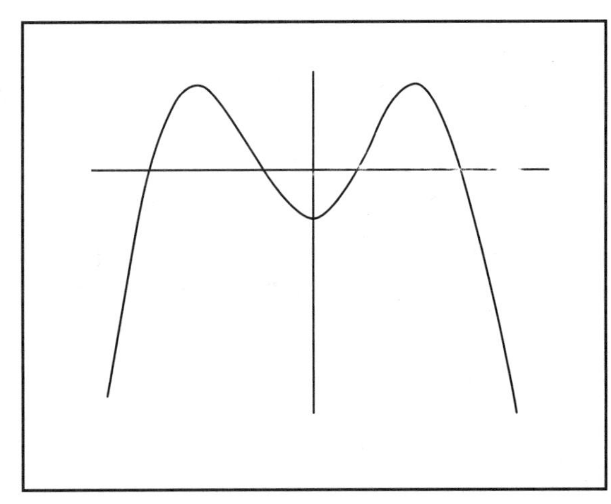

Figure 1

Xmin:

Xmax:

Ymin:

Ymax:

4. Graph the equation $y = 112x - 4x^2$. Find appropriate viewing window settings on your calculator to obtain a picture like that shown in Figure 2. List your **Xmin, Xmax, Ymin,** and **Ymax** values.

Xmin:

Xmax:

Ymin:

Ymax:

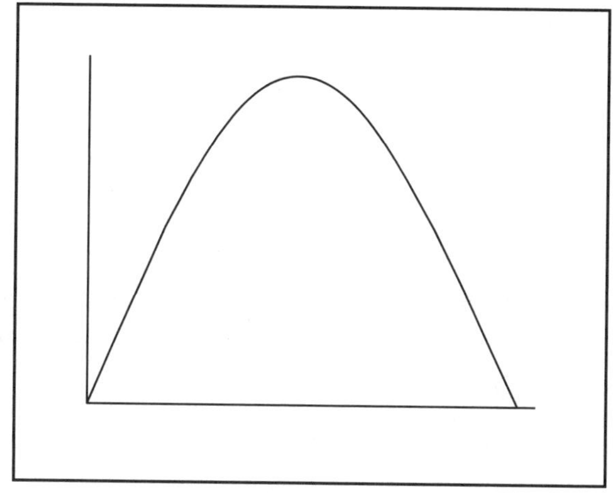

Figure 2

PART II: USING SOME BUILT-IN ROUTINES (DO NOT "ZOOM AND TRACE")

There are several built-in routines available on the Texas Instruments graphing calculators that you will be expected to use on occasion. In particular, you should be acquainted with your machine's root/zero finder, intersect feature, and maximum/minimum finder.

5. Graph the equation $y = 2 - 0.4x^2 - 0.25x^3$ in the standard viewing window and make a sketch. Use the root/zero finding feature of your calculator to find the x-intercept. Give the value of the x-intercept accurate to four decimal places.

Remark: The x-coordinate you found is the approximate solution to the equation $2 - 0.4x^2 - 0.25x^3 = 0$. (Check this)

Estimate:

6. Graph the following two equations in the standard viewing window and make a sketch.

$$y = 0.5x^2 + 3x - 4$$
$$y = 2x + 1$$

Use the intersect routine on your machine to find the points of intersection of the two graphs. Give the coordinates accurate to four decimal places.

Remark: Notice that the x-coordinates of the points of intersection are the approximate solutions to the equation

$$0.5x^2 + 3x - 4 = 2x + 1.$$

(Check this)

Estimates:

7. Graph the equation $y = 0.2x^3 - 4.1x - 1$ in the standard viewing window. You should notice that the graph has one "peak" and one "valley", both located in the fourth quadrant. First, use the maximum finder on your machine to find the coordinates of the peak. Next, use the minimum finder to find the coordinates of the valley. Record the coordinates of both points, accurate to four decimal places.

Peak:

Valley:

APPENDIX II

Dividing Polynomials

OBJECTIVES

1. Perform Long Division of Polynomials
2. Perform Synthetic Division

In this section we perform division of polynomials. Long division of polynomials, like division of real numbers, relies on multiplication and subtraction skills. Synthetic division is introduced as a tool for carrying out division more efficiently for certain situations. We begin with a review of the long division process.

Performing Long Division of Polynomials

An *algorithm* is a step-by-step systematic procedure used to perform a computation. For instance, we use the long division algorithm to divide 372 by 5 to obtain the *quotient* and the *remainder,* as shown below:

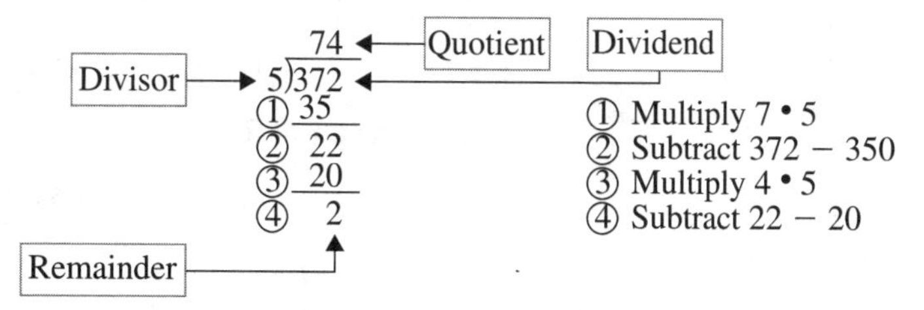

The result of the division can be expressed either in the *multiplicative form* as

$$372 = (5)(74) + 2$$

$$\text{Dividend} = (\text{Divisor})(\text{Quotient}) + \text{Remainder}$$

or in the *fractional form* as

$$\frac{372}{5} = 74 + \frac{2}{5}$$

$$\frac{\text{Dividend}}{\text{Divisor}} = \text{Quotient} + \frac{\text{Remainder}}{\text{Divisor}}$$

The long division of polynomials follows the same pattern, as illustrated in the following example.

EXAMPLE 1 Divide the polynomial

$$F = 2x^3 - 3x^2 + 5x + 2 \text{ by } D = x + 3.$$

Express the result in both multiplicative and fractional forms.

Solution We note that both the divisor

$$x + 3$$

and the dividend

$$2x^3 - 3x^2 + 5x + 2$$

are arranged in descending powers of x. (If they weren't, we would rewrite them so that they were before beginning the process.)
Then we proceed as follows:

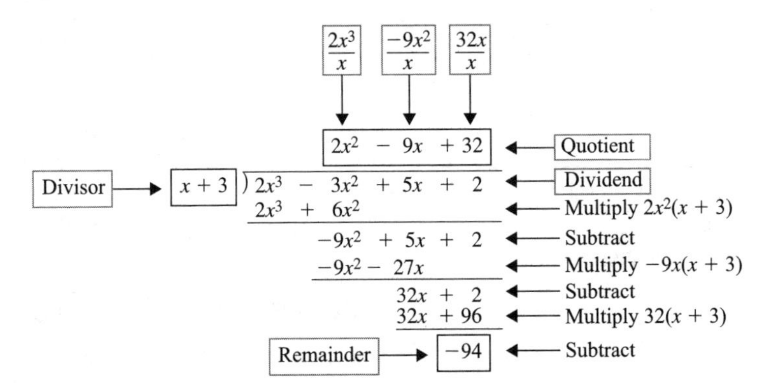

As in the previous numerical example, we can write the result either in multiplicative form:

$$2x^3 - 3x^2 + 5x + 2 = (x + 3)(2x^2 - 9x + 32) + (-94)$$

or in fractional form:

$$\frac{2x^3 - 3x^2 + 5x + 2}{x + 3} = 2x^2 - 9x + 32 + \frac{(-94)}{x + 3}.$$

❖

❖ Practice Exercise 1

Divide the polynomial $4x^3 - 5x^2 + 3$ by $x + 2$

Write the result in multiplicative form.
At times the division process results in a remainder of zero as shown in the next example.

EXAMPLE 2 Divide the polynomial $x^3 + 6x - 3x^2 - 4$ by $x - 1$.

Express the result in multiplicative form.

Solution First we note that the divisor and dividend must be rearranged in descending powers of x, then we apply the division process as follows:

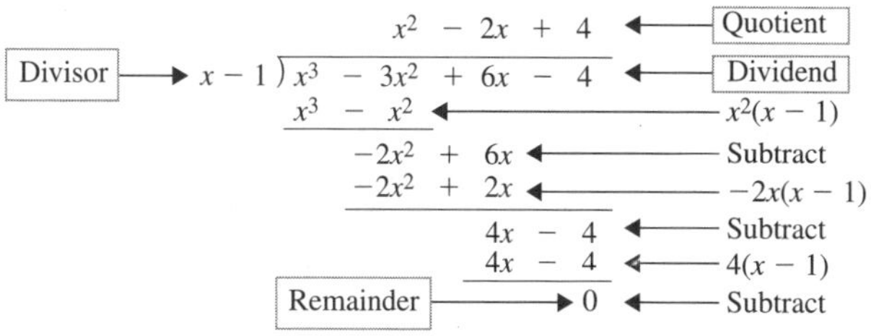

$$
\begin{array}{r}
x^2 - 2x + 4 \quad \leftarrow \text{Quotient} \\
\text{Divisor} \rightarrow x - 1 \enclose{longdiv}{x^3 - 3x^2 + 6x - 4} \quad \leftarrow \text{Dividend} \\
\underline{x^3 - x^2} \quad \leftarrow x^2(x-1) \\
-2x^2 + 6x \quad \leftarrow \text{Subtract} \\
\underline{-2x^2 + 2x} \quad \leftarrow -2x(x-1) \\
4x - 4 \quad \leftarrow \text{Subtract} \\
\underline{4x - 4} \quad \leftarrow 4(x-1) \\
\text{Remainder} \rightarrow 0 \quad \leftarrow \text{Subtract}
\end{array}
$$

The quotient is $x^2 - 2x + 4$, and the remainder is 0, so that in multiplicative form we have

$$x^3 - 3x^2 + 6x - 4 = (x - 1)(x^2 - 2x + 4).$$ ❖

Notice that when we get a remainder of zero, we obtain a factorization and the given polynomial is said to be *divisible by the divisor.* Thus, as shown in Example 2 above, $x^3 - 3x^2 + 6x - 4$ is divisible by $x - 1$.

❖ **Practice Exercise 2**

Use long division to show that $7x^3 - 5x^2 + 4x - 156$ is divisible by $x - 3$
Express the result in multiplicative form.

Example 2 and Practice Exercise 2 illustrate the following property which is stated as follows:

The Division Algorithm Result

If F and D are polynomials such that D is non zero, then there exist unique polynomials Q and R such that

$$F = D \cdot Q + R$$

where the degree of R is less than the degree of D [R may be 0]. The expression D is called the **divisor,** F is the **dividend,** Q is the **quotient,** and R is the **remainder.**

The result of the division algorithm can also be expressed in the fractional form

$$\frac{F}{D} = Q + \frac{R}{D}.$$

which carries implicit restrictions on the divisor D, since D cannot equal 0.

Performing Synthetic Division

Consider the long division process used to divide the polynomial

$$F = 3x^3 - x^2 + 2x - 18 \text{ by } x - 2.$$

First, notice that the divisor and the dividend are arranged in descending powers of x, then the process is organized as follows:

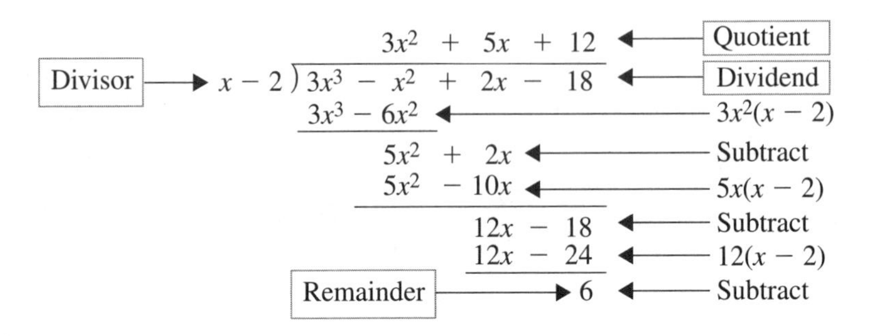

Thus the quotient $Q = 3x^2 + 5x + 12$ and the remainder $R = 6$. Now we can simplify the task of performing such a division by using an algorithm called *synthetic division,* which is an efficient way of dividing a polynomial by a binomial of the form $x - c$ where c is a constant.

Being able to divide a polynomial F by a linear binomial of the form $x - c$ quickly and accurately will be of great help in studying polynomial functions in Chapter 3. This kind of division can be carried out more efficiently by a method called **synthetic division.**

In synthetic division, we work with the coefficient(s) of the dividend and divisor to produce a result that gives the coefficients of the quotient and remainder. The process is explained and illustrated in the following table.

The synthetic division method can be used only when the divisor is of the form $x - c$.

In the illustration earlier we used long division to divide the polynomial

$$F = 3x^3 - x^2 + 2x - 18$$

by

$$D = x - 2.$$

Now we perform the same division by using the synthetic division method as follows: Notice how the long division process used earlier compares with the synthetic division method shown above:

	Explanation of Process	Result
Step 1	Arrange the terms of the polynomial F in descending powers of x, and use the 0 coefficient for any missing powers of x.	$F = 3x^3 - x^2 + 2x - 18$
Step 2	Consider the divisor in the form $x - c$. Write down the value of c (here, $c = 2$), draw a vertical line, and then list the coefficients of the dividend F.	2 | 3 −1 2 −18
Step 3	Leave space below the row of coefficients, draw a horizontal line, and copy the leading coefficient of F below the line.	2 | 3 −1 2 −18 3
Step 4	Multiply 3, the leading coefficient, by 2, the value for c, and place the product, 6, above the horizontal line under the second coefficient, −1. Then add −1 to this product and place the result, 5, below the horizontal line. Now multiply 5 by 2 and place the product, 10, above the horizontal line under the third coefficient, 2. Then add 2 to this product and place the result, 12, below the horizontal line. Continue in this way, multiplying 12 by 2 and placing the product, 24, above the horizontal line under the next coefficient, −18. Then add −18 and 24 to obtain 6, placing 6 below the horizontal line. Isolate the very last sum, 6, by drawing a short vertical line.	2 | 3 −1 2 −18 6 3 5 2 | 3 −1 2 −18 6 10 3 5 12 2 | 3 −1 2 −18 6 10 24 3 5 12 | 6
Step 5	In the last row, the numbers 3, 5, and 12 are the coefficients of the quotient Q and the last number, 6, is the remainder R.	$Q = 3x^2 + 5x + 12$ $6 = R$

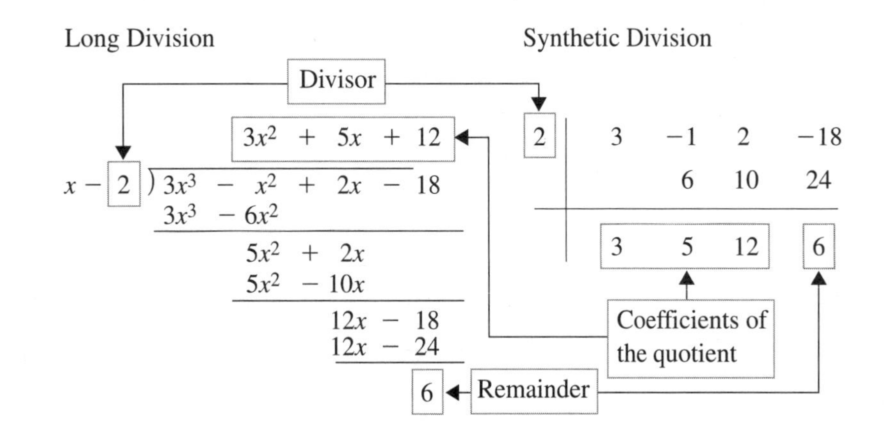

EXAMPLE 3 Use synthetic division to obtain the quotient and the remainder if the polynomial

$$F = 2x^4 - 3x^2 + 5x + 7 \text{ is divided by } D = x + 3.$$

Solution The terms of F are given in descending powers of x, but we need to use 0 to denote the coefficient of the missing x^3 term. So the coefficients of F are listed as 2, 0, -3, 5, and 7.

The divisor has the form $D = x + 3 = x - (-3) = x - c$, so $c = -3$.

By synthetic division, we get

$$\begin{array}{r|rrrrr} -3 & 2 & 0 & -3 & 5 & 7 \\ & & -6 & 18 & -45 & 120 \\ \hline & 2 & -6 & 15 & -40 & 127 \end{array}$$

Hence the quotient is $Q = 2x^3 - 6x^2 + 15x - 40$ and the remainder is $R = 127$. 

❖ Practice Exercise 3

Use synthetic division to obtain the quotient and remainder if the polynomial F is divided by a binomial D.

(a) $F = 3x^5 - 4x^3 + 7x - 2$; $D = x - 2$
(b) $F = x^3 + 8$; $D = x + 2$

❖ Answers to Practice Exercises

1. $4x^3 - 5x^2 + 7x + 3 = (x + 2)(4x^2 - 13x + 33) + (-63)$
2. $7x^3 - 5x^2 + 4x - 156 = (7x^2 + 16x + 52)(x - 3) + 0$
3. (a) $Q = 3x^4 + 6x^3 + 8x^2 + 16x + 39$; $R = 76$
   (b) $Q = x^2 - 2x + 4$; $R = 0$

PROBLEM SET

In problems 1–8, use long division to divide F by D. Identify the quotient Q and the remainder R. Also express each result in the multiplicative form.

1. $F = 4x^3 - 2x^2 + 7x - 1; D = x + 1$
2. $F = 8x^3 - x^2 - x + 5; D = x - 2$
3. $F = 3x^4 - 2x^3 + 4x^2 + 3x + 7; D = x - 3$
4. $F = 2x^5 + 3x^4 - x^2 + x - 4; D = x + 2$
5. $F = 6x^4 + 10x^2 + 7; D = 2x + 3$
6. $F = 15x^4 + 5x^3 + 6x^2 - 2; D = 5x - 3$
7. $F = -3x^3 - x^2 - 3x - 1; D = 3x + 1$
8. $F = -6x^3 + 25x^2 - 9; D = 3x - 2$

In problems 9–16, use synthetic division to find the quotient Q and the remainder R.

	Dividend F is:	Divisor $D(x)$ is:
9.	$2x^3 + 17x^2 - 11x - 21$	$x - 3$
10.	$x^3 - 2x^2 - 1$	$x + 2$
11.	$4x^4 - 3x^3 + 2x^2 + 5$	$x - 1$
12.	$10x^4 - 30x^3 + 63$	$x - 6$
13.	$-5x^4$	$x + 2$
14.	$3x^6 - 7$	$x + 3$
15.	$3 + x^2 - 5x^4$	$x + 2$
16.	$8 - 2x^3 + 3x^4 - 2x^5$	$x + 3$

In problems 17–20, use synthetic division to write F in the form $F = Q \cdot D + R$.

17. $F = 2x^3 - 5x^2 + 5x + 11; D = x - 1$
18. $F = 4x^4 - 2x^3 - 19x^2 + 5; D = x + 3$
19. $F = 2x^4 + 3x - 6; D = x + 2$
20. $F = 3x^5 - 2x^2 + 7; D = x - 3$

In problems 21–24, use synthetic division to find k so that by dividing F by D, the remainder $R = 0$.

21. $F = 3x^3 + 4x^2 + kx - 20; D = x - 2$
22. $F = kx^3 - 25x^2 + 47x + 30; D = x - 10$
23. $F = 2x^3 - 6x^2 + kx - 20; D = x - 3$
24. $F = x^4 - kx^2 + 6; D = x + 1$

Complex Numbers

OBJECTIVES

1. Define Complex Numbers
2. Add, Subtract, Multiply, and Divide Complex Numbers

When restricting our attention to real numbers only, we find that equations like $x^2 = -1$ and $x^2 = -7$ do not have solutions. In order to solve equations like these, it is necessary to expand our number system to allow for the square roots of negative numbers. We will look at such a number system, which mathematicians call the **complex number system**. This system will also include the real number system, with which we are already somewhat familiar, so the arithmetic within this number system will in many ways be similar to the ordinary real number arithmetic we comfortably employ.

Imaginary and Complex Numbers

The equation $x^2 = -1$ has no real number solutions, as there are no real numbers whose squares are negative. However, using the principle of square roots with this equation, we see that

$$x = \sqrt{-1} \quad \text{or} \quad x = -\sqrt{-1}$$

Substituting into the given equation, we see that these two results do, in fact, satisfy the equation (we leave it as an exercise to the reader to do this). So, it seems reasonable to say that the equation $x^2 = -1$ does have solutions, but the solutions are not real numbers; rather, they involve the square root of a negative number. Consequently, an introduction follows to what mathematicians call the **imaginary unit** and the collection of **imaginary numbers**.

The Imaginary Unit and Imaginary Numbers

- The **imaginary unit** i is a number such that $i^2 = -1$ (i.e., $i = \sqrt{-1}$)
- **Imaginary numbers** are numbers of the form $\sqrt{k}$, where $k < 0$; that is, imaginary numbers are square roots of negative numbers

Using standard simplification properties with square roots and this i notation, we can simplify imaginary numbers and write them in terms of the imaginary unit. For instance, consider the imaginary number $\sqrt{-24}$. Notice that

$$\sqrt{-24} = \sqrt{4 \cdot 6 \cdot -1} = \sqrt{4} \cdot \sqrt{6} \cdot \sqrt{-1} = 2\sqrt{6} \cdot i$$

It is commonplace to write complex numbers in terms of i, and we will do so for the remainder of this discussion. Consequently, we will refer to all numbers that we can write as bi, where b is a real number, as imaginary numbers.

Now, consider the quadratic equation $x^2 + 2x + 5 = 0$. Using the quadratic formula, we see that the numbers

$$x = \frac{-2 + \sqrt{-16}}{2} \text{ and } x = \frac{-2 - \sqrt{-16}}{2}$$

are solutions, but are not real numbers. Rewriting these in terms of i, we have

$$x = \frac{-2 + \sqrt{16} \cdot i}{2} \quad \text{and} \quad x = \frac{-2 - \sqrt{16} \cdot i}{2}$$

or, simplifying a bit,

$$x = -1 + 2i \quad \text{and} \quad x = -1 - 2i$$

The numbers $-1 + 2i$ and $-1 - 2i$ are arithmetic combinations of a real number and an imaginary number. We will call numbers with this form **complex numbers**.

Complex Numbers

Complex numbers are numbers that can be written in the form $a + bi$, where a and b are real numbers. Mathematicians call this form the **standard form** of a complex number. The number a is called the **real part** of the complex number, while b is called the **imaginary part** of the number.

From this definition, we see that any real number a is also a complex number, since $a = a + 0i$. Similarly, any imaginary number bi is also a complex number, as $bi = 0 + bi$. Consequently, the collection of all complex numbers includes all real numbers as well as all imaginary numbers.

Complex Number Arithmetic

Since the collection of complex numbers contains the collection of real numbers, it is quite natural to define basic arithmetic operations of addition, subtraction, multiplication and division for complex numbers in such a way that our familiar arithmetic with real numbers still holds. Specifically, when adding, subtracting or multiplying complex numbers, we proceed just as if we were working with polynomials, where we identify and collect like terms, as illustrated in the next two examples.

EXAMPLE 1 Adding and Subtracting Complex Numbers

Perform the indicated operations and write the results in standard form.

(a) $(3 + 6i) + (5 - 8i)$ (b) $(4 - i) - (6 + 3i)$

Solution As mentioned, we proceed in a manner similar to adding and subtracting binomials.

(a) $(3 + 6i) + (5 - 8i) = (3 + 5) + (6i - 8i) = 8 - 2i$

(b) $(4 - i) - (6 + 3i) = (4 - 6) + (-i - 3i) = -2 - 4i$ ❖

EXAMPLE 2 Multiplying Complex Numbers

Find the following products and write the results in standard form.

(a) $(1 + 3i)(-7 + 10i)$ (b) $12i(6 + 5i)$

Solution We multiply in the same manner as we would multiply polynomials, keeping in mind that $i^2 = -1$.

$$\begin{aligned}\text{(a)}\quad (1 + 3i)(-7 + 10i) &= (1)(-7) + (3i)(-7) + (1)(10i) + (3i)(10i)\\ &= -7 - 21i + 10i + 30i^2\\ &= -7 - 11i + 30(-1)\\ &= -37 - 11i\end{aligned}$$

$$\begin{aligned}\text{(b)}\quad 12i(6 + 5i) &= (12i)(6) + (12i)(5i)\\ &= 72i + 60i^2\\ &= -60 + 72i\end{aligned}$$ ❖

In working with the complex number system, we understand exponentiation with integer exponents in the same way as when we are working with real numbers. When looking at positive integer powers of i, an interesting pattern develops.

$$\begin{aligned} i^1 &= i = \sqrt{-1} & i^5 &= i^4 \cdot i = i\\ i^2 &= -1 & i^6 &= i^4 \cdot i^2 = -1\\ i^3 &= i^2 \cdot i = -i & i^7 &= i^4 \cdot i^3 = -i\\ i^4 &= \left(i^2\right)^2 = (-1)^2 = 1 & i^8 &= \left(i^4\right)^2 = 1\end{aligned}$$

The positive integer powers of i "cycle through" the four values i, -1, $-i$ and 1. This makes it rather easy to simplify higher positive integer powers of i just by using basic properties of exponents. For instance

$$i^{35} = i^{32} \cdot i^3 = (i^4)^8 \cdot (-i) = 1^8 \cdot (-i) = -i$$

You should note that some familiar properties of exponents from the real number system do not necessarily always hold in the complex number system. For instance, if x and y are nonnegative real numbers, then we know that $\sqrt{xy} = \sqrt{x} \cdot \sqrt{y}$ (or, using equivalent exponential notation, $(xy)^{1/2} = x^{1/2} \cdot y^{1/2}$). However, if both x and y are negative numbers, that property will not hold. For example,

$$\sqrt{-4} \cdot \sqrt{-25} = 2i \cdot 5i = 10i^2 = -10$$

However,

$$\sqrt{(-4)(-25)} = \sqrt{100} = 10$$

So,

$$\sqrt{(-4)(-25)} \neq \sqrt{-4} \cdot \sqrt{-25}$$

In general, the properties of real numbers and positive integer exponents do hold within the complex number system, but properties involving fractional exponents may not.

Now, consider the two complex numbers $5 + 9i$ and $5 - 9i$. We say that these numbers are **complex conjugates** of each other. Notice what happens when we multiply these:

$$(5 + 9i)(5 - 9i) = 25 + 45i - 45i - 81i^2 = 25 + 81 = 106$$

In fact, given any complex conjugate pair $a + bi$ and $a - bi$, we see that

$$(a + bi)(a - bi) = a^2 + abi - abi + b^2 = a^2 + b^2$$

This property of conjugate pairs is important for us because we will use it when we perform division with complex numbers.

Since $i = \sqrt{-1}$, expressions of the form $\dfrac{18 + 3i}{2 + 7i}$ (indicating division of one complex number by another) actually have a radical in the denominator, so we proceed as if we are rationalizing the denominator, except this time we will use the complex conjugate of the denominator:

$$\begin{aligned}\frac{18 + 3i}{2 + 7i} &= \frac{18 + 3i}{2 + 7i} \cdot \frac{2 - 7i}{2 - 7i} \\ &= \frac{36 + 6i - 126i - 21i^2}{2^2 + 7^2} \\ &= \frac{57 - 120i}{53} \\ &= \frac{57}{53} - \frac{120}{53}i\end{aligned}$$

In general, then, we have the following summary of addition, subtraction, multiplication and division in the complex number system.

Arithmetic in the Complex Number System

Suppose that $a + bi$ and $c + di$ are complex numbers. Then

- $(a + bi) + (c + di) = (a + c) + (b + d)i$
- $(a + bi) - (c + di) = (a - c) + (b - d)i$
- $(a + bi)(c + di) = (ac - bd) + (bc + ad)i$
- $\dfrac{a + bi}{c + di} = \dfrac{a + bi}{c + di} \cdot \dfrac{c - di}{c - di} = \dfrac{(a + bi)(c + di)}{c^2 + d^2}$, provided $c + di \neq 0$

Appendix III Exercises

For these exercises, assume that $i = \sqrt{-1}$.

1. Write each expression in the form $a + bi$.

(a) $\sqrt{-16}$

(b) $3 + \sqrt{-8}$

(c) $7 - 3\sqrt{-9}$

(d) $\dfrac{2 - \sqrt{-4}}{6}$

(e) $i^6 + 5$

2. Write each expression in the form $a + bi$.

(a) $\sqrt{-64}$

(b) $-6 - \sqrt{-12}$

(c) $\sqrt{-14} + \sqrt{-21}$

(d) $2i^2 - 6i + 9$

(c) $\dfrac{\sqrt{-36} + \sqrt{-81}}{i^5}$

3. Perform the indicated operations and simplify. Write each result in the form $a + bi$.

(a) $(3 + 2i) + (7 - 3i)$

(b) $(4 - i) - (7 - 3i)$

(c) $(-3 - 6i) - (3 + 7i)$

(d) $(18 + 2i) + (-6 - 5i)$

4. Perform the indicated operations and simplify. Write each result in the form $a + bi$.

(a) $(-3i)(-6i)$

(b) $\sqrt{-36}\sqrt{-9}$

(c) $2i(5 + 3i)$

(d) $-7i(6 + i)$

5. Perform the indicated operations and simplify. Write each result in the form $a + bi$.

(a) $(2 + 8i)(3 + 5i)$

(b) $(-2 - 5i)(-9 + 4i)$

(c) $(3 + 4i)^2$

(d) $(-5 - 2i)^2$

6. Perform the indicated operations and simplify. Write each result in the form $a + bi$.

(a) $\dfrac{7}{4+i}$

(b) $\dfrac{6}{3-2i}$

(c) $\dfrac{9}{7i}$

(d) $\dfrac{8i+16}{8i}$

7. Perform the indicated operations and simplify. Write each result in the form $a + bi$.

(a) $\dfrac{3i}{4+2i}$

(b) $\dfrac{2+7i}{5i}$

(c) $\dfrac{2+i}{5+4i}$

(d) $\dfrac{3-6i}{12+9i}$

(d) $\dfrac{4-8i}{2-9i}$

8. Simplify each of the following.

(a) i^7

(b) i^{11}

(c) i^{52}

(d) $(-i)^9$

(e) $(-i)^{34}$

9. Simplify each of the following.

(a) $(3i)^4$

(b) $(-2i)^5$

(c) $5i^5 + 4i^3$

(d) $\dfrac{i^5 + i^6 + i^8}{(1-i)^4}$

(e) $(1 - i)^3(1 + i)^3$

APPENDIX IV

Formulas and Facts

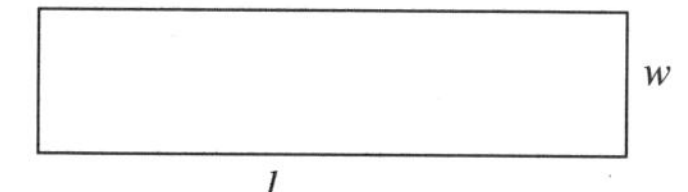

$A = lw$
$P = 2l + 2w$

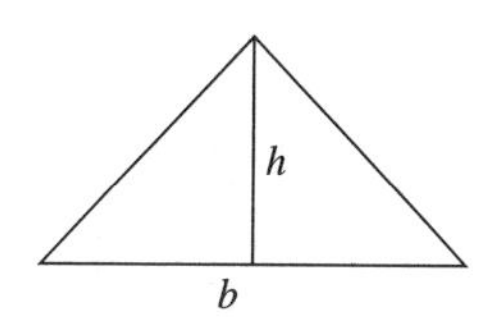

$A = \frac{1}{2}bh$

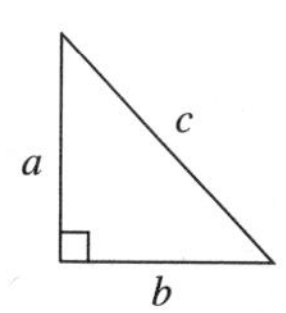

$a^2 + b^2 = c^2$
(Pythagorean Theorem)

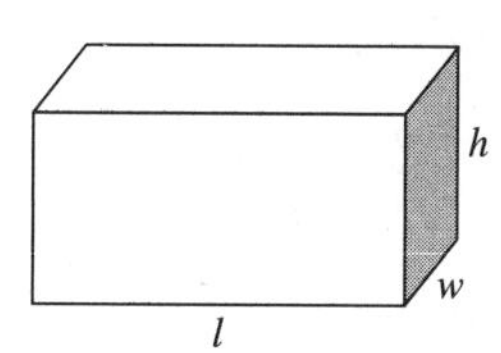

$V = lwh$
$S = 2lw + 2wh + 2lh$

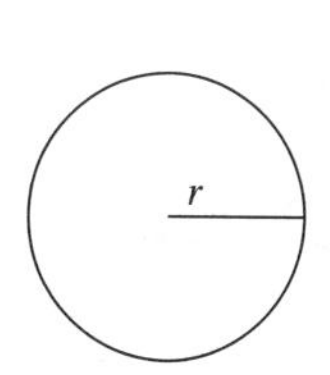

$A = \pi r^2$
$C = 2\pi r$

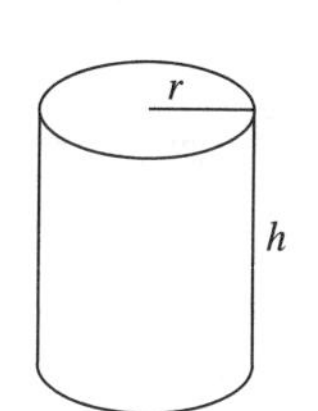

$V = \pi r^2 h$

$$ax^2 + bx + c = 0 \Rightarrow x = \frac{-b \pm \sqrt{b^2 - 4ac}}{2a}$$ (Quadratic Formula)

Interval on Number Line	Interval Notation	Inequality
 a (closed) to b (closed), x	$[a, b]$	$a \leq x \leq b$
 a (open) to b (closed), x	$(a, b]$	$a < x \leq b$
 a (closed) to b (open), x	$[a, b)$	$a \leq x < b$
 a (open) to b (open), x	(a, b)	$a < x < b$
 ← to b (closed), x	$(-\infty, b]$	$x \leq b$ or $-\infty < x \leq b$
a (closed) →, x	$[a, \infty)$	$x \geq a$ or $a \leq x < \infty$
 ← to b (open), x	$(-\infty, b)$	$x < b$ or $-\infty < x < b$
 a (open) →, x	(a, ∞)	$x > a$ or $a < x < \infty$

Exponents

$a^m \cdot a^n = a^{m+n}$

$(ab)^n = a^n b^n$

$$\frac{a^m}{a^n} = a^{m-n}$$

$(a^m)^n = a^{mn}$

$$\left(\frac{a}{b}\right)^n = \frac{a^n}{b^n}$$

$$a^{-n} = \frac{1}{a^n} \text{ and } \frac{1}{a^{-n}} = a^n$$

$$\sqrt[n]{a} = a^{1/n}$$

$$\sqrt[n]{\frac{a}{b}} = \frac{\sqrt[n]{a}}{\sqrt[n]{b}}$$

$$\sqrt[n]{a^m} = \left(\sqrt[n]{a}\right)^m = a^{m/n}$$

$$\sqrt[n]{\sqrt[m]{a}} = \sqrt[nm]{a}$$

Logarithms

$\log_b MN = \log_b M + \log_b N$

$$\log_b \frac{M}{N} = \log_b M - \log_b N$$

$\log_b N^y = y \cdot \log_b N$

$b^{\log_b N} = N$

$$\log_b N = \frac{\log_a N}{\log_a b}$$

APPENDIX V Answers to Selected Problems

APPENDIX II PROBLEM SET

1. $Q = 4x^2 - 6x + 13, R = -14$
$4x^3 - 2x^2 + 7x - 1 = (4x^2 - 6x + 13)(x + 1) - 14$

3. $Q = 3x^3 + 7x^2 + 25x + 78, R = 241$
$3x^4 - 2x^3 + 4x^2 + 3x + 7 = (3x^3 + 7x^2 + 25x + 78)(x - 3) + 241$

5. $Q = 6x^3 - \frac{9}{2}x^2 + \frac{37x}{9} - \frac{111}{4};\ R = -\frac{305}{4}$

6. $6x^4 + 10x^2 + 7 = \left(6x^3 - \frac{9}{2}x^2 + \frac{37x}{4} - \frac{111}{4}\right)(2x + 3) - \frac{305}{4}$

7. $Q = -x^2 - 1; R = 0$
$-3x^3 - x^2 - 3x - 1 = (-x^2 - 1)(3x + 1) + 0$

9. $Q = 2x^2 + 23x + 58; R = 153$

11. $Q = 4x^3 + x^2 + 3x + 3; R = 8$

13. $Q = -5x^3 + 10x^2 - 20x + 40; R = -80$

15. $Q = -5x^3 + 10x^2 - 19x + 38; R = -73$

17. $2x^3 - 5x^2 + 5x + 11 = (2x^2 - 3x + 2)(x - 3) + 13$

19. $2x^4 + 3x - 6 = (2x^3 - 4x^2 + 8x - 13)(x + 2) + 20$

21. -10 **23.** $-\frac{20}{3}$

APPENDIX III PROBLEM SET

1. **(a)** $4i$ **(b)** $3 + 2\sqrt{2i}$ **(c)** $7 - 9i$ **(d)** $\frac{1}{3} - \frac{1}{3}i$ **(e)** 4

3. **(a)** $10 - i$ **(b)** $-3 - 4i$ **(c)** $-6 - 13i$ **(d)** $12 - 3i$

5. **(a)** $-34 + 34i$ **(b)** $38 + 37i$ **(c)** $-7 + 24i$ **(d)** $21 + 20i$

7. **(a)** $\frac{3}{10} - \frac{3}{5}i$ **(b)** $\frac{7}{5} - \frac{3}{5}i$ **(c)** $\frac{14}{41} - \frac{3}{41}i$ **(d)** $-\frac{2}{25} - \frac{11}{25}i$ **(e)** $\frac{16}{17} + \frac{4}{17}i$

9. **(a)** 81 **(b)** $-32i$ **(c)** i **(d)** $-\frac{1}{4}i$ **(e)** 8